D1190354

Earthquake Design Practice for Buildings

Earthquake Design Practice for Buildings

Third edition

Edmund Booth

Published by ICE Publishing, One Great George Street, Westminster, London SW1P 3AA.

Full details of ICE Publishing sales representatives and distributors can be found at: www.icevirtuallibrary.com/info/printbooksales

First published 1988

Other titles by ICE Publishing:

Designers' Guide to Eurocode 8: Design of bridges for earthquake resistance.
B Kolias, MN Fardis and A Pecker. ISBN 978-0-7277-5735-7
ICE Manual of Structural Design: Buildings.
J Bull (ed.). ISBN 978-0-7277-4144-8
Structural Dynamics for Engineers Second Edition.
HA Buchholdt and E Moossavi Nejad. ISBN 978-0-7277-4176-9
Structural Systems: Behaviour and Design.
L Stavridis. ISBN 978-0-7277-4105-9
Progressive collapse of structures.
U Starossek. ISBN 978-0-7277-3610-9

www.icevirtuallibrary.com

A catalogue record for this book is available from the British Library

ISBN 978-0-7277-5794-4

© Thomas Telford Limited 2014

Table 5.3 derived from BS EN 1998-1:2004+A1:2013. British Standards can be obtained in PDF or hard copy formats from the BSI online shop: www.bsigroup.com/Shop or by contacting BSI Customer Services for hardcopies only: Tel: +44 (0)20 8996 9001, Email: cservices@bsigroup.com

ICE Publishing is a division of Thomas Telford Ltd, a wholly-owned subsidiary of the Institution of Civil Engineers (ICE).

Commissioning Editor: Jennifer Saines
Production Editor: Vikarn Chowdhary
Market Development Executive: Catherine de Gatacre

Typeset by Academic + Technical, Bristol
Index created by Indexing Specialists (UK) Ltd, Hove, East Sussex
Printed and bound by CPI Group (UK) Ltd, Croydon CR0 4YY

Contents

Introduction to the third edition

Many things have changed since David Key wrote his introduction to the first edition in 1987, but his approach as outlined above remains essentially valid. The major changes in seismic engineering can be listed as follows.

1 Publication of a European seismic code of practice and significant developments in codes elsewhere, particularly the USA.
2 A growing realisation that, in many cases, preventing collapse is not the only performance goal for earthquake engineers; minimising repair costs and preserving functionality are also important.
3 A vast increase in the number, availability and quality of earthquake ground motion recordings, and a better understanding of the influence of soils and earthquake characteristics on ground motion.
4 A greater appreciation of the factors that need to be accounted for in the seismic design of steel structures, and better knowledge of the seismic response of concrete, masonry and timber.
5 Transformation of non-linear time history analysis from a specialist research method to a potentially useful (and actually used) tool for practising engineers, and increasing use of non-linear static (pushover) techniques of analysis.
6 Development of practical methods for assessing and improving the seismic resistance of existing structures (but there is still a long way to go!).
7 Much greater use and experience of seismically isolated structures and buildings with other special measures for improving seismic performance, although though they still represent only a tiny minority of structures actually built.

The revised text reflects these changes, based on the author's 30 years' experience of consultancy work in earthquake engineering. Of David Key's original text, only his introduction above remains, but it is hoped that this new edition of the book remains true to David's original concept of a practical guide for engineers grappling, perhaps for the first time, with the problems of designing buildings in earthquake country.

Preface

Scope of the book

This book is directed towards practising engineers and advanced students who have a sound general knowledge of structural design but who may be unfamiliar with the problems of providing earthquake resistance. Earthquake engineering is a vast subject and the intention of this book is not to provide a fully comprehensive treatment. Rather, it is hoped to give its readers an understanding of those aspects of the subject that are important when designing buildings in earthquake country, with reference to more detailed sources where necessary. The scope of this book is restricted to buildings, but many of the principles discussed apply more generally to other forms of construction such as bridges, tanks and telecommunication towers.

Although earthquakes do not respect national boundaries, the practice of earthquake engineering does vary significantly between regions, and this is reflected in the differing formats and requirements of national seismic codes. This book is intended to be more general than to describe the approach of just one code, although it reflects the experience of the author, particularly of the European seismic code, Eurocode 8, and of US codes. Rather than sticking to an explanation of code requirements, the book attempts to explain the science and engineering which (in most cases!) underpins them.

Outline

Earthquakes regularly occur which test buildings much more severely than their designers might have expected, and earthquake engineers should (and do) make use of this chance (found more rarely in other disciplines) to find out whether current theories actually work out in practice. Throughout the book therefore, there is a discussion of lessons that can be drawn from the way buildings have performed in past earthquakes. That was a cornerstone of David Key's concept for the original (1987) edition of this book.

The first chapter steps back from the engineering technology of the rest of the book, and considers the wider factors that contribute to the way earthquakes affect the built environment and the people that live in it. Chapter 2 is a brief introduction to engineering seismology, covering such matters as quantifying the magnitude of earthquakes and the ground motions they produce. Chapter 3 outlines the important principles of structural dynamics applicable to seismic analysis, while chapter 4 discusses the analysis of soils (a crucial issue for a load case where the soil provides the dual and conflicting roles of both supporting and also exciting the structures founded on it). Chapter 5 presents the fundamentally important issue of the conceptual design of buildings; if the concept is wrong, it is unlikely that the seismic performance will be satisfactory. Chapter 6 gives an overview of some seismic codes of practice. Chapter 7 discusses the design of foundations, while chapters 8 to 11 discuss issues specific to seismic design in the four main materials used for building structures – concrete, steel, masonry and timber. So far, the book has concentrated on the primary structure of a building, but its contents are also very vulnerable and their failure may cause as great or even greater damage and loss. Chapter 12 therefore discusses building contents and cladding. Chapter 13 discusses seismic isolation, the technique of mounting buildings on flexible bearings to detune them from the earthquake motions. Existing buildings without adequate seismic resistance pose a huge safety and economic threat in many parts of the world, and the final chapter discusses the difficult issues of their assessment and strengthening.

Foreword

It was during a visit to Mexico City with Edmund Booth in the weeks following the 1985 earthquake disaster that I began to understand the enormity of the challenge facing the earthquake engineering profession. This was no academic subject, as many might perceive it to be, but a matter of life and death for those people and communities unfortunate enough to be caught up in the destruction. Buildings just a few years old were crushed flat as pancakes, toppled like dominos or shattered like crazy paving. Next door, buildings that looked as though they should have collapsed under their own weight years ago appeared undamaged.

The Mexico earthquake illustrated for me the breadth of the engineering challenge: with minimal time to react, everything is shaken, everyone is tested, and within a few minutes, the economic and social structures on which the community has hitherto depended lie in ruins. The Mexico earthquake was not the first disaster of recent times to test buildings constructed with some understanding of modern design principles and, of course, neither was it to be the last.

Sadly, over the intervening decades we have seen the story repeated again and again around the world. Major earthquake disasters continue to afflict rich and poor nations. In this new, third edition of 'Earthquake design practice for buildings', Edmund contrasts the 2010 Haiti disaster with the 2011 Tohoku Japan earthquake and the two earthquakes that struck Christchurch, New Zealand in the same years. The lessons for structural engineers seem all too familiar. Buildings constructed without adequate consideration of the principles of earthquake design and subjected to strong ground motions will be damaged or destroyed, often with fatal consequences. We know exactly why they are damaged yet we seem powerless to prevent future disasters.

An important part of the problem is scale. As fast as our understanding of earthquake hazard and risk advances, the global problem escalates. Since 1985 world population has increased by nearly 50%. Urbanisation is occurring at breakneck speed around the world. Development takes place on increasingly vulnerable sites. Medium and high rise concrete buildings remain the design solution of choice in developing countries and emerging economies struggling to cope with population growth and changing patterns of living.

Yet technical solutions that would greatly improve the resilience of buildings and infrastructure in earthquake prone regions are well understood by experts. The biggest challenge is one of communication; how to stimulate a proper dialogue within civil society over the level of earthquake risk that vulnerable communities are exposed to and how this risk is managed alongside other extreme risks. Should communities seek to mitigate earthquake risk through new building codes and more rigorous building inspection, through improvements to the existing building stock, better disaster management, the installation of warning systems (such as for tsunami), or even through insurance?

For all these reasons, it is vital that earthquake engineers share their knowledge as widely as possible. In this important contribution to the subject, Edmund has expertly blended rich technical insight and broad contextual discussion. If this work stimulates new dialogue and understanding amongst earthquake engineering specialists and their fellow professionals and clients around the world, it has the potential to make a real difference to the mitigation of future earthquake disasters.

Scott Steedman CBE FREng
October 2013

Acknowledgements

Earthquake engineering has many diverse branches, which is one of the reasons for its fascination; the branches range from geological matters through the complex behaviour of materials and soils under cyclic loading to non-linear dynamic analysis, with many other subjects encountered on the way. There are a number of excellent compendia on earthquake engineering, where each of the chapters is contributed by a leading expert on one of these aspects. A book such as this by a single author suffers from the drawback that one person cannot possibly have as great an in-depth knowledge of all these branches. Offsetting the limitations imposed by a single authorship is the consistency of approach brought by an author who has had to integrate all the aspects of earthquake engineering into his work as a practising engineer.

And for the author there is another advantage. Most of the chapters in this book have been reviewed by experts in the relevant field; their extensive comments provided the author with much thought provoking and challenging material. Contacting these experts gave him the opportunity to renew old friendships and forge new ones. The author is particularly grateful to two very old friends, Jack Pappin and Richard Fenwick, for providing comprehensive reviews of chapters 4 and 7 (Jack) and chapter 8 (Richard); they also generously allowed some of the text in these chapters to be based on material originally prepared by them. Ahmed Elghazouli, Sin-Tsuen Tong, Agostino Marioni, Svetlana Brzev, Suikai Lu, Peter Watt, Scott Steedman, Tiziana Rossetto, David Wald, Robin Spence, Gopal Madabhushi, Ziggy Lubkowski, Keith Fuller, Helen Crowley, Ichiro Nagashima, Michel Bruneau, RS Narayanan, Timothy Sullivan, Martin Williams, Athol Carr, Massimo Fragiacomo and Maurizio Follesa all provided reviews, advice and/or comments, for which the author is very grateful, although of course the opinions expressed and the errors that remain are his responsibility.

Thanks are also due to Pilate Moyo for providing the author with space and facilities at the University of Cape Town, where this book was completed, and to Mark van Ryneveld and Francois Viruly of UCT for stimulating discussions on chapter 1.

Finally, thanks are due to the editorial staff at ICE Publishing, and in particular to Jennifer Saines.

Notation

Notes

1 The units shown for the parameters are mainly to indicate their dimensions. Units of length, mass and time are given in metres (m), tonnes (t) and seconds (s) respectively; in dynamic equations, this leads to forces being in kilonewtons (kN). Other dynamically consistent systems of units would also be possible (for example mm, kg, s, N). Stresses pose a special problem; the dynamically consistent unit of stress in the m,t,s system of units used here is kPa (kN/m^2). However, while soil stresses are generally expressed in kPa, structural stresses are generally given in MPa (MN/m^2), and this has been done here. When using formulae with dimensionless constants, care is needed to choose consistent units; when the constants are not dimensionless, their value depends on the system of units adopted.

2 Notation not given in this table is defined at the point of occurrence in the text.

Symbol	Description
A_c	Cross-sectional area of the concrete section: m^2
A_s	Cross-sectional area of the reinforcing steel: m^2
A_{sh}	Shear area of the concrete section: m^2
A_w	Web cross-sectional area: m^2
a_g	Peak ground acceleration: m/s^2
a_{gR}	Design peak ground acceleration for the reference return period: m/s^2
b_w	Width of beam web: m
C_d	Deflection amplification factor
c_u	Undrained shear strength of soil: kPa; Dimensionless coefficient in the US code ASCE 7 relating to the upper limit on calculated period of a building
D	Diameter of a circular column: m
d	Effective depth to main reinforcement in a concrete beam: m; Diameter of bolt or other fastener joining timber members: m
d_b	Diameter of reinforcing steel in concrete: m
d_{bl}	Diameter of longitudinal bar: m
d_r	Relative displacement between points of attachment of an extended non-structural element: m
E	Youngs modulus: MPa
E_{cd}	Design value of concrete compressive modulus: MPa

E_{cm}	Mean value of concrete modulus: MPa
E_{sd}	Design value of reinforcing steel modulus: MPa
E_{sm}	Mean value of reinforcing steel modulus: MPa
e	Maximum out of straightness of a strut: m; Length of the shear link in an eccentrically braced frame: m
$\boldsymbol{F}$	Force: kN
$\boldsymbol{F}_a$	Horizontal force on non-structural element: kN
$\boldsymbol{F}_{elastic}$	Seismic force developing in an elastic (unyielding) system: kN
$\boldsymbol{F}_i, \boldsymbol{F}_j$	Forces at levels i and j: kN
$\boldsymbol{F}_{plastic}$	Seismic force developing in a plastic (yielding) system: kN
$\boldsymbol{F}_{ult}$	Force at ultimate deflection: kN
$\boldsymbol{F}_y$	Yield force: kN
f'_c	Cylinder strength of concrete: MPa
f'_{cd}	Design compressure strength of concrete: MPa
f_{ck}	Characteristic compressure strength of concrete: MPa
f_{ctm}	Mean value of axial tensile strength of concrete: MPa
f_{cu}	Cube strength of concrete: MPa
f_{ult}	Ultimate strength of steel: MPa
f_{yh}	Yield strength of steel: MPa
G	Shear modulus: MPa or kPa
$\boldsymbol{G}_0$	Shear modulus of soil at small strains: kN/m^2
G_{cd}	Design value of concrete shear modulus: MPa
G_{cm}	Mean value of concrete shear modulus: MPa
$\boldsymbol{G}_s$	Shear modulus of soil at large shear strain: kPa
g	Acceleration due to gravity: m/s^2
H	Building height: m
H_e	Effective height of substitute structure: m
h	Minimum cross-sectional dimension of beam: m; Greater clear height of an opening in a masonry wall: m
h_b	Depth of beam: m
h_{ef}	Effective height of a masonry wall: m
h_s	Clear storey height of shear wall between lateral restraints: m
h_w	Overall height of shear wall: m; Cross-sectional depth of beam: m
I	Moment of inertia: m^4
I_{cr}	Cracked moment of inertia of a concrete section: m^4
I_E	Seismic importance factor
I_{eft}	Effective moment of inertia of a concrete section: m^4
I_g	Gross moment of inertia of concrete section, based on uncracked properties: m^4

k	Spring stiffness: kN/m; Dimensionless exponent in Equation 6.2 for distribution of seismic forces with height
k_e	Secant stiffness of a non-linear system at a given deflection: kN/m (see Figure 3.26)
k_i	Initial stiffness: kN/m
L	Length of a masonry wall: m
L^*	Critical span of beam corresponding to formation of plastic hinges within span under lateral loading: m
L'	Clear span of beam: m
L_b	Span of beam between column centre lines: m
L_i	Structural property defined in Equation 3.12: t
L_{pl}	Effective plastic hinge length: m
L_v	Bending moment to shear force ratio at the critical section of a plastic hinge forming in a concrete member
l	Effective unrestrained length of a beam or column: m
l_w	Width of shear wall: m
M	Magnitude of earthquake: t
M_A, M_B	Plastic hinge moments forming at either end of a beam: kNm
M_{cr}	Moment at which a concrete section cracks: kNm
M_i	Structural property defined in Equation 3.13: t
M_P	Plastic flexural strength of a beam: kNm
M_{RK}	Characteristic flexural strength: kNm
M_s	Surface wave magnitude of earthquake
M_w	Moment magnitude of earthquake
m	Mass: t
m_a	Mass of non-structural element: t
m_b	Body wave magnitude of earthquake
m_e	Effective mass: t
m_i, m_j	Mass at level i and j: t
N	Number of storeys in a building
N_{SPT}	Blow count per 300 mm in the standard penetration test
$N_{1(60)}$	Corrected standard penetration test blow count
P	Axial load in a column: kN
P_1	Probability of exceedence in 1 year
P_y	Probability of exceedence in y years Tensile yield force of a column: kN
q	'Behaviour' or force reduction factor for structural systems in Eurocode 8
q_a	'Behaviour' or force reduction factor for non-structural elements in Eurocode 8
R	'Response modification' or force reduction factor for structural systems in the US code IBC; Radius of a friction pendulum isolation bearing: m

R_{sh}	Reduction factor on shear stiffness of concrete element
r_d	Reduction factor for soil stress or at depth
r_y	Radius of gyration of a beam or column about its minor axis: m
S	Soil amplification factor in Eurocode 8
S_1	Spectral acceleration at 1 s period: m/s^2
S_a	Spectral acceleration: m/s^2
S_{ai}	Spectral acceleration corresponding to the period of mode i: m/s^2
$S_{a\mu}$	Ductility modified spectral acceleration for a ductility demand μ: m/s^2
S_d	Spectral displacement: m
S_S	Spectral acceleration at short period: m/s^2
S_v	Spectral velocity: m/s
T	Return period: years; Structural period: s
T_1, T_2, T_3	Periods of first, second, third modes of building: s
T_a	Fundamental vibration period of non-structural element: s; Empirically determined vibration period of a building: s
T_B, T_C, T_D	Periods defining the shape of the design response spectrum in Eurocode 8: s
T_e	Effective period of a non-linear system at a given displacement: s
t_{ef}	Thickness of a masonry wall: m
t_f	Thickness of flange of steel section: m
u_d	Design seismic displacement: m
u_e	Equivalent displacement of substitute structure in capacity spectrum method: m
$u_{elastic}$	Seismic displacement of elastic (unyielding) system: m
u_{max}	Maximum deflection: m
$u_{plastic}$	Seismic displacement of a plastic (yielding) system: m
u_{ult}	Displacement at ultimate capacity: m
u_y	Displacement at yield: m
V_1, V_2, V_3	Seismic shears at base of building corresponding to first, second, third modes: kN
V_{base}	Base shear: kN
V'_{base}	Modified base shear as in ASCE 7: kN
V_c	Shear in a column: kN
V_{Ed}	Design value of shear force in wall in the seismic design situation: kN
V_p	Shear capacity of the shear link in an eccentrically braced frame: kN
$V_{Rd,c}$	Design shear resistance of the wall without shear reinforcement: kN

V_u	Shear force in a plastic hinge under ultimate conditions: kN
V_{wall}	Shear in a shear wall: kN
v	Masonry shear strength under zero compressive load: MPa
v_d	Design in-plane shear strength of masonry: MPa
W_a	Weight of non-structural element: kN
w	Vertical loading per unit length on the beam: kN/m
x	Height above fixed base: m
z	Effective depth of flexural steel at base of wall: m
z_i, z_j	Heights above base of levels i and j: m
ε_{su}	Ultimate strain of steel
ε_{yk}	Characteristic yield strain of steel
ϕ'	Effective angle of friction in non-cohesive soil
$\phi_i(x)$	Modal deflection at height x in mode i
φ	Curvature of a beam: radians/m
φ_{av}	Average curvature of a plastic hinge: radians/m
φ_p	Curvature of a plastic hinge at rotation θ_p: radians/m
φ_u	Ultimate curvature of a plastic hinge: radians/m
φ_y	Curvature of a plastic hinge at first yield: radians/m
γ	Shear strain
γ_a	Importance factor for non-structural element, in Eurocode 8
γ_I	Importance factor
γ_m	Partial factor on material strength
η	Correction factor to adjust response for damping other than 5%
μ	Displacement ductility; Coefficient of friction
ν	Reduction factor in Eurocode 8 to convert design displacements at ultimate limit state to serviceability limit state
θ_p	Plastic rotation of a plastic hinge: radians
θ_u	Ultimate rotation of a plastic hinge: radians
θ_y	Rotation of a plastic hinge at yield: radians
ρ	Ratio of tension reinforcing steel area to cross-sectional area of concrete member; Ratio of force demand on an element to capacity of the element; Reliability factor in ASCE7
σ_t	Maximum stress in concrete due to bending, assuming an uncracked section: MPa
σ_v	Vertical stress in masonry due to permanent loads: MPa
σ_{vo}	Total vertical stress in soil at the level of interest due to gravity loads: kPa

σ'_{vo}	Effective vertical stress in soil at the level of interest due to gravity loads: kPa
τ_e	Effective shear stress in soil under design earthquake loading: kPa
ξ	Percentage of critical damping (viscous damping ratio)
ξ_{equiv}	Equivalent viscous damping ratio of substitute structure in capacity spectra method
Ω	Minimum ratio of resistance moment to design moment at plastic hinge position; Ratio of strength provided to design strength of superstructure element most affecting foundation forces

Earthquake Design Practice for Buildings
ISBN 978-0-7277-5794-4

ICE Publishing: All rights reserved
http://dx.doi.org/10.1680/edpb.57944.001

Chapter 1
The nature of earthquake risk

A stark divide separates the developed world, in which societies are highly resilient to earthquakes, from the developing world, in which even relatively small earthquakes cause large death tolls and have huge economic impacts.

England *et al.* (2011)

The tools of the trade of the earthquake engineer form the subject matter of the rest of this book. This chapter looks beyond the technical issues involved to consider the much broader range of issues which contribute to a society's success or failure reducing the impact of earthquakes.

1.1.   Introduction: technical solutions are not sufficient

This is a book written by a structural engineer for fellow construction professionals and students. The subsequent chapters address the many and diverse technical issues that have to be grappled with when designing in earthquake country. The challenge of trying to come to grips with these issues is continually fascinating and demanding, and has absorbed more than half of the author's career. Being able to understand and analyse the way in which an earthquake stresses a building, choosing buildable structural forms and materials to cope with these stresses and then translating these concepts into practical, affordable and perhaps even beautiful spaces that people want to use and live in is a vitally important task. The aim of the book is to provide an introduction to the technical issues that help make this achievable.

However, the problem of reducing the impact of earthquakes on human communities goes wider than merely getting the technical issues right, important though that is. The stark divide referred to at the head of this chapter would not be significantly changed if copies of this book, appropriately translated, were freely available in the earthquake hotspots of the world. Fostering societies that are resilient to earthquakes requires something wider than just solving the engineering aspects. While engineers should concentrate on what they are trained to do, they should also be aware of these wider vistas. The rest of this chapter attempts a brief discussion of what those vistas might include.

1.2.   Why earthquakes are different

The earthquake threat has distinct characteristics which have important consequences for the ways in which society responds to the threat, and the part that structural engineers will play in that response. These characteristics can be summarised as follows.

i.   They occur suddenly and without warning, with the potential to cause huge loss of life and destruction of wealth in a very short space of time.
ii. The time interval between destructive earthquakes affecting a particular location is often very long, perhaps many generations.
iii. An earthquake stresses much of the infrastructure that society depends on: superstructures and buried structures, building contents, transport and telecommunication systems, power supplies – all are affected.
iv. Other hazards often follow in the wake of an earthquake, notably landslides, fires and tsunamis.

Discussing these in turn:

i.   Earthquakes in most places still strike completely without notice; even in the relatively few places where sophisticated warnings systems are in place, the advance warning is at most a minute or two. By contrast, it is now usually possible to give hours or days of notice for hurricanes, floods and even volcanoes. Moreover, the sudden and swift destruction caused by an earthquake influences society's perception of seismic risk, and its response to it.
ii. Because the interval between events is often so long, knowledge residing in local populations on the types of construction that work well in resisting earthquakes does not have a chance to establish itself in seismic regions, in the way that it can (and does) for more frequent types of hazard, such as extreme winds. Even a few years after a damaging event, the urgency to 'do something' gets lost in other concerns, while knowledge and experience of good aseismic construction practice frequently gets forgotten.
iii. An earthquake is a severe test of the whole of a building – not only its internal and external structure, but also its contents. Moreover, it places the surrounding infrastructure under severe stress; electricity supplies are lost and transportation systems fail, affecting the capacity to bring in relief. Buried structures, protected from many natural hazards, are also affected; fracture of pipelines caused the loss of fire fighting ability in the 1906 San Francisco earthquake, which led to much of the damage. The dependence of large cities on its infrastructure means that they are particularly vulnerable to the multi-faceted damage an earthquake causes, and earthquake engineers have to be multi-disciplined as a result.
iv. Landslides, fire and tsunamis – the hazards that may well follow a large earthquake – need to be dealt with in different ways from those which apply to ground shaking hazard, the main subject of the rest of this book.

These unpredictable events cause the rapid and simultaneous loss of so much we need to survive, and turn the ground we rely on for our stability into a terrifying source of destruction. No wonder they hit the news headlines with such dramatic impact whenever they occur. Finding the best ways to deal with them is bound to be difficult, and narrow engineering logic cannot always provide complete solutions.

1.3.   How the toll from earthquakes varies between societies

As a rule, major earthquakes in high income societies cause very large economic losses and comparatively low death tolls, whereas the opposite is true in low income societies. Thus, the economic loss in the Tohoku Japan earthquake of 2011 is slated as the biggest ever for a single natural disaster and was about 50 times that in Haiti the previous year. By contrast, the Japanese death toll, at just under 20 000, amounted to less than 0.5% of those exposed to strong shaking, and most of the deaths were due to the tsunami which followed the earthquake. At least ten times as many people died in Haiti, and death rates in developing countries routinely exceed 5% of the exposed population.

The high cost, low death toll for developed economies explains the current emphasis in high income countries, discussed in Chapter 5, for engineering design objectives which go beyond the goal of merely achieving life safety. Whether this approach is equally appropriate for developing countries is another matter; moreover, simple measures of economic loss and death toll do not tell the whole story. While the Japanese loss in 2011 amounted to 5% of GDP, it doesn't appear to have had a significant overall effect on subsequent Japanese growth, and in fact the Nikkei index of the Tokyo stock market the day before the earthquake struck was actually 20% lower than it was two years later. By contrast, the economic loss in Haiti of around $6 billion, although about 1/50th of Japan's, probably exceeded Haiti's annual GDP. An economy already dominated by foreign aid was then swamped by around $6 billion dollars of aid. However, writing in 2013, it would be difficult to claim that Haiti has 'recovered' from its earthquake, whereas that claim perhaps could be made for Japan. For a further discussion of the situation in post-earthquake Haiti, see Ramachandran and Walz (2012).

These two earthquakes, separated in time by just over a year, are extreme examples, perhaps, but they illustrate how much the impact of an earthquake depends on the society it strikes. And there is a related point; the solutions proposed by engineers must take account of where they are to be applied. What works well in Japan is quite likely to be an inappropriate solution for Haiti.

1.4.   Preparing for earthquakes

The impact on a society of a destructive earthquake depends crucially on where the earthquake has struck, as the previous section makes clear. The next paragraphs introduce a discussion of what makes for a well prepared society.

On the purely technical side, it is of course vital to ensure that new construction incorporates the best current understanding of what works well in resisting earthquakes, and avoids what is known to work badly. Construction on land subject to the earthquake-related risks of landslides, liquefaction or tsunami should be discouraged or subject to special controls. Plans also need to be in place to upgrade existing buildings and other infrastructure that is substandard. It is important to have building regulations and standards which are kept up to date with current practice and reflect the local level of seismic hazard; having the means to ensure that these standards are observed is also essential.

These are the technical matters very much within the engineer's province, and they are vital. However, well prepared societies show other characteristics that are linked to success. They devote resources to training engineers and to carrying out research. Dissemination of knowledge among the profession through learned societies, which depend to a large extent on voluntary activities, is part of this effort. This culture of fostering professional engineering competence may be very hard to achieve in societies where the basic needs of providing shelter, healthcare and education are understandably seen as more pressing needs than preparing for an event – a damaging earthquake – which may not have been experienced for many generations. This is a problem of poverty and development, which of course applies to much more than just earthquake resistance.

Even the most technically advanced societies know that they are not immune from damage if they experience a great earthquake. So areas of high seismicity should have plans for responding to a disaster that are well rehearsed and disseminated. The plans need to address the provision of rescue services, medical care, emergency shelter, food and water supplies and so on, and need to take account of the chaotic conditions that might apply when normal services, communications systems and transport routes cannot be relied on. Training and education of the general public is also needed on how to prepare for an earthquake, for example, on how to react when the shaking starts (Figure 1.1). These are matters for administrators and other actors with backgrounds not in engineering, although of course engineers can provide important advice. Once again, disaster plans for Japan are likely to be rather different from those for Haiti.

Planning for the recovery phase after an earthquake is also important, and insurance against earthquake loss plays its part in ensuring that the necessary resources become available for

Figure 1.1 What to do in an earthquake. (The Great British Columbia Shakeout poster (www.ShakeOutbc.ca) as an example of The Great ShakeOut earthquake drills, a global drill. © Southern California Earthquake Center, USC)

reconstruction. Turkey (Bommer *et al.*, 2002) and New Zealand are among the countries that have devoted considerable attention to the role that insurance can play in mitigating the effect of earthquakes on society. Once again, engineers have a role to play here, but other professionals must take the lead.

1.5.   When the earthquake strikes

The previous section described how society can prepare itself for a destructive earthquake. This section looks at measures that have been found to help once the earthquake strikes.

A consistently reliable method of predicting an earthquake, in the sense of providing data on its size, location and time (all three parameters are needed) has never yet been achieved, and there are some grounds for supposing that the inherently chaotic nature of rock fracture that results in sudden fault movements means that it is not possible to find one. In any case, persuading the entire population of a city to evacuate on the grounds of an inevitably uncertain prediction would be difficult, particularly if previous warnings had proved to be false alarms; moreover, evacuating a large city would cause enormous problems of its own. However, once a very large earthquake has taken place, there may be a short window of opportunity to provide a warning. The seismic waves that cause damage travel through rock at around 3–6 km/s, whereas radio communications travel at the speed of light. Accordingly, if there is a network of seismometers immediately around the source of the earthquake, it can establish its size and location almost instantly, and then this information can be relayed to surrounding areas. If the earthquake appears potentially destructive, then appropriate actions can follow; for example, emergency braking can be applied to high-speed trains and other high-risk systems can be shut down. A site 50 km from the source would receive an advance warning of perhaps only 10 s, but this sort of system is being implemented in Mexico City, Istanbul, California and elsewhere (Erdick, 2006), and has been successfully used to halt high-speed 'bullet' trains in Japan. Much more cheaply, accelerometers measuring ground motions can measure the strength of the earthquake waves once they arrive and then trigger appropriate actions if the motions exceed a specified threshold. Shutoff valves on high pressure gas lines and shut-down systems for nuclear power stations are examples of this type of system that have been successfully used.

These are purely technical measures, entirely within the province of the engineer (though not primarily the structural engineer). Other measures are now described where the engineer is providing tools to aid decision making by others. A prime example is providing emergency managers with an overview of how much damage has been caused, and where it is worst. This information may be very difficult to get under the inevitably disturbed conditions of an earthquake; two high technology developments of the last decade, described next, have made significant contributions to overcoming this problem. Unlike the early warning and shutdown systems described in the previous paragraph, they can operate using systems outside the country concerned and do not rely on seismic instrumentation very near to the earthquake source.

The first of such systems produces a rapid estimation of the likely intensity and geographical distribution of the ground motion. The estimate is based on empirical models, using past experience of how shaking intensity varies with magnitude of earthquake and how it attenuates with distance from the epicentre or causative fault. Account must also be taken of local ground conditions, which can modify the ground motions intensity significantly. What is needed to produce these estimates is the size and location of the earthquake; for large events, reliable estimates of these two parameters become available within an hour or less of the event from the global earthquake monitoring system. The results of these empirical models can then be modified and corrected by actual measurements of ground motion collected from accelerometers, where available (Worden *et al.*, 2010). A further level of modification to the empirical predictions may be made from reports of damage posted by the general public on special websites; the predicted ground motions can be compared with reports of how strongly the motion was actually felt at various locations and the nature of the damage actually observed.

Given the distribution of ground shaking intensity, and a knowledge of the location and types of buildings and other infrastructure in the affected area, predictions can be made that can inform relief operations. Thus, answers are possible to questions such as where building damage is likely to be greatest or which road supply routes may have been disrupted by bridge failures or landslides. This information can be difficult to obtain quickly by other means at a relief operations control centre, but is clearly of immense value, even if inevitably it is subject to some level of uncertainty. The ShakeMap system (see Figure 1.2) is a leading example of this type of system (Wald *et al.*, 2005); it has been developed primarily for

Figure 1.2 Prediction of ground shaking intensity in Haiti, 2010 based on ShakeMap. For detailed version, see http://earthquake.usgs.gov/earthquakes/shakemap/global/shake/2010rja6/

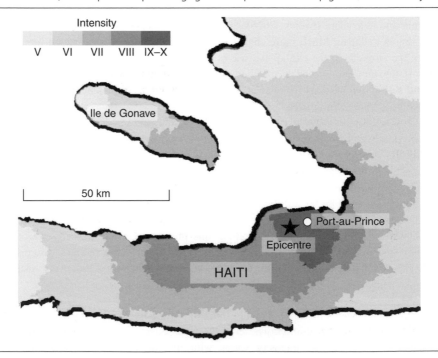

California, but ShakeMaps are routinely produced and made freely available on the US Geological Survey website for damaging earthquakes worldwide very soon after they occur.

A second system uses aerial and satellite imagery of the affected area taken after the earthquake to estimate the extent and distribution of damage. The Haiti earthquake of 2010 provided a major test of this procedure; the international community quickly became aware that a very damaging event had struck Port-au-Prince and its surroundings, but communications to and around the area were all but impossible in its immediate aftermath. It was the examination of satellite images by a volunteer team of experts spread around the world, networking by the internet, which produced the first hard data on the extent of the tragedy, and the consequent needs for the relief operations (Ghosh *et al.*, 2011).

1.6.   Reconstruction and recovery

The aftermath of a destructive earthquake provides a window of opportunity for improvements to be made, because there is often strong pressure on politicians and other decision makers to ensure that future generations are protected from a similar disaster. Importantly, this applies not just to the areas directly affected, but also the surrounding region which may fear that their turn for destruction might be next. However, these pressures are easily dissipated or diverted. For one thing, those who have not lost homes or known people who were killed or injured in the earthquake soon tend to turn their attention to other, apparently more pressing, needs. Even in the areas worst affected, 'building back better' does not always happen. Partly, this is because the immediate and urgent needs for restoring shelter, sanitation, health clinics, schools, transport links and so on as rapidly as possible crowd out the long term need to provide earthquake resistance. In this context, the UK Department for International Development (DFID) Disaster Risk Reduction policy paper recommends setting aside 10% of the disaster funds to reduce the impact of related future disasters (Rossetto, 2012).

Less excusably, many agencies involved in reconstruction do not always recognise that engineers have a role to play. Jo da Silva, in her Brunel lecture (da Silva, 2012) notes:

> At a strategic level, engineers are notably absent amongst the staff of key decision makers (the World Bank, the UN, bi-lateral donors, governments etc.). Nor do these organisations typically engage experienced engineers or engineering firms as consultants in the early stages of recovery and reconstruction, when their experience is most needed in order to prioritise reinstatement of critical infrastructure, and help develop a road-map for recovery. In the event of a pandemic or a terrorist bomb attack with numerous human casualties, failure to consult medical professionals would be unthinkable. It therefore seems absurd that engineers are not engaged in disasters where there has been considerable loss of infrastructure and thus basic services, and particularly so in an urban context.

If they really want to make an effective contribution to recovery after an earthquake, engineers need to emerge from concentrating on their comfort zone – the technical matters dealt with in the rest of this book; they must make their voice heard at the strategic level that da Silva refers to.

1.7.   An appropriate response to the earthquake threat

As noted earlier, the dramatic and sudden nature of seismic destruction means that an appropriate and balanced response is not easy for society to find, particularly in the inevitably febrile atmosphere following a destructive event. The immediate aftermath of a great earthquake provides a powerful impetus towards improving matters, but may also make it harder to take a balanced view. Following the Christchurch earthquakes of 2010 and 2011, Kelly (2012) wrote:

> Since records began in 1843, 483 people are recorded as having died in New Zealand as a result of earthquakes. The vast majority of our earthquake fatalities – 447 people – resulted from the two terrible earthquakes in Christchurch in 2011 and Napier in 1931. By comparison, more than twice as many people (1125) died on [New Zealand] roads in the three years from 2008–10.

This is partly an issue of public perception. People are prepared to trade the risks of travel by car with the convenience and utility it provides; they regard driving a car and its associated risks as an activity under their personal control, at least to some extent. By contrast, once they have arrived at their destination, they expect to be protected from external hazards, even those that are very unlikely (as was the earthquake which destroyed central Christchurch). Public pressure for 'something to be done' in the immediate aftermath of a destructive earthquake is entirely understandable; that pressure will most likely diminish with time and may disappear altogether after a few years. But to what extent should resources be devoted to ensuring the disaster is not repeated in future? These are questions whose answers will depend on the nature of the society which asks them – its priorities, state of development, political structures, attitudes to the value of human life, understanding of 'acts of God' and so on.

Given that many generations may pass before an earthquake of similar size strikes again, the questions are not easy ones to answer. Engineers have a vital role in advising on practical options, but the key decisions are likely to be taken by politicians and administrators.

1.8.   Creating earthquake-resilient communities

The earthquake that struck Haiti in 2010 gave rise to one of the worst earthquake disasters of our time in terms of human impact. It was not just that shoddy construction led to tens of thousands of people being crushed to death in their homes. The disaster was greatly compounded by the fact that Haiti for a time became practically inaccessible; the narrow roads of the city were blocked with rubble, large parts of the harbour failed, and for a short time the airport was put out of action. Moreover, many government buildings collapsed, killing the civil servants that worked in them, and putting out of action the local civil authority so badly needed to mount a rescue and recovery effort. The ruined presidential palace (Figure 1.3) became a symbol of Haiti's problems, while the recently completed US embassy, designed to resist earthquakes using the seismic isolation technology described in Chapter 13, was scarcely affected.

A fragile and poor society, beset by political problems and a range of natural hazards including frequent hurricanes, is inevitably going to respond badly to the impact of an earthquake strike.

Figure 1.3 Presidential palace, Port-au-Prince, Haiti after the 2010 Haiti earthquake (photograph courtesy of G. Madabhushi)

Addressing a situation like Haiti's, clearly needs to go much further than addressing narrow issues of seismic engineering. In fact, improving education and healthcare, fostering the development of the political system, creating a culture of new business enterprises and other items on a standard development agenda might be considered much more important than specifically addressing issues of protection against a rare earthquake. Indeed, in the long term they might also be more effective in protecting against seismic losses than attempting to introduce better earthquake-resistant construction. However, in retrospect, it is clear that some consideration of the seismic vulnerability of Haiti's capital as a system could have paid great dividends; measures such as securing at least part of the seaport and airport, reinforcing land supply routes and providing a robust disaster control centre are examples. Many international aid agencies had been established in Haiti long before 2010, but it is sobering to note that few of them had considered the potential impact of an earthquake on their broader development work, and the destruction of their offices and personnel by the earthquake further hampered relief efforts. Jo da Silva, in the Brunel lecture quoted above, suggests:

> The characteristics of a resilient community might include: good health; knowledge and education; reliable services and robust infrastructure; diverse livelihood opportunities; healthy ecosystems; the ability to organise and make decisions; and access to external assistance. There are therefore multiple ways in which engineers can contribute to building resilience.

Writing soon after the Haiti earthquake, and the ones in Japan and New Zealand that followed, Scott Steedman (2011) gave the following challenge to the engineering community.

> There has never been a better time for engineers to make the case that managing the risk of extreme events requires more than higher flood walls and better emergency planning. It requires local, regional and national authorities to see disaster mitigation as an integrated system, so that no matter how severe the situation, control over critical infrastructure – transport, power generation and key facilities – can be maintained throughout the event, or rapidly restored afterwards....

The important lesson for engineers is that hazard assessment is, by definition, an uncertain business. We need new ways of thinking through how our whole system of protection can remain under control, no matter what. Resilience is the key to deciding the extent of engineering countermeasures.... Resilience does not have to mean ever higher flood walls, or duplicated systems. Resilience means that engineers understand the consequences of failure of any single element and can communicate to governments that the system will still achieve whatever outcome is acceptable. More concrete is not the answer; more systems thinking is.

The rest of this book is about the technical tools used by structural engineers in reducing the earthquake threat to society. Theirs is a vital role, but, as this chapter has tried to argue, one that can only be fully effective if played out in a wider context.

REFERENCES

Bommer J, Spence R, Erdik M *et al.* (2002) Development of an earthquake loss model for Turkish catastrophe insurance. *Journal of Seismology* **6(3)**: 431–446.

Da Silva J (2012) Shifting agendas: response to resilience – the role of the engineer in disaster risk reduction. The 2012 Brunel Lecture. http://www.jodasilva.me/Report/Brunel_Report_FINAL%5B1%5D.pdf.

England P, Holmes J, Jackson J and Parsons B (2011) What Works and Does not Work in the Science and Social Science of Earthquake Vulnerability? *Report of an International Workshop held in the Department of Earth Sciences, University of Oxford on 28th and 29th January, 2011.* See http://comet.nerc.ac.uk/Workshop_report.html.

Erdik M (2006) *Urban earthquake rapid response and early warning systems.* 14th European Conference on Earthquake Engineering, Geneva, Switzerland, 2006.

Ghosh S, Huyck CK, Greene M *et al.* (2011) Crowdsourcing for rapid damage assessment: the Global Earth Observation Catastrophe Assessment Network (GEO-CAN). *Earthquake Spectra* **27(S1)**: S179–S198.

Kelly D (2012) *Quake safety cost must relate to risk.* The Press, Christchurch, New Zealand (15/06/12), Perspective.

Ramachandran V and Walz J (2012) *Haiti: Where Has All the Money Gone?* Center for Global Policy Development Policy Paper 004. Washington, DC, USA – see www.cgdev.org.

Rossetto T (2007) Construction Design, Building Standards and Site Selection. Chapter 12 in Benson C and Twigg J (ed.) *Tools for Mainstreaming Disaster Risk Reduction: Guidance Notes for Development Organisations.* Geneva. International Federation of Red Cross and Red Crescent Societies/the ProVention Consortium, 141–152.

Steedman S (2011) Earthquakes, again – editorial. *Ingenia* issue 47, Royal Academy of Engineering, London, UK – www.ingenia.org.uk.

Wald DJ, Worden BC, Lin K and Pankow K (2005) *ShakeMap manual: technical manual, user's guide, and software guide.* US Geological Survey, Techniques and Methods 12-A1, 132 pp. Free download from http://earthquake.usgs.gov/earthquakes/shakemap.

Worden CB, Wald DJ, Allen TI, Lin K, Garcia D and Cua G (2010) *A Revised Ground-Motion and Intensity Interpolation Scheme for ShakeMap.* Bulletin of the Seismological Society of America, Vol. 100, No. 6, pp. 3083–3096, December.

US Geological Survey (2010) Shakemap us2010rja6. See http://earthquake.usgs.gov/earthquakes/shakemap/global/shake/2010rja6/.

Earthquake Design Practice for Buildings
ISBN 978-0-7277-5794-4

ICE Publishing: All rights reserved
http://dx.doi.org/10.1680/edpb.57944.011

Chapter 2
Earthquake hazard

Earthquakes are the only remaining type of great, naturally occurring catastrophe that cannot be predicted.

McGuire (2004)

This chapter covers the following topics.

- The hazards that earthquakes give rise to.
- The principal factors in assessing ground motion.
- Influences on ground motion.
- Means of describing ground motion.
- Design ground motions.

The chapter is intended to give structural engineers unfamiliar with engineering seismology sufficient background on ground motion issues to inform their structural designs, and to give references to further sources of information.

2.1.   The hazards that earthquakes give rise to

The potential for a large earthquake to damage buildings and other structures comes in the first instance from the violent shaking of the ground. This gives rise to inertia forces, which are the primary source of potential damage that earthquake engineers must deal with, and which are the main concern of this book.

However, inertia effects in structures are not the only source of seismic hazard affecting structures, and the other main hazards associated with earthquakes are now briefly discussed.

- Structures that straddle the causative fault of an earthquake will be damaged if the fault breaks up to the surface (Figure 2.1). However, the number of buildings at risk is always much less than those affected by the ground motion; shaking damage may occur tens or even hundreds of kilometres from the fault trace, but only structures directly on the fault can be affected by its movement. For example, about 100 000 buildings collapsed or were severely damaged in the Kocaeli, Turkey earthquake of 1999, which occurred on a fault over 100 km long with displacements averaging 3–4 m, but of these buildings only about 100 directly straddled the fault break. However, long linear structures, such as viaducts or pipelines, are much more likely to be affected than buildings, because it is more likely that one part of them will straddle

11

Figure 2.1 A fault passed underneath the Shinkhang Dam, Taiwan, which moved in the earthquake of 1999, causing a vertical displacement of 9 m

9 m Vertical
fault displacement

a fault break. Design measures for pipelines crossing faults are discussed by Eurocode 8 (EC8) part 6 section 6.6 (CEN, 2004c), and by O'Rourke (2003).

■ The cyclic shear stresses induced in the soil by an earthquake may lead to soil liquefaction (the temporary loss of shear strength in loose, saturated, sandy soils), which in turn leads to foundation failure or landslides. Section 4.2 discusses liquefaction further.

■ Coastal sites may need to consider tsunamis triggered by large offshore earthquakes. Many lives were lost in tsunamis following the Indian Ocean earthquake of 2004 and the Tohoku Japan earthquake of 2011. A report by the National Tsunami Hazard Mitigation Program (NTHMP, 2001) sets out some general principles for mitigating tsunami risk while more detailed mitigation measures are contained in FEMA P646 (FEMA, 2008).

■ Fire following an earthquake has cost many lives in the past (Scawthorn, 2003).

■ Seismically induced failure of containment structures may lead to the release of noxious chemicals, flammable materials or radioactive materials. To date, the most significant release of radioactivity following an earthquake was in 2011 at the Fukushima power station in Japan (Weightman, 2011), primarily caused by the tsunami rather than the ground shaking that preceded it.

The remainder of this chapter concentrates on the primary hazard of strong ground motion, and how to describe it for the purposes of engineering design.

2.2. Earthquake basics
2.2.1 Earthquake sources
Figure 2.2 shows some of the principal terms used in describing an earthquake's location.

Figure 2.2 Describing an earthquake's location

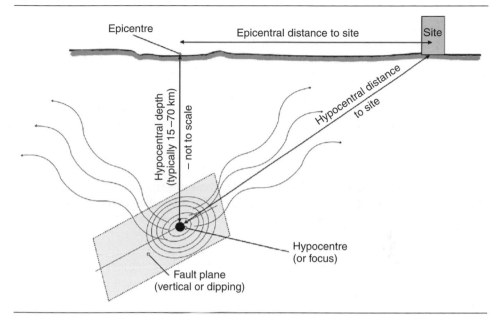

Earthquakes arise due to forces within the earth's crust tending to displace one mass of rock relative to another. When these forces reach a critical level, failure in the rock occurs along planes of weakness (fault planes) and a sudden movement occurs, which gives rise to violent motions at the earth's surface. The failure starts from a point on the fault plane called the hypocentre or focus, and propagates outwards until the forces in the rock are reduced to a level below the failure strength of the rock. The fault plane may be hundreds of kilometres long in large earthquakes, and tens of kilometres deep. In a large earthquake, the fault plane is likely to break up to the surface, but in smaller events it remains completely buried. A more complete description of the causes and types of earthquake is given by Bolt (2001).

2.2.2 Quantifying earthquakes
2.2.2.1 Earthquake magnitude and intensity
Magnitude and intensity are the two measures for quantifying an earthquake and its effects that are most commonly quoted. They are fundamentally different and it is important not to confuse them (although they often are). Earthquake magnitude is a fundamental property of the earthquake, related to its energy release on a logarithmic scale. By contrast, earthquake intensity describes the effects of the earthquake on the earth's surface, in terms of its effects on people and buildings. Unlike magnitude, the intensity of a given earthquake depends on the location at which it is measured; in general, the larger the epicentral distance (see Figure 2.2) the lower the intensity. Thus, a given magnitude of earthquake will give rise to many different intensities in the region it affects. It is important to recognise this fundamental distinction between the two measures, which are now discussed in more detail.

2.2.2.2 Earthquake magnitude

A number of different magnitude scales exist. Two common scales are the body wave magnitude M_b (suitable for measuring smaller magnitude events) and the surface wave magnitude M_s (suitable for larger events). Both are measured from sensitive instruments (seismometers), which detect ground tremors at great distances from the earthquake source. Currently, more favoured is a third scale M_w, the moment magnitude. This is directly related to the estimated energy release at the earthquake source and is suitable for all sizes of event. In broad terms, an earthquake with a magnitude less than 4 on any of the scales is unlikely to cause significant damage, while magnitudes larger than 8 are rare events affecting very large areas. Because of the logarithmic nature of the scale, a 1-point increase in magnitude represents a 30-fold increase in energy release and a 10-fold increase in fault displacement, although the violence of the accelerations at the epicentre increases much more slowly with magnitude. It is the spatial extent and long duration of large earthquakes that make great earthquakes so potentially damaging. Earthquakes larger than 9.5 are not found in practice because they would represent fault sizes larger than the dimensions of the earth's crust.

2.2.2.3 Earthquake intensity

Intensity scales are based primarily on reports of the felt and observed effects of an earthquake at a given position. As this is less precise than the measurement of magnitude, intensity values are described by Roman numerals. A number of different scales exist. Among these are the 12-point modified Mercalli intensity (MMI) scale, which is commonly used in the USA; intensity I is the lowest (not felt except by a few under especially favoured conditions), VII is the intensity at which some structural damage is likely and XII is the highest (total damage). The Japanese seismic intensity scale is similar in principle but is based on only 7 points; Bolt (2001) gives further details of both the Japanese and MMI scales. The European intensity scale (Grünthal, 1998), a development of the MSK scale, is a 12-point scale broadly similar to the MMI scale. It is more favoured in Europe, because it relates damage more precisely to the earthquake-resisting qualities of the damaged structures and to the frequency of damage.

2.2.2.4 Correlation between magnitude and intensity

Although magnitude and intensity are such different measures of an earthquake, they are of course related. Cua et al. (2010) provide a review of the relationships that have been derived between the magnitude of an earthquake, the intensity it gives rise to at a given site and the distance of the site from the earthquake's causative fault. The relationships are subject to a large amount of scatter.

Another useful type of relationship links magnitude to the area affected by at least a certain intensity level, for example, the area over which an earthquake is generally felt, corresponding to MMI IV (Table 3 in McGuire (2004) lists some of these relationships). They have been derived from earthquakes for which instrumentally determined magnitudes exist, and have subsequently been used to estimate the magnitude of earthquakes occurring in the period before seismometers existed. This can be done by deducing the felt area of the earthquake from contemporary reports in newspapers and other documents. This greatly extends the time period for which magnitude data are available.

Table 2.1 Average annual numbers of earthquakes worldwide during the twentieth century (data taken from US Geological Survey website)

Magnitude: M	Average number per year
8 and higher	1
7–7.9	15
6–6.9	134
5–5.9	1319
4–4.9	>10 000
3–3.9	>100 000

2.2.3 Occurrence of earthquakes

Table 2.1 shows the number of earthquakes that occur on average per year, as a function of magnitude. Of course, many of the earthquakes are remote from human populations and cause little if any damage.

Details of earthquakes are posted on a number of websites more or less as they occur worldwide; the British Geological Survey (www.earthquakes.bgs.ac.uk) and the US Geological Survey's National Earthquake Information Centre (earthquake.usgs.gov/regional/neic) are examples.

2.3. Earthquake probability and return periods

Almost anywhere in the world is thought to be susceptible to experiencing an earthquake of magnitude 6, which can give rise to very severe motions at its epicentre. However, in an area of low seismicity such as the UK, such an occurrence would be extremely rare, and it would be unreasonable to design against it, except perhaps for high-risk installations such as nuclear power stations. Therefore, at any rate in areas of low seismicity, something less than the 'maximum credible event' must be found for design, because a magnitude 6 event at a given site is very likely to be credible, although it may be extremely rare. The 'maximum credible' concept is more useful for sites near large active faults that break regularly within human timespans of tens or hundreds of years. Here, the worst that could occur may be much better defined, and may need checking. However, there is now a general consensus that a probabilistic rather than deterministic approach to defining earthquake hazard generally gives the most appropriate results for engineering design. The design earthquake motions are thus defined by their annual probability of exceedence P_1, or (equivalently) return period T.

P_1 is defined as the probability in any given year that ground motions of a given intensity will be exceeded. For example, in parts of California, there is a 2% annual probability that ground accelerations exceeding 0.25 g may occur, while in the UK the probability is likely to be nearer 0.01%. The return period T is then defined simply as

$$T = 1/P_1 \tag{2.1}$$

(Strictly speaking, the return period is the inverse of the annual frequency of exceedence, but the difference is small.) Often, the probability of exceedence P_y during a period of y years (i.e. greater than annual) is of most interest, where y years might represent the lifetime of a building. P_y, T and P_1 are related as follows

$$P_y = 1 - (1 - P_1)^y = 1 - (1 - 1/T)^y \qquad (2.2)$$

For example, a 475-year return period corresponds to a probability of exceedence in a 50-year building life of $1 - (1 - 1/475)^{50} = 10\%$. Therefore, if the 475-year return period peak ground acceleration (PGA) is estimated to be 25% g, there is a 10% chance that a larger PGA could be experienced at least once during any 50-year period. Generally, design return periods of the order of 500 or more years are used in earthquake-resistant design, rather than 50 years as used for many other environmental loads such as wind, because the 'tail' of the earthquake hazard distribution – the effect of rare but extremely damaging events – is generally much more significant for earthquakes than for wind. Therefore, relatively much rarer events must be considered, in order to reduce the risk of failure to levels comparable to those of other hazards. In low seismicity areas, this is particularly true, because of the almost limitless upper bound, discussed above, to the 'maximum credible event'.

2.4.   Performance objectives under earthquake loading

At any rate in principle, more than one level of performance needs to be considered in design. The US document FEMA P750 (FEMA, 2009) refers to four performance levels as listed below; FEMA P750 (NEHRP provisions) is significant because it effectively represents a draft of the earthquake provisions for the next revision of the US loading standard ASCE/SEI 7. Detailed definitions of the performance levels are given in the US standard for retrofitting existing buildings, ASCE/SEI 41-06 (ASCE, 2006), on which the following is based.

- Operational. Very light damage; all systems important to normal operation are functional.
- Immediate occupancy. Light damage; the structure substantially retains original strength and stiffness; elevators can be restarted and fire protection is operable, but some equipment and contents, although secure, may not operate.
- Life safety. Moderate damage; the structure retains some residual strength and stiffness, but the building may be beyond economic repair; many architectural, mechanical and electrical systems are damaged.
- Collapse prevention. Severe damage; building is near collapse but load-bearing elements continue to function; extensive damage to architectural, mechanical and electrical components.

The European seismic standard EC8 defines three performance levels, which are broadly similar to immediate occupancy, life safety and collapse prevention, but with different titles as follows. An 'operational' performance level is not defined. The definitions below are based on those in EC8 part 3 (CEN, 2004b).

- Damage limitation. Light structural damage; the cost of non-structural damage is small compared to the value of the structure.

- Significant damage. No global or local collapse of structure, which retains its structural integrity and a residual lateral strength and stiffness. Non-structural components are damaged but partitions and infills have not failed out of plane.
- Near collapse. Heavy damage; building is near collapse; little or no residual lateral strength.

Clearly, as the performance goal becomes less stringent (e.g. changing from operational to collapse prevention), the return period of the design earthquake motions can become longer. FEMA P750 (FEMA, 2009) sets indicative return periods for typical new buildings shown in Table 2.2, which also shows the corresponding return periods in EC8.

Definitions like these form the basis for 'performance based design', in which a performance goal under earthquake loading is associated with a probability of exceedence. In fact, limit state design, as practised in Europe for many decades, works to just the same general principle; see for example Table C.2 in EN 1990 (CEN, 2002), which defines target exceedence probabilities for the ultimate, serviceability and fatigue limit states. The specific features associated with seismic performance-based design, as developed in the USA over the past 20 years, is first the recognition of the increased importance of achieving satisfactory performance levels for non-structural elements (building contents, and so on) and of limiting repair costs, and second, the development of methods of analysis for checking that the performance levels are achieved in practice.

2.5. Representation of ground motion

2.5.1 Earthquake time histories

The earthquake intensity described previously gives a broad measure of the damaging power of an earthquake at a given location, but more precise (and less subjective) measures are usually required by engineers for the purposes of design calculations. The most precise description is given by a 'time history' of the motions at a given point. Time histories are measured by strong motion recording instruments set into action by the earthquake itself when the ground acceleration exceeds a preset threshold. Digitised records of earthquakes are freely available from a number of sources: the PEER strong motion database in California (peer.berkeley.edu/peer_ground_motion_database) and the European strong motion database (www.isesd.hi.is/ESD_Local/frameset.htm) are examples.

Figure 2.3 shows plots of horizontal acceleration, velocity and displacement against time for a Californian earthquake. The acceleration plot is a record obtained by a strong motion instrument and the other two plots have been obtained from it by integration. For structural computations, earthquake records with digitised values at intervals of 0.005 or 0.01 s are commonly used, as this will capture all the structural modes of vibration likely to be of interest.

2.5.2 Earthquake response spectra

The time history plots, although they contain a great deal of information, have two disadvantages. First, it is difficult to judge what the frequency content is, and hence how damaging the motion is for a structure with a particular natural period of vibration. Second, they are specific for a given time and place, and would not be exactly repeated even if a similar earthquake re-occurred. The earthquake response spectrum provides major advantages in both respects,

Table 2.2 Performance objectives and associated return periods in US and European seismic standards (data taken from FEMA P750 (FEMA, 2009) and CEN (2004b))

Performance goal	Return period			
	Ordinary buildings		Essential buildings (e.g. hospitals)	
	FEMA P750	EC8	FEMA P750	EC8
Operational	Not defined	Not considered	'Frequent'[a]	Not considered
Immediate occupancy/damage limitation	'Frequent'[a]	95 years (10% in 10 years)	400–900 years[b]	70–125 years[c]
Life safety/significant damage	400–900 years[b]	475 years (10% in 50 years)	2475 years (2% in 50 years)[d]	1000–1500 years[c]
Collapse prevention/near collapse	2475 years (2% in 50 years)[d]	2475 years (2% in 50 years)	Not defined	Not defined

[a] The return period corresponding to 'frequent' is not defined in FEMA P750
[b] The FEMA P750 life safety ground motions for these cases are defined as two-thirds of the 2475-year motions. The associated return period depends on the level of seismicity and will be lower for areas of high seismicity than for low to moderate seismicity
[c] The return periods for essential buildings are not explicitly stated by EC8; they depend on the parameters γ₁ and ν, which are used to factor the design ground motions. The associated return period depends on the level of seismicity; for damage limitation, it will be shorter for areas of high seismicity than for low to moderate seismicity, while for significant damage, the opposite is the case
[d] In high seismicity areas of the USA, the collapse prevention ground motions for ordinary buildings correspond to a 'maximum considered earthquake' calculated by deterministic methods; similarly for life safety in essential buildings

Figure 2.3 Time history plots for a record from the Northridge, California earthquake of 1994: (a) acceleration; (b) velocity; (c) displacement

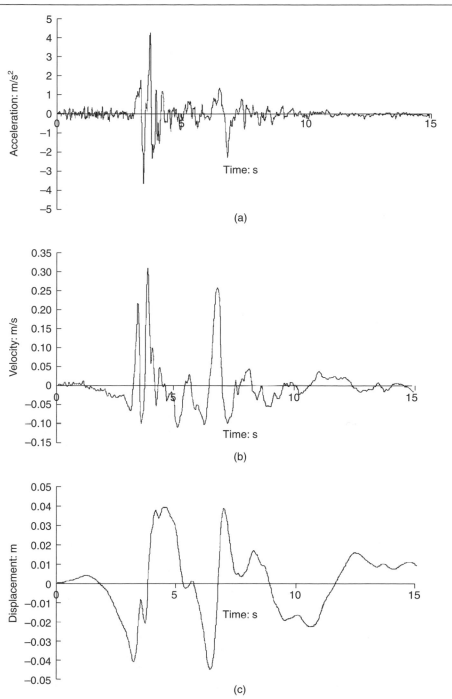

Figure 2.4 Response spectra for the Northridge, California earthquake ground motions of Figure 2.3

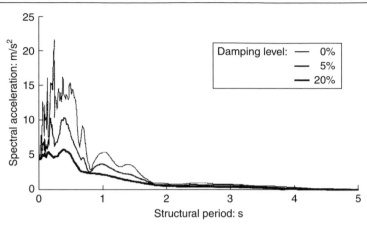

and represents a most useful tool for a design engineer to characterise earthquake motions. It represents the peak response of a linear elastic, single degree of freedom spring–mass–damper system to an earthquake, plotted against the structure's natural period. Continuous plots are drawn for each value of damping selected so that a response spectrum is represented by a family of curves, as illustrated in Figure 2.4.

Response spectra are discussed in more depth in Chapter 3. However, it is immediately clear from the peaks in the spectra shown in Figure 2.4 that the Northridge ground motions would have been particularly damaging to structures with a natural period in the range 0.1–0.4 s (representing low-rise construction up to approximately five storeys) with damping levels between 2% and 5% (a typical range for buildings responding at around their yield capacity). Increasing the damping level to 20% would have been highly effective in reducing the response in low-rise construction; this level of damping is a possibility with the addition of special damping elements, as discussed in Chapter 13. These observations would have been difficult to make from the time history traces of Figure 2.3.

A further advantage of a response spectrum is that by averaging and enveloping the spectra of several related time histories, thus smoothing the individual peaks and troughs, a representation of ground motion may be obtained that is more general than can be obtained from a particular time history (Figure 2.5). Such smoothed spectra are almost always the basis for spectra used in design.

Representing ground motions by their response spectra therefore overcomes two of the disadvantages of time history representations. Response spectra give direct information about structural response, and they can be generalised to cover the effects of a range of possible earthquakes at a given site. There is a corresponding disadvantage, though; a smoothed spectrum is quite unlike the spectrum of a real earthquake, and designing to a smoothed spectrum

Figure 2.5 Smoothed design spectrum

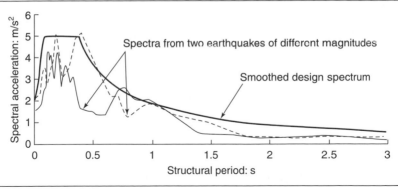

may imply designing for the simultaneous occurrence of two or more earthquakes, which in practice would never occur at the same time.

Other forms of spectral representation of ground motion are possible, in particular Fourier spectra and power spectral density plots. These have more specialist uses for the earthquake engineer, and the reader is referred to Mohraz and Sadek (2001).

2.5.3 Damaging capacity of ground motions

PGA is often quoted as the single figure representing the severity of a particular earthquake motion. Usually it is quoted as a percentage of the acceleration due to gravity; multiplying by 9.8 converts to m/s^2. In itself, PGA is rather unreliable as an indicator of damaging capacity because it may occur as only the briefest of transient values and (crucially) gives no information on the frequency content of the motions.

Peak velocity gives some information about longer period content within the ground motion, and appears to be a rather better predictor of structural damage (Bommer and Alarcon, 2006; Worden et al., 2012), although the correlation is still quite weak. A response spectrum, as discussed above, gives even more information about the effect of the motions on structures with varying structural periods and damping levels, although the duration of the motions, and hence number of loading cycles giving rise to fatigue effects, is not included.

Earthquake intensity, described in Section 2.2.2, is directly related to structural damage, and is often used by the insurance industry to estimate the likely overall damage to a large collection of buildings, for example, in a city (Spence et al., 2008). However, it is not really suitable for estimating the damage to a specific building; a single parameter such as intensity cannot by itself describe the amplitude, duration and frequency content of earthquake motions that together determine the extent to which a particular building is damaged.

2.5.4 Vertical, torsional and rocking components

For most structures the horizontal translation of earthquake ground motion has by far the greatest effect. However, four other components – one vertical translation and three rotations – also exist and may need to be explicitly taken into account in some cases.

2.5.4.1 Vertical motions

Vertical motions affect long span structures, because they may significantly change the bending and shear forces due to gravity loads. In particular, the effect of vertical motions on shear in reinforced concrete beams, and on compression in prestressed concrete beams, may need to be checked, because both failure modes are brittle and are particularly undesirable under seismic loading. Vertical motions may also reduce the effect of gravity loads in maintaining overall stability against lateral loads, and so should be allowed for in calculations of overall stability – for example, in retaining walls. Also, masonry structures and some soils rely for their shear strength on frictional resistance set up by gravity forces, so a reduction in gravity stress weakens shear resistance.

In the past, the customary assumption was that the peak vertical acceleration was two-thirds of the peak horizontal acceleration and had a similar spectral distribution. However, this is a crude approximation; on firm ground near the epicentre of earthquakes, the vertical motions can be much greater than the horizontal ones in the short period range, while they become relatively insignificant far from the epicentre. The frequency contents of vertical and horizontal motions are also quite different. Seismic standards such as EC8 (CEN, 2004a) and ASCE 4-98 (ASCE, 1998) propose vertical spectra that allow for this.

The shape of the vertical spectra at soft soil sites may differ from that of the corresponding horizontal spectra for an additional reason. Vertical motions, and hence their spectra, are much less affected than horizontal motions by the presence of soft soils overlying bedrock (see Section 2.6). This is because the vertical compressive stiffness of a soft soil is usually much greater than its shear stiffness, and so vertical compressive waves pass through more or less unmodified, whereas shear waves are amplified.

2.5.4.2 Rotational ground motions

Rotational ground motions are likely to be even less important than vertical motions for building structures and are seldom allowed for. Rocking motions (i.e. rotation about a horizontal axis) may affect very tall slender structures, and EC8 part 6 (CEN, 2004c) requires this to be accounted for in tall masts and chimneys; EC8 supplies a suitable rocking spectrum. Torsional ground motions (i.e. rotation about a vertical axis) in themselves are very unlikely to cause significant effects. However, the couple torsional-horizontal response of torsionally unbalanced structures can be very damaging. As these are triggered by horizontal (translational) rather than rotational motions, the latter are not important when analysing the torsional response of structures.

2.6. Site effects

It has long been observed that buildings (particularly tall ones) founded on poor soil generally perform much worse in earthquakes than those founded on hard soil or rock. A notable example occurred during the Mexican earthquake of 1985. This magnitude 8.1 event had its epicentre just off the Pacific Coast, and by the time the earthquake motions had travelled around 300 km eastwards to Mexico City, they had attenuated to such an extent that they were not particularly damaging. However, the centre of Mexico City is built on 30 m of very soft clay, which acted as a resonator for the motions, and amplified them in the same way as would a wobbling jelly. A high proportion of buildings, particularly

Figure 2.6 Five per cent damped spectra for Mexico City, 1985

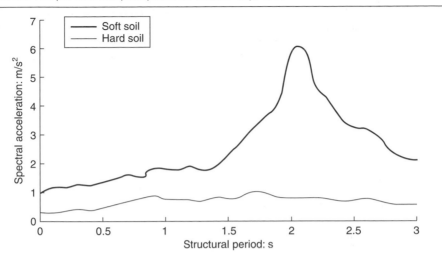

of 10–20 storeys, were severely damaged in this soft soil region, while very few were damaged on firm ground.

Figure 2.6 compares the response spectrum of motions on hard ground at Mexico City with those on the soft soil. The figure clearly shows the transformation of essentially harmless motions into ones that were particularly damaging to structures with a period of around 2 s.

Mexico City is an extreme example; the surface soils there are unusually soft, possess little damping but relatively high strength and there is a large contrast in stiffness with the material below the soft clays. However, some degree of amplification can almost always be expected in the presence of soft soils, and it must be accounted for in design. Figure 2.7 shows the

Figure 2.7 Elastic ground response spectra in Eurocode 8

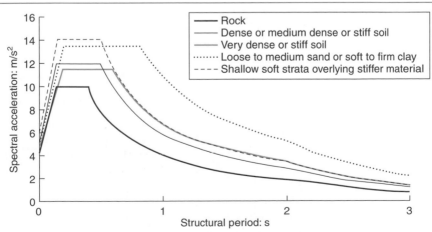

design response spectra for different soil types given in EC8 (CEN, 2004a). It can be seen that there can be a particularly significant effect for structural periods around 0.7–1 s, which corresponds to buildings of approximately five to 10 storeys. As people tend to settle on soft alluvial basins, because such basins are fertile and can support large populations, the problem is clearly a widespread one for human populations.

Section 4.3 provides further discussion of site effects.

2.7.   Quantifying the risk from earthquakes

Before addressing the topic of what ground motions a structural engineer needs to consider for the design of a particular structure, which is the subject of the next section, the means of quantifying seismic risk must first be addressed. These are seismological issues that are only touched on in what follows. Musson (2012) gives a much more comprehensive (and highly readable) account by a distinguished seismologist of the issues involved.

2.7.1   World seismicity

Much of the earth's seismic activity is concentrated into two active zones (Figure 2.8). These are the 'ring of fire' around the edge of the Pacific Ocean and a broad Alpide belt, which stretches across the Mediterranean and southern Europe to the Himalayas and into eastern China. Between them, these two areas account for approximately 97% of the world's largest earthquakes (80% in the ring of fire and 17% in the Alpide belt). These two belts, and the other areas of high seismic activity around the world, occur along the junctions between the tectonic plates, which together form the strong but brittle outer skin of the earth called the lithosphere. However, there is also significant distributed earthquake activity remote from the plate boundaries that cannot be neglected. England *et al.* (2011) have concluded that 'the greatest risk to human life from earthquakes lies not at plate boundaries, but in the

Figure 2.8 World seismicity, 1900–2010 (modified from US Geological Survey worldwide seismicity map, Department of the Interior/USGS, USA; material in the public domain)

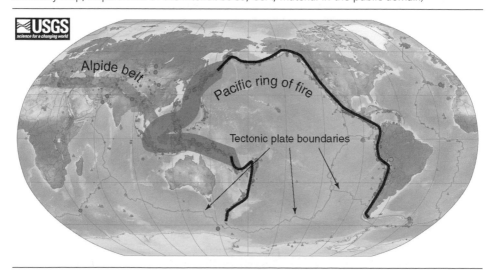

continental interiors, where growing populations are exposed to earthquake risk from distributed networks of faults that are poorly characterized'.

2.7.2 Probabilistic estimates of seismic hazard

For design purposes, as discussed in Section 2.4, earthquake engineers need to know not only the intensity of seismic ground motions at a site, but also an associated return period; this often needs to be around 500 years or longer. The process of obtaining this information is called probabilistic hazard assessment. It is carried out by performing a statistical analysis of the locations and magnitudes of earthquakes in the region surrounding a site, combined with knowledge of how ground motions attenuate with distance. The methods involved are not dealt with here, but are described in a definitive monograph by McGuire (2004).

One common method of representing seismic hazard is by a response spectrum with a specified return period, which can be used directly for design purposes, as explained in Chapter 3. A response spectrum plotted so that the return period of the spectral values is equal for all structural periods is called a uniform hazard spectrum (UHS). One issue with the UHS is that many structures – for example tall buildings – have significant responses at more than one period. The UHS provides the required return period for the response at any single period, but the joint probability of the responses it gives at two or more periods occurring simultaneously is less – in some cases, much less. A common example is when the short period response is due to a smaller magnitude local earthquake, and the long period response is due to a larger magnitude distant earthquake. It is very unlikely that both earthquakes will occur at the same time and so design to the UHS would be conservative. Abrahamson (2006) discusses a method of disaggregating the UHS into two or more 'controlling scenarios'; for instance, these would correspond to the small local earthquake and the large distant one in the example given previously. Structural design would analyse for response to all the identified controlling scenarios and take the worst case. An ATC report (NEHRP Consultants Joint Venture, 2011) provides further advice.

Another representation of hazard takes the form of a plot of a ground motion parameter such as PGA against return period (Figure 2.9).

Probabilistic seismic hazard assessments depend on the availability of reliable catalogues of earthquakes. As the return periods of interest are usually hundreds or even thousands of years, instrumental records of earthquakes, which are not available for earthquakes occurring before the start of the twentieth century, may be insufficient, particularly in areas of low seismicity where the hazard may be dominated by very rare events. To fill this gap, reports of earthquakes in newspapers, monastic records and other historical sources can be useful. The severity and extent of the damage can be used to estimate the location and magnitude of the earthquakes that caused it (see Section 2.2.2.4). Such historical reports go back to periods as early as 1200 BC in some parts of the world and form an important part of our knowledge of seismicity. Geological studies also provide important clues to past activity, especially when earthquake sources are shallow and faults can be identified on the earth's surface.

In assessing the seismicity of a site, all the available information needs to be considered. Once this has been done, it should be remembered that any prediction of seismicity remains

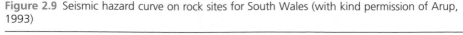

Figure 2.9 Seismic hazard curve on rock sites for South Wales (with kind permission of Arup, 1993)

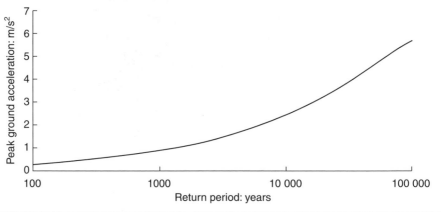

an estimate with a substantial degree of uncertainty. This is illustrated by Figure 2.10, which shows estimates of the seismic hazard in Haiti made before and after the earthquake of 2010, which devastated the capital, Port-au-Prince. The pre-earthquake estimate is for a 475-year return period PGA in the capital of approximately 16% g; this more than doubles in the estimate made after the earthquake.

2.7.3 Published sources of ground motion hazard

Performing a seismic hazard assessment for a site is a lengthy exercise requiring specialist expertise. In many cases, however, this exercise is not required because the site will be covered by a seismic hazard map in a seismic code of practice. For example, the 32 member countries of the European Committee for Standardisation provide seismic hazard maps within their respective national annexes to EC8 (CEN, 2004a), and countries such as the USA, New Zealand, China and Japan provide similar maps. These maps are all relatively recent but the maps in some codes are less reliable because they are out of date, although the level of hazard indicated may remain a statutory minimum obligation for design purposes.

When hazard information from a code is unobtainable or potentially unreliable, recourse may be made to the many published seismic hazard maps that are freely available from the web. In 1999, the Global Seismic Hazard Assessment Program (GSHAP) published maps of 475-year PGA for the whole world on http://seismo.ethz.ch/GSHAP, although some of this information is rather outdated; see for example Figure 2.10. An international exercise arising from GSHAP, called SESAME, published a seismic hazard map for Europe and the Mediterranean (European Seismological Commission, 2003); the map is generally considered more reliable than GSHAP for those parts. In 2009, the Global Earthquake Model (GEM) Foundation (www.globalquakemodel.org) was established, and in a major, well-funded, international effort is developing state-of-the-art tools and models for a global approach to determining earthquake hazard and risk. GEM intends to provide free access to comprehensive measures of seismic hazard for the whole world, including PGA at various return periods. For Europe, the SHARE project (www.share-eu.org), which is part of GEM, provides free access to a very

Figure 2.10 Changing seismic hazard estimates for Haiti. Both estimates are for the 475-year return period PGA on rock ((a) courtesy of GSHAP; (b) modified from Frankel *et al.*, 2011, U.S. Department of the Interior/USGS, U.S., material in the public domain)

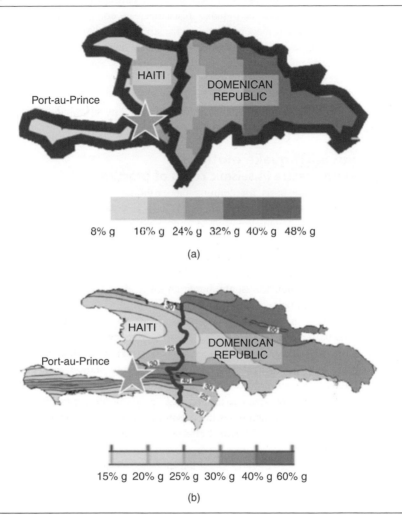

wide range of seismic hazard parameters for Europe on the website www.efehr.org. In addition to these global initiatives, journals such as *Earthquake Spectra* contain many papers on regional seismic hazard studies (for example, Frankel *et al.*, 2011, cited in Figure 2.10), which can provide useful information.

2.7.4 Areas of low seismicity

The public is, naturally enough, greatly concerned with areas of high seismicity. However, in many areas of low seismicity the problem can be dismissed too readily. Few areas of the world are free from earthquakes altogether; even a low seismicity area such as the UK can have a significant risk at very long return periods, as Figure 2.9 suggests.

A normal well-constructed building, designed for moderate wind forces, should be able to resist minor ground shaking reasonably well. However, structures that are unusually flexible or exceptionally brittle may be sensitive to small nearby earthquakes or more distant larger earthquakes. The importance of earthquake resistance, in comparison to other external hazards such as wind or flood, becomes greater for structures of great importance or when the consequences of failure are especially serious, such as nuclear reactors. This is because very long return periods must be considered in design and this has a proportionately greater effect for earthquakes than for other external hazards.

Further discussion of the issues of design for low and very low seismicity areas is given by Booth and Skipp (2008) and PD6698 (BSI, 2009).

2.8. Design earthquake motions

2.8.1 Response spectra in seismic codes of practice

Response spectra form the basis for defining the earthquake motions in current codes of practice. In most cases, these will form the main basis for design, although they may be supplemented by the special studies described in the next section.

Code spectra always take the form of smoothed spectra intended to envelope the range of conditions for which they are specified (Figure 2.5), and are supplied for a range of standard soil profiles (Figure 2.7). They are usually based on UHS, described earlier in Section 2.7.2. PGA are usually supplied in seismic zoning maps attached to the codes; the PGA values usually relate to values on rock and give a measure of the underlying seismicity of the region, without including the influence of local soil deposits. The global seismic hazard maps produced by GSHAP and SESAME (see Section 2.7.3) are of this form. The standard procedure to define a code response spectrum is therefore to choose an appropriate PGA for the site and a standard soil profile most appropriate for the site. The design spectrum is then found by multiplying the standard normalised shape by an appropriate PGA. This is the system used by EC8, which also gives the option of allowing for the presence of topographical effects (Section 4.3.4) as well as soil effects.

By contrast, US seismic codes after 1997 no longer based design spectra on PGA. Instead, two parameters are used, namely peak spectral acceleration at short period (S_s) and spectral acceleration at 1 s period (S_1). This two-parameter system allows spectral shapes to reflect varying regional geology and seismology much better than the use of a single parameter, namely PGA. Future revisions of EC8, and perhaps other codes, may well follow the Americans and adopt a multiparameter system for specifying hazard.

2.8.2 Response spectra from special studies

In some circumstances, special studies may be carried out to define a response spectrum for a particular site. Such studies are often performed for high-risk installations such as nuclear power stations, and are specified in some codes (including EC8 and the US code ASCE 7) for sites where the soils may be unusually prone to amplifying ground motions, such as those at Mexico City.

A number of possibilities exist. One such is when the fundamental seismicity of a site is uncertain, either because of the lack of a reliable seismic zoning map for the region or because such zoning maps that do exist relate to different return periods from those required. In this case, a site-specific hazard assessment (Section 2.7.2) would be carried out, in order to obtain a value for the PGA (for EC8) or S_s and S_1 (for US codes – see previous section) on rock at the site. Provided that the soil profile at the site is well known and well characterised in the code, a design spectrum can then be obtained directly from the code; this procedure allows both regional seismicity and site effects to be accounted for.

Alternatively, the regional seismicity may be well specified, but the effects of the overlying soil may not be adequately allowed for in codes of practice. In this case, the earthquake motions found in a notional rocky outcrop near the site would be established from the code, and these could then be modified to allow for the overlying soils, using the analytical methods described in Chapter 4.

Of course, if neither regional seismicity nor the overlying effects of the soils were well established, both procedures would be necessary. That is, the rock motions at the site would first have to be found, and then the effect of overlying soils allowed for.

2.8.3 Earthquake time histories for design purposes

For most projects, an earthquake response spectrum is likely to form the initial basis for design ground motions. However, response spectra provide no information about the duration of the motion. This is important when significant changes in soil or structural properties take place with time under repeated cyclic loading. To account for this, and to allow more generally for non-linear effects, earthquake time histories are required, in order to carry out a 'time history' analysis. There are a number of ways of deriving design time histories for a particular site that correspond to the design spectrum specified in a code, and these methods are described briefly below; a more extensive and authoritative review is provided by NEHRP Consultants Joint Venture (2011).

The most direct way is to select the records of real earthquakes with a magnitude and distance appropriate to the seismic hazard at the site. For sites where there is significant amplification of the ground motions in the top layers of soil, it is usually most satisfactory to choose appropriate records taken for rock sites, and then modify them analytically for the soil, as described in Chapter 4. Codes of practice such as EC8 require that at least three records be used (although in practice many more are usually needed to obtain reliable results), and that the average of the response spectra of the individual records should envelope the design response spectrum for the site given in the code. It is likely that some or all of the records would need to be factored in order to achieve this at all structural periods of interest. Even with such factoring, it may be difficult to find a set of records that envelopes the code spectrum without being unduly conservative at some structural periods.

A second technique, based on stochastic methods, is to generate a set of 'artificial' time histories for which the average spectrum envelopes a specified design spectrum. A number of software packages exist to perform this task (for example, SIMQKE, 1976) and it is easily possible to produce a set of three or four statistically independent records that meet the requirements of a code such as EC8 or (more stringently and completely) the ASCE standard ASCE 4-98 (ASCE, 1998). However, such methods are not currently favoured; the problem is that the records they produce tend to look quite unlike those of real earthquakes, and in particular tend to have many more damaging cycles. This becomes important for cases in which every cycle of loading causes cumulative damage or movement; examples are the onset of liquefaction in soils, the sliding or overturning of retained walls and low cycle fatigue effects generally in structures. Also, these methods do not model the special characteristics of ground motions near the causative fault of an earthquake (near-fault effects), described in Section 4.3.5.

More realistic time histories may be obtained by choosing representative time histories, as described in the previous paragraph, and then using the freeware RSPMatch (Hancock *et al.*, 2006) to modify the records, so that they match the design spectrum more closely. The program may be freely run from http://nees.org/resources/tools/.

A third way has been developed, which is particularly useful for low seismicity countries such as the UK, where there are almost no ground motion records of damaging earthquakes at all. In this method, sensitive instruments are used to record the accelerations at the chosen site of many small earthquakes. As the number of earthquakes occurring increases exponentially with decreasing magnitude, a sufficient number of records can be acquired over a period of a few years even in the UK. The small earthquakes represent breaks over small areas of a fault, which are accurately located. By combining a large number of these records using a 'Green's function' technique, the effect of the break of a much larger area of fault, and hence much larger earthquake can be simulated. The technique is further discussed by Erdik and Durukal (2003), who also present other methods of deriving time histories directly from geophysical parameters.

REFERENCES

Abrahamson N (2006) Seismic hazard assessment: problems with current practice and future developments. Keynote address: *First European Conference on Earthquake Engineering and Seismology, Geneva, Switzerland.*

ASCE (1998) ASCE 4-98: Seismic analysis of safety-related nuclear structures and commentary. American Society of Civil Engineers, Reston, VA, USA. A revision is expected in 2014.

ASCE (2006) ASCE/SEI 41-06: Seismic rehabilitation of existing buildings. American Society of Civil Engineers, Reston, VA, USA.

Bolt B (2001) The nature of earthquake ground motion. In *The seismic design handbook (2nd edition)* (Naeim F (ed)). Kluwer Academic Publishers, Boston, MA, USA.

Bommer JJ and Alarcon JE (2006) The prediction and use of peak ground velocity. *Journal of Earthquake Engineering* **10**(1): 1–31.

Booth E and Skipp B (2008) *Establishing the necessity for seismic design in the UK*. Research report for the Institution of Civil Engineers, London, UK.

BSI (2009) PD6698: 2009. Recommendations for the design of structures for earthquake resistance to BS EN 1998. BSI, Chiswick, UK.

CEN (2002) EN 1990: 2002: Eurocode – basis of design. European Committee for Standardisation, Brussels, Belgium.

CEN (2004a) EN 1998-1: 2004: Design of structures for earthquake resistance. Part 1: General rules, seismic actions and rules for buildings. European Committee for Standardisation, Brussels, Belgium.

CEN (2004b) EN 1998-3: 2004: Design of structures for earthquake resistance. Part 3: Seismic assessment and retrofitting of existing buildings. European Committee for Standardisation, Brussels, Belgium.

CEN (2004c) EN 1998-4: 2004. Design of structures for earthquake resistance. Part 6: Towers, masts and chimneys. European Committee for Standardisation, Brussels, Belgium.

CEN (2004d) EN 1998-5: 2004: Design of structures for earthquake resistance. Part 5: Foundations, retaining structures and geotechnical aspects. European Committee for Standardisation, Brussels, Belgium.

Cua G, Wald D, Allen T *et al.* (2010) *"Best Practices" for Using Macroseismic Intensity and Ground Motion Intensity Conversion Equations for Hazard and Loss Models in GEM1.* GEM Technical Report 2010-4. See www.globalquakemodel.org/system/files/doc/GEM-TechnicalReport_2010-4.pdf.

England P, Holmes J, Jackson J and Parsons B (2011) *What works and does not work in the science and social science of earthquake vulnerability?* Report of an International Workshop held in the Department of Earth Sciences, University of Oxford, 28 and 29 January, 2011.

Erdik M and Durukal E (2003) Simulation modelling of strong ground motions. In *Earthquake Engineering Handbook.* Chen W-F and Scawthorn C (eds). CRC Press, Boca Raton, FL, USA.

European Seismological Commission (2003) *European–Mediterranean Seismic Hazard Map.* Giardini D, Jiménez MJ and Grünthal G (eds). See http://wija.ija.csic.es/gt/earthquakes/.

FEMA (Federal Emergency Management Agency) (2008) P646: *Guidelines for Design of Structures for Vertical Evacuation from Tsunamis.* Federal Emergency Management Agency, Washington, DC. See www.fema.gov/library/viewRecord.do?id=3463.

FEMA (2009) P750: *NEHRP recommended provisions for seismic regulations for new buildings and other structures.* Prepared by the Buildings Seismic Safety Council for the Federal Emergency Management Agency, Washington, DC, USA.

Frankel A, Harmsen S, Mueller C, Calais E and Haase J (2011) *Documentation for initial seismic hazard maps for Haiti: U.S. Geological Survey Open-File Report 2010-1067.* U.S. Department of the Interior/USGS, U.S. http://pubs.usgs.gov/of/2010/1067/ [accessed 27/09/2013].

Grünthal G (ed.) (1998) *European Macroseismic Scale, 1998.* Cahiers du Centre Européen de Géodynamique et de Séismologie, Vol 7. European Seismological Commission, Luxembourg.

Hancock J, Watson-Lamprey J, Abrahamson N *et al.* (2006) An improved method of matching response spectra of recorded earthquake ground motion using wavelets. *Journal of Earthquake Engineering* **10** (Suppl. 01): 67–89.

McGuire R (2004) *Seismic hazard and risk analysis.* Earthquake Engineering Research Institute, Oakland, CA, USA.

Mohraz B and Sadek F (2001) Earthquake ground motions and response spectra. In *The seismic design handbook.* Naeim F (ed.). Kluwer Academic Publishers, Boston, USA.

Musson R (2012) *The million death quake – the science of predicting earth's deadliest natural disaster.* Palgrave Macmillan, New York, NY, USA.

NEHRP Consultants Joint Venture (2011) *NIST GCR 11-917-15 – selecting and scaling earthquake ground motions for performing response-history analyses.* US National Institute for Standards and Technology, Gaithersburg, MD, USA. See www.nehrp.gov/pdf/nistgcr11-917-15.pdf.

NTHMP (National Tsunami Hazard Mitigation Program) (2001) *Designing for tsunamis – seven principles for planning and designing for tsunami hazards.* NTHMP. See http://nthmp.tsunami.gov.

O'Rourke MJ (2003) Buried pipelines. In *Earthquake Engineering Handbook.* Chen W-F and Scawthorn C (eds). CRC Press, Boca Raton, FL, USA.

Scawthorn C (2003) Fire following earthquakes. In *Earthquake Engineering Handbook*. Chen W-F and Scawthorn C (eds). CRC Press, Boca Raton, FL, USA.

SIMQKE (1976) *I: A program for artificial motion generation*. Department of Civil Engineering, Massachusetts Institute of Technology (distributed by NISEE, University of California, Berkeley, http://nisee.berkeley.edu/).

Spence R, So E, Jenny S *et al.* (2008) The Global Earthquake Vulnerability Estimation System (GEVES): an approach for earthquake risk assessment for insurance applications. *Bulletin of Earthquake Engineering* **6**: 463–483.

Weightman M (2011) *Japanese earthquake and tsunami: Implications for the UK nuclear industry.* Final report by HM Chief Inspector of Nuclear Installations. Health and Safety Executive, Bootle, UK. See www.hse.gov.uk/nuclear/fukushima/final-report.pdf.

Worden CB, Gerstenberger MC, Rhoades DA and Wald DJ (2012) Probabilistic relationships between ground-motion parameters and Modified Mercalli Intensity in California. *Bulletin of the Seismological Society of America* **102**(1): 204–221.

Earthquake Design Practice for Buildings
ISBN 978-0-7277-5794-4

ICE Publishing: All rights reserved
http://dx.doi.org/10.1680/edpb.57944.033

Chapter 3
The calculation of structural response

> The myth, then, was that refinement of the analysis process improved the end result.
>
> Priestley (2003)

This chapter covers the following topics.

- Stiffness, mass, damping and resonance.
- Response spectrum analysis.
- Linear and non-linear time history analysis.
- Equivalent static analysis.
- Non-linear static (pushover) analysis.
- Capacity design.
- Displacement and force-based analyses.

Seismic engineers need to have a good understanding of the dynamic, non-linear response of structures to earthquake loading, and this chapter aims to provide a mainly qualitative introduction to the subject and to the range of analytical techniques that are available for calculating structural response.

3.1.   Introduction

Earthquake loading poses the structural analyst with one of the most challenging problems in engineering. A violent and essential unpredictable dynamic ground motion imposes extreme cyclic loads on engineering materials whose response under such conditions is complex and incompletely understood. If this is the case, engineering designers for whom this book is written may wonder whether there is any point in their getting to grips with the complex underlying theory of dynamic seismic analysis. In fact, current methods of analysis provide important insights into the way that structures respond to earthquakes, and hence the ways in which designers can control this response. Moreover, a basic understanding of analytical principles is essential to enable an informed and critical use to be made of computer-generated results, which form the basis for most seismic analysis and design. Therefore in the author's view earthquake engineers must make the effort to understand the basics of dynamic seismic analysis. To obtain a thorough understanding of the subject, reference needs to be made to some of the classic texts on the subject, for example Clough and Penzien (1993) and Chopra (2012), which are, however, lengthy and require considerable effort.

This chapter does not attempt to reproduce this literature; its aim is to give the reader unfamiliar with the subject an outline of the most important principles and analytical methods and some help with the jargon employed. It is worth repeating that despite its sophistication and fascination, analysis for seismic response gives results that are almost always beset with very large uncertainties, both in the input motions and the structural response to them. The analysis is only a stage in the design process, and pages of computer output or complex mathematics should never be used as a replacement for sound engineering judgement.

3.2. Basic principles of seismic analysis

Seismic forces in a structure do not arise from externally applied loads. They are therefore different from more familiar effects such as wind loads, which are caused by external pressures and suctions on a structure. Instead, response is the result of cyclic motions at the base of the structure causing accelerations and hence inertia forces. The response is therefore essentially dynamic in nature and the dynamic properties of the structure, and in particular natural period and damping, are crucial in determining that response. Any seismic analysis, if it is to be at all realistic, must allow for this dynamic character, even if it is only in a simplified way.

The dynamic nature of the response is clearly a complicating factor, but there is a further analytical difficulty. Most engineered buildings are designed to withstand extreme earthquakes by yielding substantially, and the designer must therefore have some understanding of the non-linear dynamic response of structures under extreme cyclic excitation. In principle, this poses very complex analytical problems. In practice, a combination of highly simplified analytical methods and appropriate design and detailing are often sufficient to secure satisfactory behaviour. However, it is essential to understand the basis and limitations of such techniques. Subsequent sections describe some basic principles of dynamic analysis, initially based on essentially linear elastic methods using earthquake response spectra. The implications for the analyst/designer of post-yield response are then introduced, and the final section reviews the main analytical techniques used for buildings, including the displacement-based methods of analysis and design, which are gaining increasing prominence.

3.2.1 Resonance

Everyone has experienced the phenomenon of resonance in some form, for example the juddering that only occurs at a particular speed when driving a car with an unbalanced wheel. Resonance takes place when the period of excitation (in this example, the time for one revolution of the unbalanced wheel) matches the natural period of the structure (the period of the car body bouncing on its suspension).

Figure 3.1 shows the familiar curves for the steady-state response of a simple system subject to constant sinusoidal ground motion. The response is shown here in terms of peak acceleration of the system, divided by the peak ground acceleration (PGA) to give a normalised response; it shows a marked peak when the system period matches that of the input motion, causing resonance.

By contrast, very rigid systems with low periods track the ground motion closely. The normalised response therefore tends to unity as the system period tends to zero, or (equivalently) as the ground motion period becomes very long in comparison to the period of the structure.

Very flexible springs, on the other hand, act to isolate their masses from the input motion and so response tends to zero when the period of the structure is very long compared to that of the ground motion. In other words, response becomes small for very long-period structures or for very short period motions. This is the principle behind, for example, isolation mounts for rotating machinery and also (as discussed later) seismic isolation systems for earthquake-resistant buildings.

Figure 3.1 describes the steady-state response to constant amplitude single-period motions. By contrast, earthquakes are transient phenomena and the associated ground motions contain a range of periods. Nevertheless, certain periods tend to predominate, depending chiefly on the magnitude of the earthquake and the soil conditions at the site. The match between these predominant periods and the periods of a particular structure is crucial in determining its response. Figure 3.2 shows a response spectrum for a typical earthquake. Response spectra have already been introduced in Section 2.5.2 and are discussed in more detail in Section 3.3; the similarity in broad outline between Figures 3.1 and 3.2 is, however, immediately apparent. Thus, at zero period, the normalised response is unity. As the structural period increases, the trend (despite the spikiness for low levels of damping) is to increase to a maximum and then reduce to a level eventually approaching zero.

Predominant ground motion periods at a firm soil or rock site are typically in the range 0.2–0.4 s while periods can reach 2 s or more on very soft ground. As building structures have fundamental periods of approximately $0.1 N$ (where N is the number of storeys), it can be seen that resonant amplification may well take place in common ranges of building height.

3.2.2 Damping

When the cyclic excitation on a structure ceases, its response tends to die away. This is the phenomenon known as damping. Figures 3.1 and 3.2 show that the level of damping has an influence on response that may be as important as structural period.

If the damping is assumed to be 'viscous', that is, the damping force varies as the velocity of the system relative to the ground, the mathematics become reasonably easy to solve. For this reason, the assumption of viscous damping is often adopted in analysis, although practical mechanisms of damping in buildings often follow somewhat different patterns, as discussed later. Viscous damping is usually expressed in terms of the percentage of critical damping ξ, where $\xi = 100\%$ (critical damping) is the lowest level at which a system disturbed from rest returns to equilibrium without oscillation (Figure 3.3). The percentage reduction between successive peaks in a cycle is approximately $2\pi\xi\%$, for small values of ξ; thus in Figure 3.3, the 5% damped system reduces from an initial displacement of 1 to $(1-0.05 \times 2\pi)$ or about 0.7 after one cycle. Note that 5–7% damping represents the upper bound of damping found in most building structures responding at or around their yield point, while 20% damping represents an achievable level with the introduction of specially engineered damping devices.

It can be shown that for a sinusoidal excitation, ξ is related to the ratio of energy dissipated by damping per cycle to the peak elastic strain energy stored (Figure 3.4), a useful result for appreciating the physical significance of ξ.

Figure 3.1 Steady-state response to sinusoidal ground motion

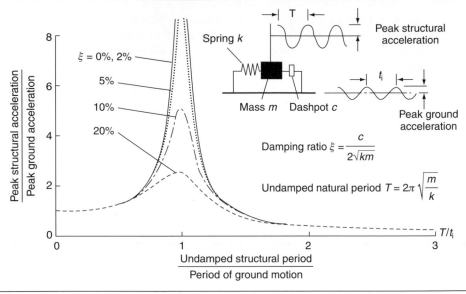

A well-known text book result is that peak steady-state response at resonance under single-period excitation, as shown in Figure 3.1, is approximately $1/(2\xi)$ times the input motion. Therefore, the resonant response becomes infinite as the damping falls to zero. For the transient condition of an earthquake excitation Figure 3.2 shows a lower level of amplification

Figure 3.2 Acceleration response spectrum for El Centro earthquake of 1940

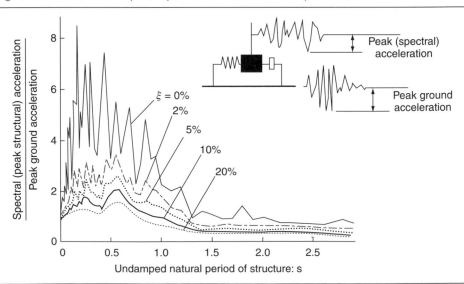

Figure 3.3 Effect of viscous damping level on the decay of free vibrations

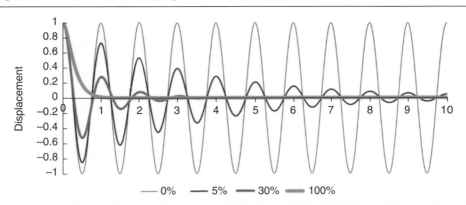

at resonance; typical ratios of peak response to input are 2.5–3 for $\xi = 5\%$ (compared with 10 in Figure 3.1 for constant sinusoidal excitation) and 5–8 for $\xi = 0\%$ (compared with infinity for Figure 3.1). At or near resonance, therefore, earthquake response is less sensitive to the damping level than steady-state sinusoidal response. However, comparison of Figures 3.1

Figure 3.4 Energy stored and dissipated in (a) undamped and (b) damped systems

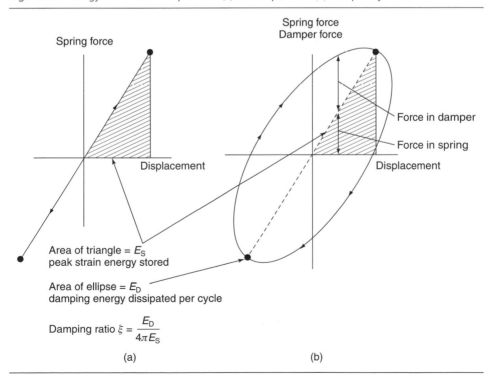

and 3.2 shows that under earthquake loading, response is more dependent on damping at periods away from resonance than is the case for single-period excitations.

Damping in buildings arises from a variety of causes, including aerodynamic drag (usually small), friction in connections and cladding (typically approximately 1% at amplitudes well below that corresponding to yield), damping associated with the soil and foundations (important in modes of vibration involving large soil deformation), and bond slip and cracking in reinforced concrete. These sources of damping predominate when stresses are generally below yield. Plastic yielding gives rise to a different source of energy dissipation. Here, the damping energy is dissipated plastically as the structure cycles through hysteresis loops (Figure 3.4), rather than as a result of viscous drag; hence the damping is referred to as 'hysteretic' rather than 'viscous'. An important difference between the two is that in hysteretic damping the dissipated energy is proportional to peak displacement, while in viscous damping it is proportional to the displacement squared. Care is therefore needed when using viscous damping in the analysis of yielding structures; when the viscous damping value used is appropriate for elastic response, it may overestimate the energy dissipated at high amplitude, and so be unconservative.

3.2.3 Determining structural periods of buildings

As already discussed, the natural period and damping of a structure are the crucial parameters in determining its response to an earthquake ground motion. In the next two sub-sections, the determination of these two structural parameters is discussed.

The period of an undamped mass supported on a spring is given by

$$T = 2\pi\sqrt{m/k} \tag{3.1}$$

as shown in Figure 3.1. Doubling the mass therefore increases the period by approximately 40%, and the same is true if the stiffness is halved. More complex structures can have their natural periods determined from their mass and stiffness. A useful approximation for buildings with a regular distribution of mass and stiffness is

$$T \cong 2\sqrt{\delta} \text{ (seconds)} \tag{3.2}$$

where δ is the lateral deflection in metres of the top of the building when subjected to its gravity loads acting horizontally; see for example equation 4.9 of part 1 of the European seismic code, Eurocode 8 (EC8) (CEN, 2004). Many structural analysis programs exist that produce more exact answers, although they are always worth checking by simple means, including those discussed in the next paragraph.

Theoretically derived periods should always be treated with some caution. While the mass of a building structure may be reasonably easy to determine, its stiffness is usually much more uncertain. Non-structural elements such as cladding and partitions tend to add stiffness and thus decrease natural periods. Moreover, the stiffness and hence the period depend on the amplitude of response primarily because when the structure starts to yield, stiffness decreases; in the case of concrete structures, the reduction starts well before the effective yield point. An interesting paper by Ellis (1980) suggests that simple empirical formulae based on building height provide more accurate predictions of fundamental period than even quite sophisticated analyses. Most codes of practice including the European standard EC8 (CEN, 2004) and the US standard ASCE 7 (ASCE, 2010) provide empirical formulae of this kind. Moreover, the US standard requires that if the empirically derived period results in substantially higher seismic forces than those corresponding to an analytically derived period (i.e. brings the structure closer to resonance), then the forces based on the analytical period must be increased.

3.2.4   Determining damping level in buildings

Figures 3.1 and 3.2 show that the level of damping greatly influences response. However, unlike period, damping can only be determined empirically and measurements in buildings show the level varies over a large range in practice. It is found to be highly dependent on amplitude; for low to moderate levels of excitation (applicable to serviceability considerations) damping levels are generally in the range 1–2% of critical, while for levels of excitation with stresses approaching yield, damping may reach 3–10%. Concrete and masonry buildings tend to be at the higher end of the range, and steel at the lower end. Smith *et al.* (2010) provide a review of damping levels in tall buildings, based on field measurements. Figures for design purposes are given in ASCE 4 (ASCE, 1998).

A near-universal assumption is that damping in earthquake-excited buildings is 5%. Two provisos should be borne in mind when using this figure

1   It is only appropriate to severe earthquakes and would normally be unconservative for moderate events where yielding does not occur. Therefore, 5% damping may well be unconservative when checking for serviceability conditions.
2   In a ductile structure responding to the 'no-collapse' or 'life-safety' level event, the hysteretic damping is usually much greater than the viscous damping. Therefore, a change in viscous damping has only a small effect on response. For example, the maximum response of a structure achieving a displacement ductility of 4 typically only reduces by less than 10% if the viscous damping doubles from 2.5% to 5%. This is well within the margin of uncertainty and therefore differences in viscous damping between different types of building are neglected by codes such as EC8 when considering the response of buildings to events in which they are expected to yield substantially.

3.3.   Linear response spectrum analysis
3.3.1   Earthquake response spectra

Calculating the earthquake response, even of a simple structure idealised as a linear spring/mass/dashpot system (Figure 3.2) is complex. Response spectrum analysis (RSA) provides a much simpler method, often possible without the use of computers. This is done by restricting the calculation to the maximum response of the system during the earthquake, without having to calculate behaviour at other times. As the maximum response is usually the quantity of greatest engineering interest, this is both useful and convenient. The method relies on the previous calculation (using computers) of the maximum response of a series of simple systems with a range of periods from short to long and with various levels of damping.

The maxima (called spectral values) are then plotted against the natural period of the system to produce the response spectrum shown in Figure 3.2. Spectra can be plotted for spectral acceleration, velocity or displacement. Note that the damping is assumed to be viscous and not hysteretic.

It is easy to show that the spectral (i.e. maximum) response of all idealised linear systems with the same period and percentage of critical damping is the same for a given earthquake motion. Thus, a 10 t mass with 5% damping and 1 s period deflects and accelerates just as much as a 10 kg mass with the same damping and period when subjected to, say, the motions recorded during the El Centro earthquake of 1940. The response spectrum therefore becomes a powerful and versatile design tool. Knowing the mass, damping and period of a structure (providing it can be idealised as a simple linear spring/mass/dashpot system) and given an acceleration response spectrum, the following quantities of interest to the designer can be derived.

$$F = mS_a \tag{3.3}$$

where F is the peak spring force, m is the mass and S_a is the spectral acceleration.

$$S_d = F/k \tag{3.4}$$

where S_d is the spectral deflection, F is peak spring force and k is the spring stiffness.

Combining Equations 3.1, 3.3 and 3.4 gives

$$S_d = mS_a/(4\pi^2 M/T^2)$$
$$= S_a T^2/(4\pi^2) \tag{3.5}$$

Together, Equations 3.3 and 3.5 show that with an earthquake response spectrum, two of the quantities of most use to earthquake engineers – namely maximum force and deflection in a given earthquake – can be derived for a simple structure, provided its mass, natural period and damping are known. Figure 3.5 illustrates this with an example.

It should be noted that Equation 3.3 relates to the maximum spring force in a system, neglecting the damping force. Figure 3.3 shows that the total peak force due to spring and damper peaks just before the point of maximum displacement. Therefore, S_a is slightly less than the true peak acceleration during an earthquake, and strictly speaking is defined as the 'pseudo-spectral acceleration' calculated from the peak deflection such that (by rearranging Equation 3.5)

$$S_a = S_d(4\pi^2/T^2) \tag{3.6}$$

Note that the quantity (mass × pseudo-spectral acceleration) represents the peak spring force within the system. As the damping forces are in most cases fictitious quantities representing energy loss, it is the peak spring forces that are of most interest when assessing the structure's

Figure 3.5 Deriving peak force and deflection from a response spectrum

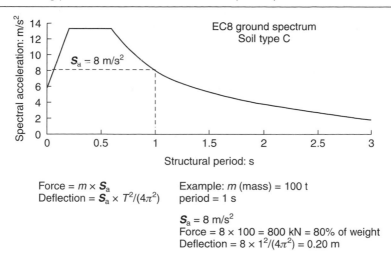

Force = $m \times S_a$

Deflection = $S_a \times T^2/(4\pi^2)$

Example: m (mass) = 100 t

period = 1 s

S_a = 8 m/s^2

Force = 8 × 100 = 800 kN = 80% of weight

Deflection = 8 × 1^2/(4π^2) = 0.20 m

requirement for strength. In any case, for low levels of damping ($\xi < 20\%$) the difference between pseudo-spectral and true spectral acceleration is very small. This is because the velocity is small at the time the true spectral acceleration peak occurs, and so the damping force (which is proportional to velocity) is also small. Usually, therefore, acceleration response spectra refer to pseudo-spectral accelerations. Of course, for undamped systems, there is no difference between pseudo and true quantities.

3.3.2 Smoothed design spectra

Each 'time history' of earthquake motions produces its own unique response spectrum, with a shape reflecting the frequency content of the motions. As explained in Chapter 2, in design, a smoothed enveloped spectrum is used (Figure 2.5), which irons out the spikes in response and effectively encompasses a range of different possible motions assessed for a particular site.

One other point to note is that the response is assumed to be linear, so doubling the input motions doubles the response. Response spectra can therefore be normalised – for example, with respect to PGA. If the shape of the spectrum is known and is independent of the ground motion intensity, the spectrum for a particular site can be found by factoring the normalised spectrum by the appropriate hazard value, for example 20% g PGA. The design spectra in EC8 are presented in this way, although the method has limitations for soft soil sites, as was discussed in Section 2.6.

3.3.3 Absolute and relative values

One common source of confusion in earthquake engineering relates to the fact that not only the structure, but also the ground, moves. Therefore, should motion be quoted relative to the ground or in absolute terms? It is particularly important to remember that spectral accelerations are always quoted as absolute values, whereas spectral velocities and displacements are relative values, being the difference in motion between the mass and the ground. It may help to remember that the forces in the spring and dashpot result, respectively, from the relative deflection and velocity, but the absolute acceleration of the mass equals the spring plus dashpot force divided by the mass.

3.3.4 Displacement spectra

All the response spectra presented so far have shown accelerations. These are of fundamental importance to the earthquake engineer because they relate to the maximum inertia (i.e. mass times acceleration) forces that develop during an earthquake and hence to the strength that a structure needs to resist those forces safely. However, spectra can also be drawn for maximum displacement; increasingly, displacement spectra are being regarded by earthquake engineers as of equal or greater importance than acceleration spectra. It might be expected that there would be a close relationship between the displacement and acceleration spectra of a given earthquake, and Equations 3.5 and 3.6 demonstrate that this is indeed the case. Given a displacement spectrum, a (pseudo-)acceleration can immediately be derived, and vice versa; the acceleration spectrum equals the displacement spectrum times $4\pi^2$ divided by the period squared. Figure 3.6 shows the acceleration and displacement spectra for a Californian earthquake. Displacement spectra have provided the basis for many of the insights connected with 'displacement-based design' (DBD) methods, discussed further in Section 3.8.

As the period of a structure becomes much longer than the predominant periods of the ground motions, the structure becomes increasingly isolated from the ground motions, and the ground moves independently of the structure, which remains nearly motionless. The spectral displacement at a very long period would therefore be expected to approach the peak

Figure 3.6 Displacement and acceleration spectra

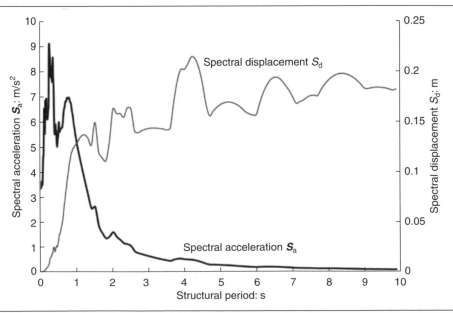

Figure 3.7 Pseudo-spectral velocity spectra and single-sided Fourier acceleration spectrum

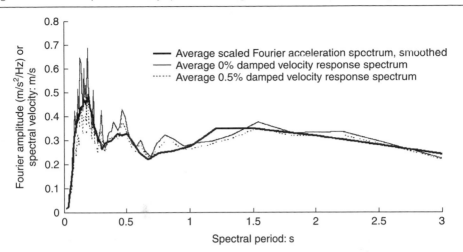

displacement of the ground (PGD). The PGD of the ground motions in Figure 3.6 was 0.15 m, and it can be seen that spectral displacement for periods exceeding 5 s is around this value.

3.3.5 Velocity and Fourier spectra

Velocity spectra can also readily be derived. For convenience, the 'pseudo-spectral' velocity is shown, which by analogy with Equation 3.6 is defined as

$$S_v = S_d(2\pi/T) \tag{3.7}$$

This slightly overestimates the true peak velocity, although for low levels of damping, the discrepancy is small. Velocity is rarely a quantity of direct interest to earthquake engineers, and the primary importance of the velocity spectrum is that for zero damping, it can be shown to be a fairly close upper bound to the single-sided Fourier acceleration spectrum of the relevant earthquake (Figure 3.7). Fourier spectra can be used to derive power spectral densities, see for example ASCE 4 (ASCE, 1998: equation 2-4.1), which are used in probabilistic analysis methods and are also specified in ASCE 4 when checking the adequacy of time histories for design purposes.

3.3.6 Capacity displacement spectra

Equation 3.5 shows that the three quantities – spectral acceleration, spectral displacement and structural period – are uniquely related for a specified level of damping; given two of them, the third is always known. So far, acceleration spectra (spectral acceleration plotted against structural period) and displacement spectra (spectral displacement against structural period) have been discussed. There is, however, a third possibility, namely plotting spectral acceleration directly against spectral displacement, and the result is called a capacity displacement spectrum (Figure 3.8). The 'capacity' of the title refers to the fact that by multiplying the

Figure 3.8 Capacity displacement spectrum

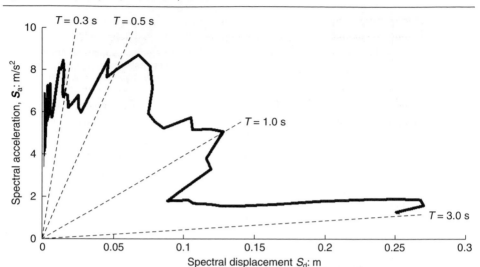

vertical (acceleration) axis by mass, a peak spring force is obtained, and the 'non-linear static' method of analysis relates this force to the capacity of a structure. Equation 3.5 can be rearranged to the form

$$T = 2\pi\sqrt{(S_d/S_a)}$$
(3.8)

Constant ratios of S_d/S_a therefore represent constant values of structural period, as shown in Figure 3.8. The use of capacity spectra is discussed later, in Section 3.7.9.4.

3.3.7   Systems with multiple degrees of freedom

Almost all practical structures are much more complex than the 'single degree of freedom' (SDOF) spring/mass/dashpot systems shown in Figures 3.1 and 3.2, which have been discussed so far. However, many structures can be idealised as SDOF systems. A water tower with a rigid tank full of water supported by a relatively light frame is an example. The seismic response of many buildings is dominated by their fundamental sway mode and this fact can be used to create an SDOF idealisation. Irvine (1986), in his excellent text on dynamics, shows how many systems encountered in engineering practice can be treated in this way.

More complex structures need to be analysed by considering not only the fundamental mode but also the higher natural modes of vibration, which are a characteristic of the stiffness and mass distribution of the structure. These natural mode shapes, which are a structural property independent of the forcing vibration, are shown in Figure 3.9 for a typical 10-storey building. Many computer programs exist to perform this calculation. A computer model of the structure must be established, just as would be required for a normal static analysis, using for example beam elements. However, in addition the mass of the structure must be specified

Figure 3.9 Mode shapes and periods of a 10-storey frame building

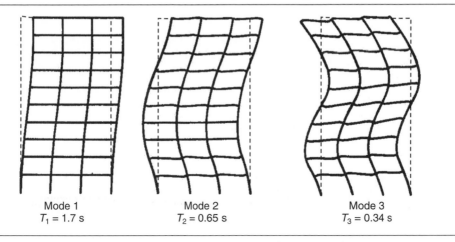

Mode 1	Mode 2	Mode 3
$T_1 = 1.7$ s	$T_2 = 0.65$ s	$T_3 = 0.34$ s

by adding mass elements to the model. With this mass and stiffness information, calculation of mode shapes and periods is a standard calculation performed by many structural analysis packages.

It turns out that the response of a linear structure can generally be calculated by considering the response in each of its modes separately (Figure 3.11) and then combining the separate modal responses. This is possible because each mode of vibration has an associated unique

Figure 3.10 Mode shapes of the first three modes of a cantilever beam

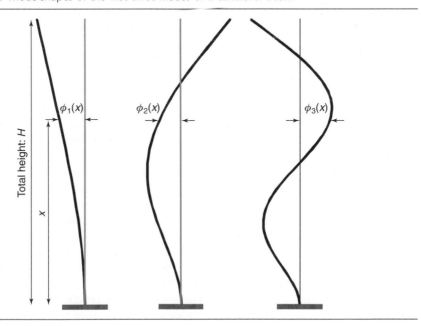

Figure 3.11 Modal RSA of a MDOF structure

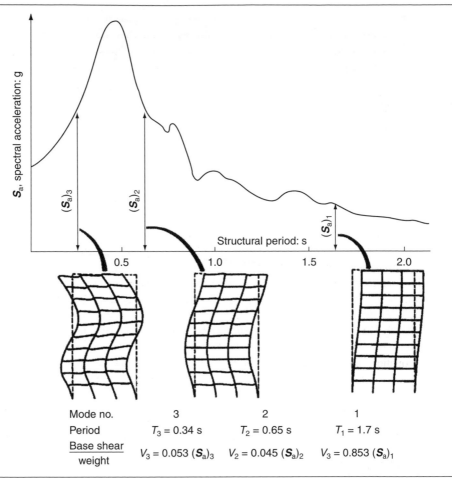

period and also a unique mode shape; therefore one parameter (e.g. the top deflection) is sufficient to define the entire deformation of the structure in that mode. In effect, therefore, each mode is a SDOF system. The basic form of Equations 3.3 and 3.5 still holds, but the equations must be modified as follows. For the base shear in each mode, the total mass in Equation 3.3 must be replaced by the appropriate 'effective' mass, which is always less than the total mass. For deflections and accelerations at any point in the system in each mode, the spectral values S_a and S_d must be multiplied by a structural constant and the value of the mode shape at the point under consideration. Clough and Penzien (1993: p. 617ff) give the values for a cantilever beam as follows (see Figure 3.10).

Base shear in mode $i = S_{ai}(L_i^2/M_i)$ (3.9)

Equation 3.9 is the equivalent for multiple degrees of freedom (MDOF) systems of Equation 3.3, which applies to SDOF systems. As explained below, the term L_i^2/M_i has the dimensions

Table 3.1 Modal multiplying factors for a uniform cantilever shear beam

Mode		1	2	3	4
Base shear factor $= (L_i^2/M_i)/$ total mass (Equation 3.9)		82%	8.0%	3.6%	1.2%
Acceleration factor $= (L_i/M_i)\phi_i(x)$	Top	127%	40%	27%	16%
(Equation 3.10)	Mid height	90%	−28%	−19%	−11%
Deflection factor $= (L_i/M_i)\phi_i(x)$	Top	127%	40%	27%	16%
(Equation 3.11)	Mid height	90%	−28%	−19%	−11%

of mass, and is always less than the total mass of the structure. A very useful result is that $\Sigma L_i^2/M_i$ summed over all modes is equal to the total mass. (This is strictly only true for lumped mass models without rotational inertia. For discussion of a case in which the sum of modal masses may have given a false impression that sufficient modes were captured, see Canterbury Earthquakes Royal Commission (2012: Vol. 2, section 3.6, conclusion 1(b).)

Therefore, sufficient modes must be considered in analysis to 'capture' an adequate proportion of the total mass. Codes often require that sufficiency is indicated when 90% of the mass is captured; for the example in Table 3.1, it can be seen that the first three modes capture 93.6% (i.e. 82% + 8% + 3.6%) of the total mass.

Similar equations apply for acceleration and displacement. These quantities obviously vary with height and so must include the mode shape, $\phi_i(x)$ (see Figure 3.11).

$$\text{Acceleration at level } x \text{ in mode } i = S_{ai}\{(L_i/M_i)\phi_i(x)\} \tag{3.10}$$

$$\text{Displacement at level } x \text{ in mode } i = S_{ai}\{(L_i/M_i)\phi_i(x)\}\{T^2/4\pi^2\} \tag{3.11}$$

where S_{ai} is the spectral acceleration corresponding to the ith mode period, $\phi_i(x)$ is the modal deflection at height x in mode I and L_i, and M_i are structural properties defined in Equations 3.12 and 3.13 below.

Equation 3.10 relates the peak acceleration at any level of the structure in a particular mode to that of its SDOF equivalent. The term $\{(L_i/M_i)\phi_i(x)\}$ is a dimensionless parameter, which Table 3.1 shows can be either greater or less than one. Thus, the acceleration at the top of a building swaying in its first mode is 27% greater than for its SDOF equivalent, but at mid height it is 10% less.

Similar remarks apply to the peak relative displacement, which is given by Equation 3.11, the MDOF equivalent of Equation 3.4.

For distributed two-dimensional systems, L_i and M_i are calculated from Equations 3.12 and 3.13.

$$L_i = \int_0^H m(x)\phi_i(x)\,dx \tag{3.12}$$

$$M_i = \int_0^H m(x)\{\phi_i(x)\}^2 \, \mathrm{d}x \tag{3.13}$$

where $m(x)$ is the mass per unit length at height x and $\phi_i(x)$ is the modal deflection at height x in mode i.

As there are as many modes as degrees of freedom, changing to a modal analysis at first sight does not appear to help much. However, it turns out that the effective masses of the higher modes (i.e. the term (L_i^2/M_i) in Equation 3.9) are low in the case of many practical structures. Therefore, a good approximation to response can usually be obtained from considering only the first few modes of vibration (and often only the lowest in each horizontal direction). For example, Table 3.1 shows the multiplying factors for a uniform cantilever shear beam; they would be typical for a regular building with an unbraced frame. The base shear in the first mode is 82% of that for a lumped mass/spring system with the same mass and period, while the ratio drops to 8% in the second mode and only 1.2% in the fourth. The acceleration and deflection at the top of the cantilever are 27% greater than for the equivalent SDOF system in the first mode, but substantially less in other modes.

The results of such an analysis give the maximum response of the structure for each mode of vibration. Although it is rigorously correct to add the response in each mode at any time to get the total response, the maximum modal responses, calculated from RSA, do not occur simultaneously. Therefore a simple numerical addition of maximum modal responses usually results in a significant overestimate of the real maximum. The square root of the sum of the squares (SRSS) combination of modal responses (whereby each modal response is squared and the square root of the sum of all such squared responses is calculated) usually gives an acceptable estimate of the true overall maximum, but it is only an estimate. Circumstances in which the SRSS estimate may be unconservative are when there is significant response in two or more modes with very similar natural periods (the more sophisticated complete quadratic combination (CQC) method allows for this) and when there is significant response in modes with periods lower than the predominant periods of the earthquake motions (an effect not allowed for in CQC). In the latter case, a safe approximation is to add very short period responses; a less conservative method is given in Section 3.7.2.1 of ASCE 4 (ASCE, 1998). Most commercial computer programs that provide RSA include SRSS and CQC combination methods; these methods are discussed further in ASCE 4.

The effect of these combination methods is that the fundamental mode is likely to contribute most of the base shear (unless of course the spectral accelerations of other modes are very much higher, which may be the case in very tall buildings). This explains the previous assertion that, in many cases, a building can be treated as an SDOF system corresponding to the fundamental mode. However, when the first mode is well out of resonance with the earthquake motion but the second and third mode periods are close to resonance (a common situation for buildings of more than 20 storeys), shears and deflections at higher levels are likely to be strongly influenced by higher modes, as occurs in the case shown in Figure 3.12. Hence there is usually a need to carry out a multi-mode analysis, rather than just a fundamental-mode analysis, for tall buildings.

Figure 3.12 Modal contributions to shear force in a typical frame building

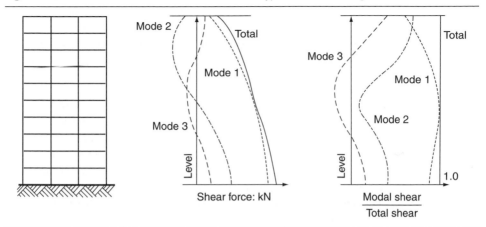

There is an important consequence of the differing contributions of different modes to shear over the height of the structure, namely that the maximum shear force at any level is unlikely to occur simultaneously with the maxima at other levels or with maximum bending moments. Shear and bending moment diagrams obtained from a RSA are therefore enveloped maxima and are not an equilibrium set of actions. In particular, maximum shear force does not equal the rate of change of maximum bending moment, as would be the case in a conventional static analysis.

The previous paragraphs have given a simplified account of multi-mode RSA, one of the most common types of dynamic analysis in engineering design practice. The advantages and disadvantages of the method are described further in Section 3.7.4.

3.4. Non-linear response to earthquakes
3.4.1 Ductility demand and supply
Discussion has so far been in terms of linear-elastic response. However, most structures are designed to yield in the event of an earthquake and so post-yield response is often of crucial importance. As a result, the ductility (or lack of it) that a structure possesses becomes a vital consideration.

For the engineering designer, it is helpful to distinguish between ductility demand and supply. The ductility demand an earthquake makes on a structure is defined as the maximum ductility that the structure experiences during that earthquake. Ductility demand is a function of both the structure and the earthquake; thus in general, the demand decreases as the yield strength of the structure increases, and the demand increases as the intensity of the motions increases.

The ductility supply is, by contrast, a property only of the structure; it is defined as the maximum ductility a structure can sustain without fracture or other unacceptable consequences. For a new structure, it is in principle entirely under the control of the designer, who can set the ductility supply to the level needed.

The objective of the designer, therefore, is to ensure that ductility supply exceeds demand by a sufficient margin to ensure safe performance in the design earthquake. A major purpose of seismic analysis is to establish the level of ductility demand in a structure. This must then be matched by the equally important design measures needed to ensure the existence of an adequate ductility supply. Such measures are at the heart of most current seismic codes of practice, as discussed in later chapters.

3.4.2   Definition and measurement of ductility

'Ductility' can be defined as the ability of a structure to withstand repeated cycles into the post-elastic range without significant loss of strength. One quantified measure is shown in Figure 3.13. This defines 'displacement ductility' for a simple yielding system as the ratio of plastic deflection to yield deflection. Displacement ductility defined in this way is needed in the simple analytical techniques commonly used to allow for yielding behaviour in buildings. However, as discussed later, it is unreliable as a predictor of seismic performance, and other measures of ductility are also needed.

Note that Figure 3.13 shows ductility supply, not demand, because the plastic deformation shown corresponds to an ultimate value. By contrast, Figures 3.15 and 3.16, discussed later, show ductility demand, because they relate to the maximum deformation a structure experiences during an earthquake. It is of course the designer's task to ensure that this never exceeds the supply – that is, the ultimate deformation.

3.4.3   Ductility modified response spectrum analysis

Just as a linear response spectrum gives the maximum response of a linear SDOF system to a given earthquake, so ductility modified spectra can be developed for the response of a ductile SDOF. In constructing ductility modified spectra, the yield strength of the SDOF is chosen so that the ductility demand during a given earthquake is limited to a given value μ. Families of ductility modified curves can then be drawn corresponding to different global displacement

Figure 3.13 Quantifying displacement ductility of a simple system

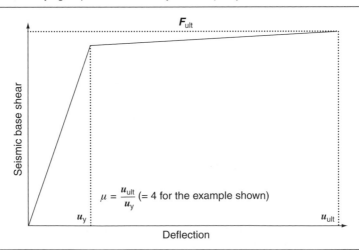

Figure 3.14 Ductility modified response spectra for El Centro

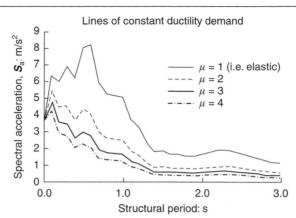

ductility demands μ (see Figure 3.14). For each value of μ, the yield strength has been set such that the peak displacement equals μ times the yield displacement.

Figure 3.14 is the non-linear equivalent of Figure 3.2, differing only in showing the response for different maximum ductility demands rather than different levels of hysteretic damping. The similarity between the two figures is not a coincidence; the increasing ductility demands in Figure 3.14 represent increasing amounts of damping, although hysteretic damping is involved, rather than the viscous damping of Figure 3.2. Note particularly that high ductility demands are ineffective in reducing response in very stiff structures with structural periods close to zero, in just the same way that applies to viscous damping. This is because both viscous and hysteretic damping arise from internal structural deformations (represented as spring tension or compression in an SDOF idealisation). These structural deformations are relatively small compared with the ground movements in very stiff structures; hence the associated damping and consequent reduction in response are also relatively small.

For an elastic perfectly plastic SDOF system, Equation 3.4 becomes modified as follows.

$$S_{d\mu} = \mu \frac{F_y}{k} = \mu m S_{a\mu} \frac{T^2}{4\pi^2 m} = \mu \frac{S_{a\mu} T^2}{4\pi^2} \tag{3.14}$$

Here, k is the pre-yield stiffness of the spring and F_y is the yield force in the spring. T is the period of the structure before it yields. The logic behind Equation 3.14 is as follows. The deformation at yield is F_y/k, and (by definition) the maximum deformation $S_{d\mu}$ under the earthquake is $\mu F_y/k$, because the yield strength of the structure has been set to achieve a global ductility demand μ. Equation 3.14 then follows directly. Note that it is exact (unlike the relationships shown in Figure 3.15). The important implication of Equation 3.14 is that in ductile structures, displacements are μ times greater than their elastic equivalents with the same level of stress. The reason is that plastic strains increase the displacements in the yielding structure.

Construction of ductility modified spectra directly from the earthquake record is in principle straightforward using appropriate software, and a number of programs exist to do this.

Figure 3.15 Forces and deflections in plastic and elastic systems: FLEXIBLE structures (DEFLECTIONS preserved)

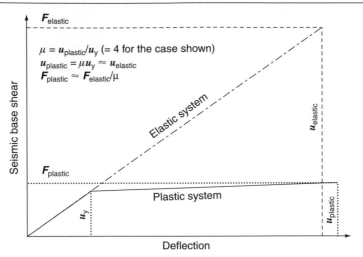

However, an approximate ductility modified spectrum can be estimated much more directly from the elastic spectrum, as is now explained.

For structures with very long initial periods (i.e. very flexible structures), whether ductile or elastic, the maximum displacement equals the peak ground displacement; essentially the structure is so floppy that the structure stays where it is and the ground moves beneath it. Therefore, displacements of very flexible ductile and elastic structures are equal. The same result – equality of displacements in elastic and yielding structures – holds approximately for most structures when the initial period is greater than the predominant period of the ground motions (approximately 0.1–0.3 s for firm ground sites, 1 s or more for very soft sites).

Therefore, for medium to long period structures, it is quite easy to see that the acceleration in the plastic structure is a factor of about μ lower than its elastic equivalent. This is because the plastic structure suffers a force μ times lower than it would have done if it had remained elastic, but it experiences the same maximum deformation $u_{plastic}$ (see Figure 3.15). However, it will almost certainly undergo some permanent deformation, and probably 'damage' as well.

This result does not apply to stiff structures with periods lower than the predominant ones of the earthquake motion. For rigid structures ($T = 0$), the force in both yielding and elastic systems must be the same, and equal to the structural mass times PGA. It follows that the deformation in the plastic (yielding) structure is μ times greater than its elastic equivalent (Figure 3.16). Therefore, ductility (hysteretic damping) is no advantage to a very rigid structure, in just the same way that viscous damping (Figure 3.2) does not reduce the response of rigid systems. These results are summarised in Table 3.2. Using the results, it can be seen that an approximate ductility modified acceleration response spectrum can relatively easily be

Figure 3.16 Forces and deflections in plastic and elastic systems: RIGID structures (FORCES preserved)

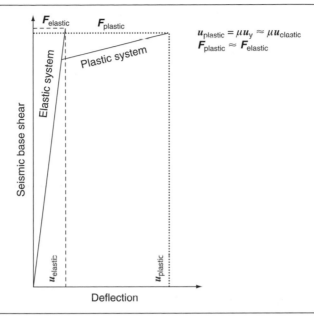

$$u_{plastic} = \mu u_y \approx \mu u_{elastic}$$
$$F_{plastic} \approx F_{elastic}$$

drawn from the corresponding elastic spectrum. For periods above the spectral peak, the ductility modified spectral acceleration $S_{a\mu}$ is obtained by dividing the elastic spectral acceleration S_a by μ, the global displacement ductility factor. For rigid systems, the ductility modified and elastic spectral accelerations are equal. For intermediate periods, the elastic spectral

Table 3.2 Comparison of forces and deflections in elastic and yielding (plastic) structures

	Elastic structure	Yielding structure
	Medium to long period structures	
Acceleration	S_a	$(S_{a\mu})/\mu$
Force	$F_{elastic}$	$(F_{elastic})/\mu$
Deformation	$u_{elastic}$	$u_{plastic} \approx u_{elastic}$
	Very short period structures	
Acceleration	S_a	$(S_{a\mu})/\mu$
Force	$F_{elastic}$	$(F_{elastic})/\mu$
Deformation	$u_{elastic}$	$u_{plastic} \approx \mu\, u_{elastic}$
	Short to medium period structures	
Acceleration	S_a	$(S_{a\mu})/N$
Force	$F_{elastic}$	$(F_{elastic})/N$
Deformation	$u_{elastic}$	$u_{plastic} \approx N\, u_{elastic}$
	where N is a factor between 1 and μ	

acceleration is divided by a factor between μ and 1. In EC8, the reduction factor increases linearly between 1 and μ as the structural period increases from 0 s to the period at the start of the spectral peak value (between 0.05 and 0.20 s, depending on the type of soil and magnitude of earthquake). Other more complex relations have been published, but the one in EC8 should be sufficient for most purposes, given the other uncertainties involved.

3.4.4   Application of ductility modified spectra to MDOF systems

Section 3.3.7 showed how a rigorously correct analysis of a linear elastic MDOF system was possible on the basis of a response spectrum constructed for a linear SDOF system. It might be thought that a similar extension of a ductility modified response spectrum from SDOF to MDOF would be possible. In other words, an MDOF structure, for which the displacement ductility demand was required to be μ, could be analysed for accelerations and forces as if it were an elastic structure, but with the elastic ground acceleration spectrum replaced by a ductility modified spectrum corresponding to the required value of μ. The yield strengths necessary to limit the ductility demand to μ would be obtained directly from such an analysis, but (from Equation 3.14), all the displacements would need to be increased by a factor μ. The assumption that the analysis of a ductile MDOF is possible in this way underpins most practical and code-based designs using RSA.

Unfortunately, there is no fundamental reason why such an analysis should apply. As explained in Section 3.3.7, the linear MDOF analysis works by effectively splitting the structure into a series of SDOF systems, each representing one of its natural modes of vibration. However, once the structure yields, the unique mode shapes on which the linear analysis relies no longer apply, and the modal periods start to increase.

When yielding is spread uniformly through the structure, so that the post-yield deformed shape is similar to the elastic first mode shape, then ductility modified RSA should give reasonable answers, but it must be remembered that the answers from such analysis are never exact. Figure 3.17 compares two analyses of the same structure subjected to the same ground motion. In the first analysis, the structure remains elastic and experiences a base shear of 6 MN. In the second analysis, which used rigorous non-linear time history methods, and not a ductility modified spectrum, the structure had the same initial stiffness but was allowed to develop a displacement ductility μ of 2.

As expected, the base shear in the yielding structure has halved to just under 3 MN, with similar reductions (i.e. a reduction factor of $1/\mu$ applying throughout the height of the building. However, the storey drifts (difference in deflections between one storey and the next), which ductility modified RSA would have predicted to be the same as in the elastic case, were in fact very different. The drifts at the bottom were about double the elastic values, and at the top about half – a fairly typical result for a structure (such as this) in which yielding is well distributed throughout the structure.

The results of ductility modified RSA will be much more in error for structures in which the yielding is concentrated at one level – for example, in a 'soft storey' structure. The use of RSA must therefore be coupled with a method, for example capacity design (Section 3.5), that ensures such failures cannot occur.

Figure 3.17 Comparison of forces and deflections in the yielding and elastic response of a typical 10-storey building

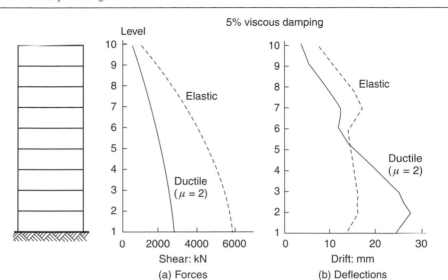

(a) Forces

(b) Deflections

3.4.5   Consequences of yielding response

There are a number of important consequences of yielding that a designer should bear in mind.

1   Member forces remain well below the level they would have reached, had the structures remained elastic (Figure 3.17(a)). The reduced response is due to the hysteretic damping associated with the yielding. For structures with an initial period greater than the predominant period of the earthquake, the lengthening of the structural period caused by yielding will also help to reduce response.

2   The post-yield deformed shape is markedly different from the elastic condition (Figure 3.17(b)). In the yielding areas (usually the lower levels of a building), deformations tend to be greater than elastic values, while in other areas they tend to be less. Therefore, the implicit assumption in most codes that the deflections remain equal to the elastic deflections corresponding to 5% viscous damping is likely to be unconservative at the lower levels of buildings.

3   Members are damaged; this can be thought of as a low-cycle fatigue effect. Hence the number as well as the magnitude of the yielding cycles is important.

4   In redundant (hyperstatic) structures such as frames, gravity moments become significantly redistributed, which may significantly affect the frame's earthquake-resisting properties.

5   The increase in ratio of deflection to restoring forces means that P–delta effects (discussed in Section 3.6.5) become relatively much more important.

Current practice deals with these five considerations in the following ways.

1   The member force reduction due to ductility is allowed for (at least in part) by reducing elastic forces by factors such as R in ASCE 7 (ASCE, 2010) or q in EC8, which depend on the available ductility. EC8 (unlike ASCE 7) takes full account of the reduced effectiveness of ductility in very stiff structures.
2   Post-yield deflections calculated from a ductility modified RSA should be treated with caution, for the reasons discussed above; however, this aspect receives no attention in EC8 or ASCE 7.
3   Low-cycle fatigue effects are generally dealt with by appropriate detailing rather than direct analysis. For example, code rules for the provision of transverse steel at a potential plastic hinge location of a reinforced concrete beam are greatly influenced by the need to prevent the flexural stiffness and strength from degrading during repeated cycles of yielding.
4   Moment redistribution can have significant effects in frames in which gravity loading produces moments that are a substantial fraction of the yield moments (Fenwick and Davidson, 1987). In codes such as EC8 and ASCE 7, the frame moment due to a simple addition of gravity and seismic actions may be redistributed to the same extent as is allowed for gravity loading alone or wind combined with gravity. However, under seismic loading, significant plastic rotations are expected, both positive and negative, which is a completely different situation from that applying to gravity and wind loading. Priestley *et al.* (2007) propose that beam support moments should be designed for the worst case of either seismic loading without gravity, or gravity loads with appropriate load factors. This implies a lower required flexural strength at supports than code requirements, but greater design span moments.
5   Codes give rules for when P–delta effects need to be considered, as discussed in Section 3.6.5 which appears later. These rules usually imply that P–delta effects can be neglected, but this may be unconservative (Fenwick *et al.*, 1992).

3.4.6   Curvature ductility

Defining ductility in terms of displacement leads to the useful analytical techniques based on ductility modified response spectra, described above, which form the basis for most current seismic standards. However, displacement ductility is much less useful as a predictor of ductility supply, and also of seismic performance. The reasons are now discussed.

Figure 3.18 shows that some parts of the structure are likely to start to yield well before the nominal yield deflection u_y is reached. Moreover, after the onset of yielding, further deformation tends to concentrate in the yielding regions, rather than in the parts of the structure that remain elastic. Therefore, local measures of ductility, for example, based on the curvature of the plastic hinges, provide a much better measure of demand than one based on the global displacement. 'Curvature ductility' is defined as the ratio of maximum curvature of a beam to the curvature at the effective yield point. It can be seen from Figure 3.18 that the curvature ductility demand on the beams at the base of the structure will be many times the displacement ductility demand, based on the deformation at roof level.

As another example of why curvature ductility is a better predictor of performance than displacement ductility, compare the two concrete building frames shown in Figure 3.19. We

Figure 3.18 Relation between curvature and displacement ductility in a building frame

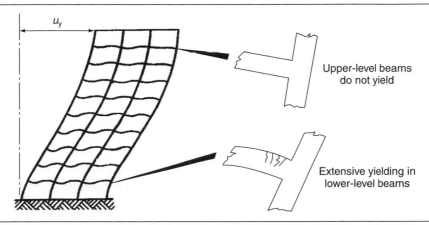

Upper-level beams do not yield

Extensive yielding in lower-level beams

will assume that both frames have the same concrete section sizes, but that frame A has considerably more reinforcement in the columns than frame B, which therefore represents the dangerous 'weak column, strong beam' system discussed in more detail in Chapter 8. It can be seen that the displacement at roof level of both frames is the same, and therefore the displacement ductility demand is also likely to be similar. However, the curvature ductility in frame B's columns is much higher than that in frame A's members, and indicates that frame B is much nearer collapse.

Figure 3.19 Comparison of curvature and displacement ductility demand in two frames with equal top displacement

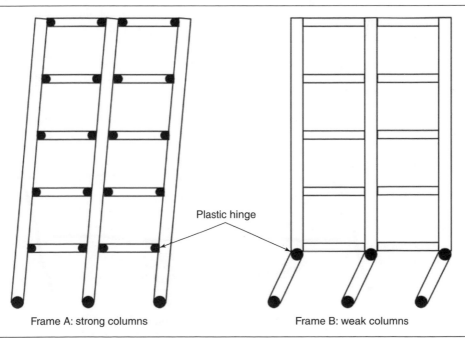

Plastic hinge

Frame A: strong columns

Frame B: weak columns

Curvature ductility is therefore a much better indicator of ductility demand on a structure than displacement ductility. It forms the basis of some of the most important of the detailing rules of EC8 for concrete structures, and is also a key parameter in the direct DBD methods outlined in Section 3.8.

3.5.   Analysis for capacity design

In most cases, code provisions are based on the premise of ductile behaviour under severe earthquake loading. Such behaviour requires that yield capacity is reached first in ductile modes of response (such as bending of well-detailed steel or concrete beams) rather than brittle modes (such as shear in poorly detailed concrete beams, buckling of slender steel struts or failure of welded connections). The analytical methods described so far, based on the ductility modified response spectrum, are not in themselves sufficient to achieve this aim. What is needed in addition is a procedure, known as capacity design, to check that the requisite hierarchy of strength is present, implying that ductile modes are weaker than brittle modes. In essence, capacity design requires that the brittle elements are designed to be strong enough to withstand the full capacity of the ductile, yielding elements – hence the term 'capacity design'.

An important concept in capacity design is that of 'overstrength'. The brittle members need to be strong enough to withstand the forces induced by yielding of the ductile members allowing a suitable margin to give a high level of confidence that the brittle elements will not reach their failure loads. The overstrength of the yielding regions must allow for various possibilities, including strain hardening in steel, the possibility that actual strength on site is greater than specified strength and (sometimes) uncertainties in analysis. Moreover, the required strength in the brittle members must be based on the actual strength provided in the ductile elements; this almost always exceeds the minimum code requirement, due for example to the rounding up of member dimensions or bar diameters for practical reasons.

A straightforward example of capacity design is to check that the shear strength of a concrete beam in a frame under sway loading exceeds the force corresponding to the development of plastic hinges (Figure 3.20).

Figure 3.20 Capacity design for shear strength in a reinforced concrete beam

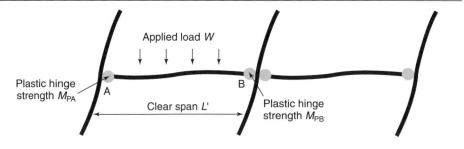

Shear strength of beam AB must exceed $(M_{PA} + M_{PB})/L' + W/2$

Figure 3.21 Capacity design for column flexural strength

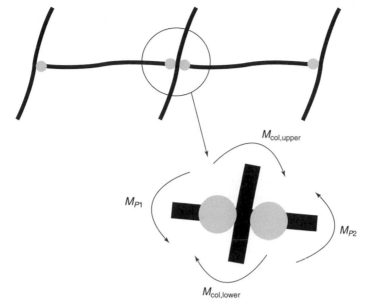

Flexural strength of column framing into joint must exceed
plastic flexural strength of beams ($M_{P1} + M_{P2}$)

The design shear strength in this example follows in a statically determinate manner from the flexural strength at the plastic hinge points. Note, however, that in this example the hinges are assumed to form at the ends of the beam, which may not be the case for relatively high levels of applied gravity load (see Section 8.2.6).

Another example relates to the columns of unbraced sway frames. Here the aim is to ensure that yielding occurs first in the beams and not the columns, in order to achieve a 'strong column/weak beam' structure (frame A in Figure 3.19) and to avoid soft or weak storeys. The capacity design procedure is to ensure that the flexural strength of the columns framing into a joint exceed the sum of the plastic yield moments at the ends of the beams (Figure 3.21). There is more uncertainty here than for the previous case; although the beam moments are known to equal the plastic strengths, the distribution of moment between columns above and below any joint is not statically determinate because the ratio of $M_{\text{col,upper}}$ to $M_{\text{col,lower}}$ in Figure 3.21 depends on the points of contraflexure in the columns. For example, if the point of flexure is just below the joint, $M_{\text{col,lower}}$ will be quite small, and the upper column must resist the full plastic strength of the beams ($M_{P1} + M_{P2}$).

Different codes treat the problem in different ways. The simplest approach is to require that the sum of the flexural strengths of the columns at each joint exceeds the sum of the beam flexural strengths by a suitable margin; this is essentially the ASCE 7 requirement, and is also a general requirement of EC8. The New Zealand concrete code NZS 3101 (New Zealand Standards, 2006) and its commentary provide the most detailed and complex procedure for concrete sway frames. This provides the greatest assurance that plastic hinges will not develop in columns during an earthquake. The simpler procedures should prevent the formation of column hinges simultaneously at the top and bottom of a storey, and hence prevent a weak storey collapse, but may not prevent hinge formation at one end of a column.

The great advantage of capacity design, and the reason why it now finds a place in all major codes, is that it is a simple procedure implemented purely from considerations of static equilibrium, without recourse to complex dynamic analysis. Nevertheless, it ensures ductile response, more or less independently both of the actual dynamic response and of the nature of the earthquake ground motions.

3.6.   Other considerations for a seismic analysis
3.6.1   Influence of non-structure
Non-structural elements such as cladding and partitions are not usually explicitly allowed for in analysis but they may have an important and not always beneficial effect on response. For example, cladding can stiffen a structure and may bring its natural period closer to resonance with the predominant period of an earthquake. However, deflections will be reduced, which is usually beneficial.

As another example, infill blockwork that is not full height may create a short column whose shear strength is less than its bending strength and that is therefore prone to brittle failure (Figure 8.3).

The designer has two alternatives. Either non-structural elements can be fully separated from the main structure, or the interaction between structure and non-structure must be allowed for in analysis. The first alternative creates a more predictable system, but may well lead to its own problems – for example, separation joints between infill masonry and structural frames are hard to detail satisfactorily to provide the weatherproofing and out-of-plane restraint that is needed. The second alternative can sometimes lead to satisfactory results – for example, the increased strength and stiffness provided to a structural frame by rigid infill masonry panels may more than offset the reduction in ductility and predictability.

3.6.2   Site effects
The nature of the soils at a site can have a dominating influence on the seismic motions at the site (Figures 2.6 and 2.7) and may also significantly affect the dynamic characteristics of structures built there, by increasing the foundation flexibility and structural damping. These important considerations are discussed in Chapter 4.

3.6.3   Effective stiffness of concrete elements
The axial and flexural stiffness of concrete elements depends not just on concrete strength and gross section area of the element, as is usually assumed, but also depends strongly on the amount of reinforcement and on the level of strain (Priestley *et al.*, 2007). This is discussed further in Section 8.4.1.

3.6.4   Torsional response

Seismic ground motions are predominantly translational, not rotational. However, where the centres of mass and stiffness of a structure do not coincide, coupled lateral–torsional response occurs (Figure 3.22). Structures with significant torsional eccentricity are found to have a much worse performance during earthquakes. Coupled lateral–torsional response cannot of course be analysed using two-dimensional models, because three-dimensional behaviour is involved. Static analysis by applying code-required forces at the centre of mass may under-estimate response because of dynamic effects; a possible example is when the period of the lateral–torsional mode of vibration matches the predominant period of the earthquake. A linear dynamic analysis (e.g. a RSA) would capture this effect but may still significantly underestimate the response after yielding, because the less stiff side tends to yield first, becoming even more flexible and hence adding to the eccentricity. A designer's first response for new structures should be to avoid significant structural eccentricity when possible. When this is not possible, or for an existing structure with torsional eccentricity, the non-linear dynamic response should be accounted for. Priestley *et al.* (2007) cites recent research findings and presents recommendations for design.

Codes of practice treat torsion in the following ways.

■ Requiring additional strength (up to 50% in Japanese codes and 20% in EC8) beyond that indicated by analysis.
■ Requiring more sophisticated analysis if eccentricity exceeds prescribed limits. For example, explicit non-linear dynamic analysis is specified by Japanese codes and three-dimensional modal RSA by EC8 and US codes.

Figure 3.22 Couple lateral–torsional response

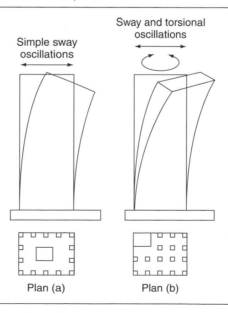

Figure 3.23 *P*–delta moments

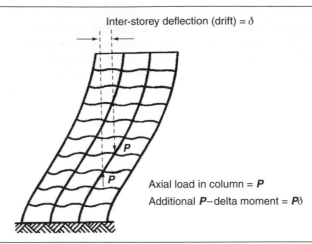

Inter-storey deflection (drift) = δ

Axial load in column = *P*

Additional *P*–delta moment = $P\delta$

- When static analysis is permitted, the US code ASCE 7 requires the distance between the point of application of lateral load and the centre of stiffness to be increased, if the eccentricity exceeds a threshold.
- Codes usually specify an 'accidental' torsion, that is, an offset of the point of application of lateral load by 5–10% of building dimension from the centre of mass. This ensures a minimum torsional stiffness and strength to deal with mass and stiffness distributions that depart from the designer's assumptions.

3.6.5 *P*–delta effects

Lateral deflections give rise to gravity-induced moments (Figure 3.23). Usually the moments are small, but when the product of gravity load ('*P*') and the lateral deflection ('delta') is a significant fraction of the seismic overturning moment, the resulting '*P*–delta' effect should be allowed for. It can be easily incorporated into a non-linear analysis, but needs special techniques to include in a linear-elastic analysis; most standard linear-analysis computer programs do not allow for it.

Codes such as EC8 and ASCE 7 allow neglect of *P*–delta effects if specified deflection limits are not exceeded; as discussed in Section 3.4.5 item 5, these limits are not always conservative.

3.7. Seismic analysis of buildings
3.7.1 Objectives

The objectives of the seismic analysis of a building structure are likely to include the following.

- To establish member strength requirements to prevent undue damage in frequent (lower intensity) earthquakes.
- To establish ductility demands in members designed to yield in rare (extreme) earthquakes, and hence ensure that appropriate measures, such as ductile detailing, are in place to develop the demand.

- To establish strength requirements in brittle members required to remain elastic in rare earthquakes.
- To calculate displacements
  - non-structural elements such as cladding and services are suitably protected
  - to prevent impact between adjacent structures
  - to check that *P*–delta effects are not significant.
- To establish the nature of dynamic design input to equipment mounted on the structure, for example machinery, storage tanks, etc.

3.7.2   Types of analysis

Various forms of linear and non-linear analysis are possible, which build on the theoretical basis set out in previous sections. Sections 3.7.3 to 3.7.5 describe the main linear analysis methods used in current practice, while Sections 3.7.6 to 3.7.8 describe the non-linear methods.

When significant ductility is assumed in design, a structure designed purely on the basis of a linear elastic analysis may well be unsafe. When no explicit non-linear analysis is performed, minimum provisions are essential to ensure satisfactory post-yield behaviour, as discussed in Section 3.5.

3.7.3   Equivalent linear static analysis

All design against earthquake effects must consider the dynamic nature of the load. However, for simple regular structures, analysis by equivalent linear static methods is often sufficient. This is permitted in most codes of practice for regular, low to medium-rise buildings and begins with an estimate of peak earthquake load calculated as a function of the parameters shown in Table 3.3.

For example, the seismic base shear F_b in EC8 part 1, section 4 is given by

$$F_b = S_d(T)m\lambda F_b \tag{3.15}$$

where $S_d(T)$ is the ductility modified spectral acceleration for a period T, PGA $a_{gR}I$, behaviour factor q, and the appropriate soil type and spectral shape type. For example, the plateau value of $S_d(T)$ equals $2.5Sa_{gR}\gamma_I/q$. $\lambda = 0.85$ for shorter period structures (around $T \le 1$ s, depending on soil type) over two storeys, or for buildings with less than three storeys. $m = \Sigma G_{ki} + \Sigma \psi_{EI}Q_{ki} =$ mass considered for seismic loading, taken as all of the permanent masses ΣG_{ki}, plus a proportion ψ_{EI} of the variable masses ΣQ_{ki}, where ψ_{EI} typically equals 0.3, and other symbols are defined in Table 3.3.

The calculated load is then applied to the structure as a set of static horizontal loads with a prescribed vertical distribution, approximating to the first-mode response of a regular building. The forces obtained from the analysis are only used to size the elements intended to yield, with other elements sized by capacity design procedures (Section 3.5).

The theoretical basis for equivalent static analysis is that the static forces are chosen to produce the same extreme deflected shape as would actually occur (momentarily) during

63

Table 3.3 Parameters to consider in a simple seismic analysis

Parameter	Associated symbols in EC8 (CEN, 2004)	Associated symbols in ASCE 7 (ASCE, 2010)
Geographical location	a_{gR} Design ground acceleration on rock or firm ground[a]	S_s and S_1 (spectral accelerations at short period and at 1 s)
Effect of foundation soils	S Soil factor T_B, T_C, T_D Corner periods	F_a, F_v short and long period site coefficients
Intended use, which influences acceptable level of damage	γ_I Importance factor	I_E Seismic importance factor
Structural form, which influences the available ductility	q Behaviour factor	R Response modification factor
Weight of structure and contents	$\Sigma\, G_{kj} + \Sigma\, \psi_{Ei}Q_{ki}$ Full characteristic dead load plus reduced characteristic live load	W Effective seismic weight (full dead plus reduced live)
First-mode period of the structure	T	T

In both EC8 and ASCE 7, structural irregularities in plan and elevation may lead to increased strength requirements
[a] In EC8, geographical location also determines the choice of type 1 or 2 spectral shapes, which accounts for whether sites are influenced by earthquakes of larger (type 1) or smaller (type 2) magnitudes

the earthquake. For a structure responding in only one mode, the velocity is zero at all points in the structure when this maximum deflection is experienced. The equivalent static force therefore equals mass times acceleration (d'Alambert force) at each point. Therefore, an exact equivalence between equivalent static and dynamic analysis is possible. However, when more than one mode is involved, different levels in the structure reach their extreme response at different moments of time and a single set of static forces can never truly represent the dynamic maxima at all levels. Equivalent static analysis can, therefore, work well for low to medium-rise buildings without significant coupled lateral–torsional modes, in which only the first mode in each direction is of significance. Tall buildings (over, say, 20 storeys), in which second and higher modes are usually important, or buildings with torsional effects are much less suitable for the method, and both EC8 and ASCE 7 require more complex methods to be used in these circumstances. It may still be useful, even here, as a 'sanity check' on later results using more sophisticated techniques.

3.7.4 Modal response spectrum analysis

As described in Section 3.3, RSA involves calculating the principal elastic modes of vibration of a structure. The maximum responses in each mode are then calculated from a response spectrum and these are summed by appropriate methods (e.g. the SRSS or CQC methods discussed earlier) to produce the overall maximum response. As is the case for equivalent static design, the results of the analysis are only used to size the yielding elements of the structure, with other elements sized by capacity design procedures. Many software packages exist to perform RSA; the main difference from analysis for gravity or wind loading is that information must be provided about the mass of the structure, as well as its stiffness.

The major advantages of modal RSA, compared with the more complex time history analysis described later, are as follows.

- The size of the problem is reduced to finding only the maximum response of a limited number of modes of the structure, rather than calculating the entire time history of responses during the earthquake. This makes the problem much more tractable in terms both of processing time and (equally significant!) size of computer output.
- Examination of the mode shapes and periods of a structure gives the designer a good feel for its dynamic response.
- The use of smoothed envelope spectra (Figure 2.6) makes the analysis independent of the characteristics of a particular earthquake record.
- RSA can very often be useful as a preliminary analysis, to check the reasonableness of results produced by other, perhaps more complex, analysis.

Offsetting these advantages are the following limitations.

- RSA is essentially linear and can make only approximate allowance for non-linear behaviour. This is perhaps the major disadvantage for the design of most buildings.
- RSA assumes that the stiffnesses in positive and negative directions of loading are equal. If they are unequal, then the structure will tend to 'ratchet' towards the more flexible direction, because deflections will not be recovered under loading in the reverse, less flexible direction. Ductility demands and P–delta effects will be greatly increased as a result. Braced steel structures with an unequal number of braces in compression and tension for a given direction of loading are one example. Codes usually prohibit this situation. Another example, not currently treated in codes, is a structure with T-section shear walls in which more T-flanges are in compression for one direction of loading than the other.
- The results are in terms of peak response only, with a loss of information on frequency content, phase and the number of damaging cycles, which have important consequences for low-cycle fatigue effects. Moreover, the peak responses do not generally occur simultaneously – for example, the maximum axial force in a column at mid height of a moment-resisting frame is likely to be dominated by the first mode, while its bending moment and shear may be more influenced by higher modes and thus may peak at different times. The same also applies to estimates of the peak responses due to simultaneous earthquake loading in more than one direction; see the discussion in Section 6.12.

- It will also be recalled (Section 3.3.7) that the global bending moments calculated by RSA are envelopes of maxima not occurring simultaneously and are not in equilibrium with the global shear force envelope.
- Variations of damping levels in the system (for example, between the structure and the supporting soils) can only be included approximately. ASCE 4 (ASCE, 1998: Section 3.1.5) discusses this topic.
- Modal analysis as a method begins to break down for damping ratios exceeding about 0.2, because the individual modes no longer act independently (Gupta, 1990).
- The method assumes that all grounded parts of the structure have the same input motion. This may not be true for extended systems, such as long pipe runs or long-span bridges. Der Kiureghian *et al*. (1997) have proposed ways of overcoming this limitation.

3.7.5   Linear time history analysis

The complete 'time history' of response to an earthquake can be obtained by calculating the response at successive discrete times, with the time step (interval between calculation times) sufficiently short to allow extrapolation from one calculation time to the next. A time step of one quarter of the period of the highest structural mode of interest should usually be sufficiently short for a linear analysis, although much shorter steps are required for non-linear analysis. This solution method in the 'time domain' is further discussed by Clough and Penzien (1993).

A linear time history analysis of this type overcomes all the disadvantages of RSA, provided non-linear behaviour is not involved. The method involves significantly greater computational effort than the corresponding RSA. With current computing power and software, the task of performing the number crunching and then handling the large amount of data produced has become a non-specialist task. More problematical is the choice of suitable input time histories to represent the ground motions at a site, as discussed in section 2.8.3.

3.7.6   Linear time history analysis in the frequency domain

Linear time history analysis can also be performed in the 'frequency domain', whereby the input motion is split into its single period harmonic components – Fourier spectrum – by means of Fourier analysis. The analysis is performed by summing the separate responses to these harmonic components; it can therefore only be used for linear responses, when superposition is valid. The output is also obtained as a set of Fourier spectra, which can then be used to compute time histories of results in the time domain. The details and theoretical basis of the technique are described by Clough and Penzien (1993).

The possibility of increased computational efficiency when using frequency domain analysis is of less importance now, because of the ready availability of computing power. It is, however, sometimes used in soil–structure interaction analyses, because the flexibility of supporting soils can best be represented by frequency-dependent springs and this requires a frequency domain analysis. It is also the basis of some probabilistic methods, which have a wider application. As noted above, non-linear analysis is not possible in the frequency domain.

3.7.7 Use of linear or non-linear analysis

By its nature, linear analysis can give no information on the distribution of post-yield strains within a structure, and only limited information on the magnitude of any post-yield strains that might develop. The best that can be hoped for is that by means of an elastic analysis the structural strength can be set to a level that will limit post-yield strains to acceptable levels. However, most structural failures during earthquakes occur as a result of elements experiencing strains beyond the limit that they can sustain. Non-linear analysis offers the possibility of calculating post-elastic strains directly, which is an enormous potential advantage. With the availability of increased computing power and more sophisticated software, non-linear methods are being increasingly used in design practice. It is noteworthy that simple non-linear time history analyses have been effectively mandatory for the seismic design of tall buildings in Japan since at least the 1980s (Fitzpatrick, 1990).

3.7.8 Non-linear time history analysis

Non-linear effects can be allowed for by stepping through an earthquake and extrapolating between calculation times, in just the same way as for a linear time history analysis. The simplest (and most tractable) analytical models consist of frame elements in which non-linear response is assumed to be concentrated in plastic hinge regions at their ends. More complex models can involve non-linear plate and shell elements. ASCE 41 (ASCE, 2006) provides information on plastic hinge properties for use in the non-linear analysis of steel and concrete structures.

Non-linear methods enable the most complete allowance to be made for the combination of dynamic response with the onset of plasticity, and variation in time-dependent parameters such as the possible loss of strength and stiffness of plastic hinge regions under repeated large cyclical strains, or the increase in pore water pressures in soils. Naturally, this extra information is bought with very considerably increased computational effort; the time steps used must be much less than those in a linear analysis. Clough and Penzien (1993) describe the solution techniques involved.

There are a number of reasonably user-friendly commercial packages available that will carry out a non-linear time history analysis and can analyse a practical size of building frame using a desktop computer. Twenty-five years ago, non-linear time history analysis was a difficult and time-consuming specialist exercise; by contrast, believable results are now reasonably easily obtained. However, there are still many pitfalls; results may be critically dependent on small variations in input parameters and sensitivity studies are likely to be needed, particularly as the non-linear cyclic response characteristics assumed in the computer model are probably only a crude and uncertain approximation of reality. At the very least, a range of different input motions must be used, because substantially different responses can be obtained from input motions with similar response spectra (see Section 2.8.3).

At the time of writing (2013), non-linear time history analysis is not generally used routinely in design practice, and is reserved for special cases such as unusually important buildings or those with novel means of earthquake protection. However, the static non-linear analyses described next are becoming increasingly favoured.

3.7.9   Non-linear static and 'displacement-based' methods

3.7.9.1   General

Non-linear static methods have recently gained wide currency, and offer the advantage of giving direct information on the magnitude and distribution of plastic strains within a structure. These are based on the ground motions represented by the design response spectrum in a code of practice, without the difficulties inherent in a non-linear time history analysis and the associated requirement to choose suitable ground motion time histories. As explained in more detail below, the method involves modelling a frame structure as an equivalent SDOF structure, whose properties have been determined by means of a 'static pushover' analysis performed on a non-linear model of the frame. The peak displacement of this equivalent SDOF structure is then determined directly from the design response spectrum, and then imposed on the frame model to determine the peak plastic strains in the frame, and their distribution.

Modelling a complex non-linear frame as a SDOF is clearly a drastic simplification, and the results can never be as 'accurate' as those obtained from more complex methods. In some cases, the method may be difficult to use. In particular, this is likely to apply if the building structure in question is subject to significant torsional response, because a SDOF idealisation cannot capture both translational and torsional response. However, displacement-based methods address the huge drawback of the response spectrum techniques that have become standard in western design practice, namely that such methods cannot properly capture the non-linear behaviour that characterises the intended response of most buildings during their design earthquake. Moreover, by promoting concentration on fundamentals without being confused by a mass of detail, the very simplicity of non-linear static analysis may enable designers to make better choices than they would when using more complex methods; read again the quote from Nigel Priestley that heads this chapter.

Non-linear static methods are often referred to as 'displacement-based' (as opposed to force-based) design methods, because peak displacements, rather than peak forces, are more obvious during the process. However, the distinction is somewhat artificial, as displacement and forces are inextricably linked in any method of analysis.

The description of the methods that follows is based on US practice – for example, as contained in ASCE 41 (ASCE, 2006), although it has gained much wider international acceptance. Annexe B of EC8 part 1 (CEN, 2004) sets out a non-linear static method based on the 'N2' procedure described by Fajfar (2000). EC8 part 2 for bridges and part 3 for the assessment and retrofit of existing buildings also provide advice on non-linear static analysis.

3.7.9.2   Static pushover analysis

The first stage in the process is to perform a 'static pushover analysis' (Figure 3.24). This involves defining a set of lateral forces, with a vertical distribution corresponding to those of the inertia forces developed in an earthquake, which are applied as a static load case to a non-linear model of the structure. All the forces are gradually increased by the same proportion, and the deflection of the top structure is plotted against the total applied shear; this is the basic pushover curve (Figure 3.25). As yielding occurs in the structure, its properties are appropriately modified – for example, plastic hinges are introduced at the ends of yielding beams.

Figure 3.24 Static pushover analysis

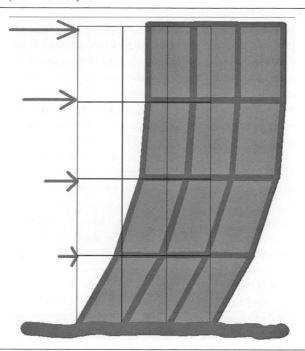

3.7.9.3 Target displacement method

Having calculated a pushover curve, two methods are available to calculate the maximum deflection at the top of the building under the design earthquake motions. The most straightforward method to use is the target displacement method of ASCE 41 (ASCE, 2006); the methods set out in the various parts of EC8 are similar. This uses the result, noted earlier, that the peak deflection of a yielding system is usually quite well predicted by that of an

Figure 3.25 Static pushover curve

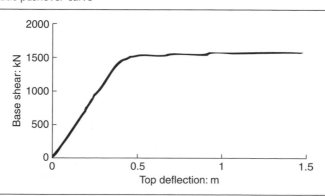

elastic system with the same period and damping. The target displacement method therefore uses Equation 3.5 to find the peak deflection of a linear SDOF structure with a period corresponding to the first mode of the building, using an elastic, 5% damped response spectrum. This is of course the 5% spectral deflection at the first mode period. If the structure remained elastic, if structural damping were 5% and if the first mode dominated the response, then the top deflection would equal this deflection increased by the modal factor, which from Table 3.1 is approximately 1.27 for a uniform cantilever. For other than short period structures, this elastic estimate often provides a good approximation for a plastically responding structure, even when the structure yields significantly (see Figure 3.15). However, it is an underestimate for short period structures (Figure 3.16), or for structures subject to strength or stiffness degradation. ASCE 41 provides a detailed formula to relate the actual top deflection to the spectral deflection, based on these principles, which take account of building period, height, hysteresis characteristics, etc.

Having calculated the top deflection of the building, the static pushover analysis is used to calculate the forces and plastic strains throughout the structure that correspond to this top deflection. This process has been automated in many computer programs.

3.7.9.4 Capacity spectrum method of ATC 40

The second method is the 'capacity spectrum' method of ATC40 (ATC, 1996). This appears much more complex, but once mastered can provide a good insight into the processes and assumptions involved. It forms one of the key theoretical bases of the direct displacement design methods referred to in Section 3.8. The method uses the pushover curve to define an equivalent viscous linear system, called the 'substitute structure' with a secant stiffness corresponding to the maximum displacement and a level of viscous damping related to the hysteretic damping in the real structure (Figure 3.26). The steps involved are (in outline) as follows. A fuller description is given by Chopra (2012).

Figure 3.26 Determining the period and damping of the substitute structure from a pushover curve

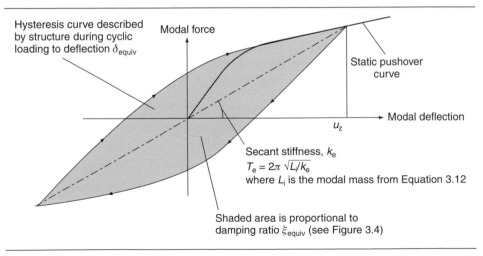

Hysteresis curve described by structure during cyclic loading to deflection δ_{equiv}

Modal force

Static pushover curve

Modal deflection

u_z

Secant stiffness, k_e

$T_e = 2\pi \sqrt{L_i/k_e}$

where L_i is the modal mass from Equation 3.12

Shaded area is proportional to damping ratio ξ_{equiv} (see Figure 3.4)

1   Transform the forces and deflections in the pushover curve to modal quantities, in order to reduce the real MDOF structure to an equivalent SDOF structure corresponding to the predominant mode of vibration. This SDOF structure is known as the 'substitute structure'. Referring to Equations 3.10 and 3.11, it can be shown that this involves division of the forces and deflections by the modal factor

$$L_i/M_i \tag{3.16}$$

2   Here, the suffix i refers to the predominant mode of deformation in the earthquake, which will usually be the first mode. A typical value of the modal factor for regular framed buildings is 1.25. Of course, the mode shape will change once the structure yields, and so the modal factors will also change slightly. However, this theoretical change is small compared to the other approximations involved in the method.

3   Calculate the modal mass, L_i (Equation 3.12). For regular framed buildings, this is approximately 60% of the total mass for the first mode; once again it will change slightly when the structure yields.

4   Make an estimate of the top deflection under the design ground motions. This can be based on the deflection the structure would have experienced had it remained elastic, which (as noted above for the target displacement method) is often a good approximation. Divide this by the modal factor (L_i/M_i) to get the equivalent modal deflection δ_{equiv}.

5   From the static pushover curve, find the slope of the secant stiffness corresponding to this equivalent modal deflection (Figure 3.26).

6   Calculate the period T_e and damping ξ_{equiv} of an equivalent linear SDOF system corresponding to this equivalent modal deflection (Figure 3.26). Note that, with increasing deflection, both T_e and ξ_{equiv} increase. Priestley et al. (2007) provide a standard formulae for ξ_{equiv}, depending on the structural form involved, and these are plotted in Figure 3.28(c).

7   From the response spectrum of the design ground motions, calculate the maximum displacement of the substitute structure u_{max} corresponding to T_e and ξ_{equiv}.

8   If u_e (the initial guess) differs from δ_{max} (the value found from the response spectrum), repeat from step 4 with a modified value of u_{max}, and iterate until satisfactory convergence is achieved.

9   The top displacement of the real structure is then given as $u_{\text{max}}(L_i/M_i)$.

This analysis can be achieved more directly by plotting the design spectrum in the form shown in Figure 3.8, that is, as spectral acceleration against spectral displacement for various damping levels. The static pushover curve can also be plotted on this curve, provided the necessary transformations are made. First it must be converted to modal quantities by dividing forces and deflections by the modal factor (L_i/M_i), as discussed before. The modal force must then be converted to a modal acceleration by dividing by the modal mass L_i.

The advantage of this method is that any point on the static pushover curve represents a particular value of structural period. A particular point on one of the capacity spectrum curves also represents a structural period (Figure 3.8). Therefore, the point at which the push-over curve intersects one of the capacity spectrum curves represents a structural period,

Figure 3.27 Capacity spectrum graphical method for determining displacements

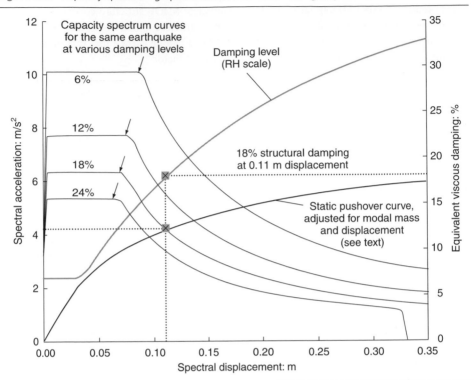

displacement and acceleration demand that is common to both pushover and spectrum curves. What is not necessarily in common, however, is the structural damping. By drawing a series of capacity spectrum curves at different damping levels, the damping level can be found where the intersection point implies a structural damping level achieved by the structure at its peak displacement. In the example of Figure 3.27, the peak displacement is found by the intersection of the 18% damped spectrum with the pushover curve; the spectral acceleration is 4.2 m/s² and the spectral displacement is 0.11 m. By checking the damping level corresponding to this displacement from the appropriate curve, it can be seen that the structural damping does in fact equal 18%.

Having found the peak or spectral displacement of the equivalent SDOF structure, it must be reconverted to a top displacement of the real structure by multiplying by the modal factor (Equation 3.16).

3.7.9.5 Interpretation of results

Both target displacement and capacity spectrum methods allow an estimate of the maximum seismic deflection to be made from a conventional response spectrum. This deflection can then be substituted back into the original static pushover analysis, and the corresponding degree of yielding in the structure can be established. For example, the rotation of plastic hinges in the

Table 3.4 Typical acceptance criteria for plastic hinge rotations in concrete frame members resisting seismic loads

| | Acceptance criteria: plastic rotation angle in radians | | |
| | Performance level | | |
	Immediate occupancy	Life safety	Collapse prevention
Beams			
Low shear, well confined	0.5–1%	1–2%	2–2.5%
High shear, well confined	0.5%	0.5–1%	1.5–2%
Low shear, poorly confined	0.5%	1%	1.0–2%
High shear, poorly confined	0.15%	0.5%	0.5–1%
Columns			
Low axial load, well confined	0.5%	1.2–1.5%	1.6–2.0%
High axial load, well confined	0.3%	1.0–1.2%	1.2–1.5%
Low axial load, poorly confined	0.5%	0.4–0.5%	0.5–0.6%
High axial load, poorly confined	0.2%	0.2%	0.2–0.3%

beams can be found. These quantified measures of local yielding are correlated with the degree of damage that the structure would experience when experiencing the calculated maximum deflection. For example, the rotation of plastic hinges at the ends of beams provides a measure of local yielding. ASCE 41 (ASCE, 2006) provides tables that relate the extent of local yielding to one of the structural performance objectives defined in Table 2.2. A typical table for reinforced concrete members forming part of a seismic resisting frame is shown in Table 3.4. Lower acceptance criteria would apply to members governed by shear failure rather than flexure, but high criteria would apply to structural members ('secondary' elements) not relied on to resist seismic loads. Annexes A and B of EC8 part 3 provide analytical expressions for acceptance criteria in steel and concrete elements, and annexe B of EC8 part 2 gives an analytical method for concrete plastic hinges forming in bridge piers.

3.8. Displacement-based design

3.8.1 Introduction to displacement-based design

For over 20 years, ductility modified RSA, coupled with capacity design checks and seismic detailing rules, has formed the main basis for seismic codes. The evidence from earthquake damage over this period is that this basis generally results in safe buildings that protect their inhabitants from injury, although it is much less effective in minimising economic damage. However, RSA is an essentially linear method that gives no information on the distribution of plasticity within a structure, or on the important changes in stiffness that take place as a structure (particularly a concrete one) undergoes cyclic loading past the point of yield.

As described in the previous section, 'displacement-based' methods of analysis are being developed as a replacement to RSA-based methods (often referred to, rather misleadingly in the author's view, as 'force-based' methods), although a complete and general consensus on how they should be implemented in codes has yet to be achieved. In direct DBD of new buildings, developed by Priestley and his co-workers, displacement is set as the principal design objective, chosen to be consistent with achieving a given performance objective (see Table 2.1), such as life safety or damage limitation. Many other DBD methodologies have also been proposed in the past two decades (see Sullivan *et al.*, 2003) but the direct DBD method is probably the most developed.

Before outlining DBD, it is helpful to set out how much it differs in concept from conventional design methods. In the latter, the initial stiffness (i.e. before the structure has yielded) is used to determine the building's structural period. The corresponding spectral acceleration, modified by a specified level of displacement ductility, is then used to calculate the design base shear. However, initial stiffness is difficult to establish with certainty, particularly in concrete structures (see Section 8.4.3). Moreover, the link between performance and displacement ductility is also very imperfect. Hence the two bases for choosing design strength – initial stiffness and displacement ductility – are both highly uncertain. In what is claimed as a major advantage of DBD, the design displacement, known because it is set by the performance objective, effectively sets the strength needed to achieve that objective. There is thus a much more direct link between the governing design parameters and the performance goal, which therefore should be satisfied with greater reliability than in conventional methods of design.

A brief explanation of DBD now follows; a fuller summary is given in Section 2.5 of Christopoulos and Filiatrault (2006), while Priestley *et al.* (2007) give an extended exposition.

3.8.2 Outline of displacement-based design

DBD is based on the capacity spectrum method of analysis described in Section 3.7.9.4, which involves modelling a building as an equivalent SDOF system, called the 'substitute structure'. Using certain simplifying assumptions, the design deflection of the substitute structure is chosen at the outset to be consistent with achieving an intended performance goal; it is generally the lesser of the following two values

1 the deflection corresponding to the maximum inter-storey drift which can be tolerated by non-structural elements
2 the deflection corresponding to the maximum tolerable strain in the most critical structural member.

In principle, the choice of design deflection is straightforward; in practice, as noted in Section 3.8.3, it poses considerable difficulties.

The yield deflection of the substitute structure is also calculated, using approximations. The ratio of design to yield displacement can then be found; this is the displacement ductility demand, which determines the effective damping level of the substitute structure. Priestley *et al.* (2007) provide equations for the relationship between displacement ductility and effective damping, which depend on the type of structure (concrete frame, concrete shear wall and so on) and its associated hysteretic energy dissipation during seismic loading (see Figure 3.28(c)).

Figure 3.28 Fundamentals of direct displacement based design (reproduced from Priestley *et al.*, 2007, courtesy of IUSS Press)

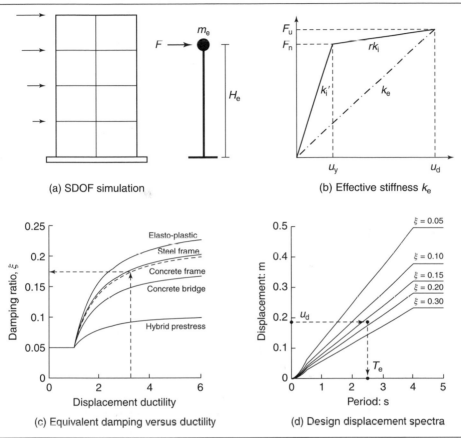

(a) SDOF simulation

(b) Effective stiffness k_e

(c) Equivalent damping versus ductility

(d) Design displacement spectra

So far, in stark contrast to conventional methods of design, the process has been independent of the design ground motions. At this stage, a design response spectrum is introduced, which will usually be obtained from a code of practice; however, in a reverse of the normal procedure, the design (spectral) displacement is used to determine the effective period of the substitute structure rather than the other way round (see Figure 3.28(d)). The base shear V_{base}, and hence the required strength of the structure, is then determined from the following relationships.

$$T_e = 2\pi\sqrt{m_e/k_e} \quad \text{(see Equation 3.1)} \tag{3.17}$$

Rearranging Equation 3.17 gives

$$k_e = 4\pi^2 m_e/T_e^2 \tag{3.18}$$

Hence, by relating force to stiffness times deflection

$$V_{\text{base}} = k_e\Delta_d = (4\pi^2 m_e/T_e^2)\Delta_d \tag{3.19}$$

Where

$k_e =$ effective secant stiffness at the design displacement (Figure 3.28(b))

$T_e =$ effective period of substitute structure at the design displacement (Figure 3.28(d))

$V_{base} =$ seismic shear in substitute structure at the design displacement, and hence design base shear for the building

$\Delta_d =$ design displacement of the substitute structure, corresponding to the chosen performance objective

$$= \sum_{i=1}^{n} m_i \Delta_i^2 \bigg/ \sum_{i=1}^{n} m_i \Delta_i \qquad (3.20)$$

$m_e =$ effective mass of the substitute structure

$$= \left(\sum_{i=1}^{n} m_i \Delta_i \right)^2 \bigg/ \sum_{i=1}^{n} m_i \Delta_i^2 \qquad (3.21)$$

Where

$\Delta_i =$ design deflection of the building at level i, corresponding to the chosen performance objective. How this is established is discussed in Section 3.8.3.

$m_i =$ mass of the building at level i

$n =$ total number of levels (storeys) in the building.

In Equations 3.20 and 3.21, note that the term $\sum_{i=1}^{n} m_i \Delta_i$ is the equivalent of L_i in Equation 3.12, and $\sum_{i=1}^{n} m_i \Delta_i^2$ is the equivalent of M_i in Equation 3.13.

At this stage, the process essentially reverts to that used in conventional design, although there are differences of detail. The base shear, found from Equation 3.19, is distributed up the structure in accordance with the inelastic deflected shape, and the resulting structural forces are used to set the strength of the plastic yielding mechanisms – for example, flexure at the ends of beams in a frame structure or tension in steel braces. The other elements are then designed for protection from yielding by a capacity design procedure.

In principle, these procedures are straightforward, and have been set out in code format by Sullivan *et al.* (2012). In practice, the summary above does not touch on a number of issues, and in particular the choice of design displacement needs further discussion.

3.8.3 Choosing the design displacement: introduction

As previously explained, a key feature of DBD is that the main parameter driving subsequent design decisions – displacement – can be linked much more directly and reliably to seismic performance than is the case for conventional design. Moreover, displacement can be used to account for the performance of both non-structural and structural elements. How this is done is now described.

3.8.4 Choosing the design displacement: limiting non-structural damage

In DBD, storey drift – the deflection of one floor relative to the floor below – is used as the proxy for non-structural damage. It is a rather crude measure – for example, drift caused by

Table 3.5 Non-structural performance levels and proposed storey drifts (data taken from Sullivan *et al.*, 2012)

Performance level	Storey drift		
	No damage	Repairable damage	No collapse
Buildings with brittle non-structural elements	0.4%	2.5%	No limit[a]
Buildings with ductile non-structural elements	0.7%	2.5%	
Buildings with non-structural elements designed to sustain building displacements	1.0%	2.5%	

These are peak storey drifts during the earthquake. Sullivan *et al.* (2012) recommend that residual drifts at the end of the earthquake are also checked against lower limits, once the initial design has been completed. This would require a non-linear time history analysis
[a] No limit is given for non-structural-related drift at collapse prevention because it is assumed that the structural related limit (Table 3.6) is the only criterion of interest.

a rigid body rotation of the structure will cause much less damage than a shearing deformation, and some non-structural elements are sensitive to accelerations rather than deflections (see Chapter 12). Also, some non-structural elements can tolerate more drift than others – or can be designed to do so. These factors need to be borne in mind when setting drift limits. Table 3.5 sets out the maximum drifts recommended by Sullivan *et al.* (2012).

Given a maximum drift and a deflected shape, the deflections corresponding to the chosen performance goal can easily be calculated at all levels of the building by factoring the deflected shape to a point at which the drift in the critical storey is reached. Priestley *et al.* (2007) propose inelastic deflected shapes for various types of building structures. The deflection of the substitute structure Δ_d is then calculated from Equation 3.20.

3.8.5 Choosing the design displacement: limiting structural damage

Relating design displacement to structural performance is less straightforward. In most cases, DBD uses the maximum plastic strain in the structure to determine structural performance. Table 3.6, adapted from Sullivan *et al.* (2012), gives suggested strain values. It is clear that increasing displacement causes an increase in plastic strain, but a relationship between the two needs to be established. The methods described below are approximate; they assume that the geometry of the structure – for example, overall beam and column dimensions, beam spans and storey heights in a frame structure – have already been chosen, but the amount of reinforcement in concrete elements or precise steel section sizes has yet to be determined.

There is a clear relationship between displacement and plastic strain for concrete cantilever structural walls, because the plasticity will usually be concentrated at a single plastic hinge at the base of the wall. Here the post-yield deflected shape can be expressed as a sum of contributions from the elastic response of the (non-yielding) upper part, plus the effect of plastic rotation of the base hinge (Figure 3.29). As maximum plastic strain and plastic hinge rotation can be linked by the methods described in Section 8.4.2, a design deflection corresponding to a given performance

Table 3.6 Structural performance levels and proposed corresponding plastic strains (data taken from Sullivan *et al.*, 2012)

Performance level	Limiting strain in plastic hinges		
	No damage	Repairable damage	No collapse
Concrete compressive strain	0.4%	$0.4\% + 1.4(\rho_v f_{yh}(\varepsilon_{su})_t)/(f'_{cc})^{a,b}$	1.5 times repairable damage limit
Longitudinal reinforcement tensile strain	1.5%	$0.6(\varepsilon_{su})_l$ or 5% if less	1.5 times repairable damage limit
Structural steel strain – class 1 sections, flexural plastic hinges[c]	1%	No limit[d]	No limit[d]
Structural steel strain – class 2 and 3 sections, flexural plastic hinges[c]	Yield strain	Yield strain	Yield strain

[a] ρ_v = volumetric ratio of transverse hoops or spirals, f_{yh} = yield stress of transverse confining reinforcement, f'_{cc} = expected concrete compressive strength, $(\varepsilon_{su})_t$ = ultimate strain of transverse (confining) reinforcement, $(\varepsilon_{su})_t$ = ultimate strain of longitudinal (flexural) reinforcement
[b] Sufficient transverse confinement steel must be provided to sustain the maximum concrete compressive strain at the no-collapse limit state (see Figure 8.15 in Chapter 8). This check would be performed at detailed design stage
[c] The class of a steel section (i.e. class 1, 2 or 3) relates to its 'compactness', which is defined in Section 9.3.3
[d] The section must satisfy the plastic hinge rotation capacity requirements of clause 5.6 of Eurocode 3 part 1-1 (CEN, 2005)

Figure 3.29 Linking deflection in a structural wall to base hinge rotation

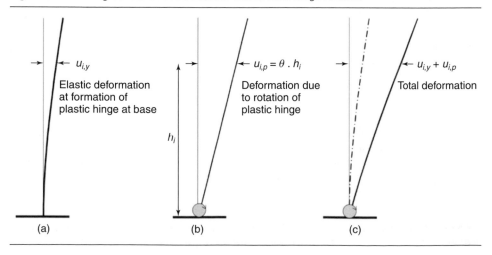

(a) $u_{i,y}$
Elastic deformation at formation of plastic hinge at base

(b) $u_{i,p} = \theta \cdot h_i$
Deformation due to rotation of plastic hinge
h_i

(c) $u_{i,y} + u_{i,p}$
Total deformation

level is relatively straightforward to establish. Alternatively, hinge rotation limits for shear walls corresponding to various limit states can be found in ASCE 41 (ASCE, 2006).

In steel and concrete frames, many hinges form and the link between strain and building deflection is much harder to establish. However, frames are of course more flexible than walls, and it is more likely that the drift limits of Table 3.5 will govern before the strain limits of Table 3.6. Sullivan *et al.* (2012) recommend that the initial design of frames is performed on the basis of drift-limited deflection and then a non-linear static or dynamic analysis is used to determine the maximum plastic hinge rotations. These rotations can be used to derive maximum plastic strains (see e.g. Section 8.4.2) or can be used to compare directly with plastic rotation limits such as those given in Table 3.4. If the plastic strain is found to be excessive, another iteration of design is needed.

3.8.6   Calculating the yield displacement

Having established the design displacement, the yield displacement must also be found, as follows. As discussed further in Section 8.4.3, Priestley *et al.* (2007) propose that the curvature of concrete and steel sections at yield is set by the section geometry, rather than the strength. Thus, for a rectangular concrete shear wall, the yield curvature φ_y can be taken as

$$\varphi_y = \frac{2.10\varepsilon_{yk}}{l_w} \tag{3.22}$$

where l_w is the length of the wall base, and ε_{yk} is the characteristic yield strain of the reinforcing steel (see Table 8.3). The deflection at yield of a cantilever shear wall can be related to this yield curvature at its base, so that the deflected shape at yield (Figure 3.29(a)) can be established from the wall geometry, without knowing its yield strength. For a multistorey cantilever wall building that is reasonably uniform with height, the yield deflection u_y of the substitute structure can be shown to be approximately equal to

$$u_y = 0.2H^2\varphi_y \tag{3.23}$$

on the assumption that the applied lateral loading increases linearly with height, where H is the total height of the wall.

Similar procedures can be used to derive the yield deflection of unbraced steel and concrete frames. Priestley *et al.* (2007) show that it can be taken with sufficient accuracy as

$$u_y = K\varepsilon_{yk}H_e\frac{L_b}{h_b} \tag{3.24}$$

where $K = 0.5$ for reinforced concrete frames, or 0.65 for steel frames with full strength rigid joints, ε_{yk} = characteristic yield strain of the main reinforcing steel, H_e = effective height of the substitute structure, approximately equal to $\frac{2}{3}H$ in multistorey regular frames, and H in single storey frames, where H is the total height of the building, L_b = span of beams between column centre lines and h_b = overall depth of beams.

In buildings in which the geometry of the walls or frames differs across the building, application of Equations 3.23 and 3.24 is less straightforward. Priestley *et al.* (2007) discuss the case of such structures with less regular geometry, and also that of structures that combine walls and frames.

Having established both design and yield displacements of the substitute structure, the ratio of the two – the displacement ductility – can be established, and hence the effective damping of the substitute structure can be determined, as explained previously (Figure 3.28(d)). Using the appropriate design response spectrum, required member strengths can then be established as explained in Section 3.8.2.

REFERENCES

ASCE (1998) ASCE 4-98: Seismic analysis for safety-related nuclear structures. American Society of Civil Engineers, Reston, VA, USA. (NB A revised version is due to be published in 2013).

ASCE (2006) ASCE/SEI 41-06: Seismic rehabilitation of buildings. American Society of Civil Engineers, Reston, VA, USA. (NB A revised version is due to be published in 2013 as ASCE/SEI 41-13).

ASCE (2010) ASCE/SEI 7-10: Minimum design loads for buildings and other structures. American Society of Civil Engineers, Reston, VA, USA.

ATC (1996) ATC-40: Seismic evaluation and retrofit of concrete buildings. Applied Technology Council, Redwood City, CA, USA.

Canterbury Earthquakes Royal Commission (2012) *Final Report Volume 2*. See http://canterbury. royalcommission.govt.nz/.

CEN (2004) EN 1993-1-1: 2004: Design of structures for earthquake resistance. Part 1: General rules, seismic actions and rules for buildings. European Committee for Standardisation, Brussels, Belgium.

CEN (2005) EN 1998-1: 2005: Design of steel structures. Part 1-1: General rules and rules for buildings. European Committee for Standardisation, Brussels, Belgium.

Chopra AK (2012) *Dynamics of Structures: Theory and Applications to Earthquake Engineering*, 4th edn. Prentice-Hall, Upper Saddle River, NJ, USA.

Christopoulos C and Filiatrault A (2006) Principles of Passive Supplemental Damping and Seismic Isolation. IUSS Press, Pavia, Italy.

Clough RW and Penzien J (1993) *Dynamics of Structures*. McGraw Hill, New York, NY, USA.

Der Kiureghian A, Keshishian P and Hakobian A (1997) *Multiple support response spectrum analysis of bridges including the site-response effect and the MSRS code*. Report no. UBC/EERC-97/02, College of Engineering to the California Department of Transportation, University of California at Berkeley, California, CA, USA.

Ellis BR (1980) Determining the natural periods of buildings. *Proceedings of the Institution of Civil Engineers* **69**: 763–776.

Fajfar P (2000) A non-linear analysis method for performance-based seismic design. *Earthquake Spectra* **16/3(August)**: 573–592.

Fenwick RC and Davidson BJ (1987) Moment redistribution in seismic resistant concrete frames. In *Pacific Conference on Earthquake Engineering, New Zealand*, Vol. 1, pp. 95–106.

Fenwick RC, Davidson BJ and Chung BT (1992) *P*–delta actions in seismic resistant structures. *Bulletin of the New Zealand Society for Earthquake Engineering* **25**: 56–69.

Fitzpatrick A (1990) Structural design of Century Tower, Tokyo. *The Structural Engineer* **70/18(September)**: 313–317.

Gupta AK (1990) *Response Spectrum Method in Seismic Analysis and Design of Structures*. Blackwell Scientific Publications, Cambridge, MA, USA.

Irvine HM (1986) *Structural Dynamics for the Practising Engineer*. Allen and Unwin, London, UK.

NZS (New Zealand Standards) (2006) NZS 3101: Part 1: The design of concrete structures. NZS, Wellington, New Zealand.

Priestley MJN, Calvi GM and Kowalsky MJ (2007) *Displacement-based Seismic Design of Structures*. IUSS Press, Pavia, Italy.

Smith RJ, Merello R and Willford MR (2010) Intrinsic and supplementary damping in tall buildings. *Proceedings of the Institution of Civil Engineers – Structures and Buildings* **163(SB2)**: 111–118.

Sullivan TJ, Calvi GM, Priestley MJN and Kowalsky MJ (2003) The limitations and performances of different displacement based design methods. *Journal of Earthquake Engineering* **7(special issue 1)**: 201–241.

Sullivan TJ, Priestley MJN and Calvi GM (eds) (2012) *DBD12: A model code for the displacement-based design of structures*. IUSS Press, Pavia, Italy.

Earthquake Design Practice for Buildings
ISBN 978-0-7277-5794-4

ICE Publishing: All rights reserved
http://dx.doi.org/10.1680/edpb.57944.083

Chapter 4
Analysis of soils and soil–structure interaction

Seismic loading is unique in that the medium (i.e. the soil) which imposes the loading on a structure also provides it with support.

This chapter covers the following topics.

- Soil properties for seismic design.
- Liquefaction.
- Site amplification, topographical effects and fault breaks.
- Slope stability.
- Soil–structure interaction.

The designer of earthquake-resistant structures needs some understanding of how soils respond during an earthquake; not only is this important for the foundation design itself, but the nature of soil overlaying bedrock has a crucial modifying influence on the seismic hazards posed by the site. This chapter gives an introduction to the subject.

4.1. Soil properties for seismic design
4.1.1 Introduction
The response of soils to earthquake excitation is highly complex and depends on a large range of factors, many of which cannot be established with any certainty. The discussion that follows is intended to highlight the important features that apply to the most standard cases; often, specialist geotechnical expertise will be needed to resolve design issues encountered in practice.

4.1.2 Soil properties for a dynamic analysis
In common with any structural system, the dynamic response of soil systems depends on inertia, stiffness and damping. These three properties are now discussed in turn; a more complete discussion is provided by Pappin (1991).

4.1.2.1 Inertia
This can easily be determined from the soil's bulk density, which for most clays and sands is in the range 1700–2100 kg/m^3. There are exceptions, however; for example, Mexico City clay has a bulk density of only 1250 kg/m^3.

4.1.2.2 Stiffness and material damping

Generally, the shear behaviour of soils will be of most concern; the behaviour in compression, characterised by the bulk stiffness, is less important. This is because the bulk stiffness of saturated materials is very high, being approximately equal to that of water divided by the soil porosity. For compression effects (for example, the transmission of P or seismic compression waves, important for vertical motions) the soil therefore acts in an essentially rigid manner with little modification due to dynamic effects. Soils with significant proportions of air may have much lower bulk stiffness, which may, therefore, need consideration. Further discussion here is confined to shear behaviour, which dominates response to horizontal seismic motion.

Figure 4.1 shows a typical cyclic response of a soil sample under variable-amplitude shear excitation. There are three important features to note when comparing the small with large shear strain response.

First, the stiffness, determined from the slope of the stress–strain curve, decreases with shear strain.

Second, the area contained within the hysteresis loop formed by the stress–strain curve increases with shear strain. As explained in Chapter 3, this area is directly related to the level of hysteretic damping. Therefore, soil damping increases with strain level, as more energy is dissipated hysteretically. It is important to note that the dissipated energy is generally much more dependent on amplitude than the rate of loading. This is in contrast with

Figure 4.1 Idealised stress–strain behaviour of a soil sample in one-dimensional shear

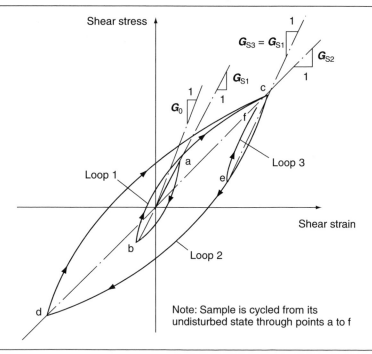

viscous damping, in which the damping resistance depends on velocity, and so for example tends to zero for very slow rates of cycling. No such reduction to zero occurs in soils. Soil damping is thus hysteretic and therefore the usual assumption of viscous damping is invalid. Analysis in the frequency domain (Section 4.4.2) is one way of accounting for this. Time history analysis with the soil modelled to account for the sort of stress–strain behaviour shown in Figure 4.1 is another possibility.

Third, after a large shear strain excursion, the hysteresis loop reverts to its original shape for a small cyclic excitation; that is, loop 3 in Figure 4.1 is similar in shape to loop 1, despite the intervening loop 2. Therefore, both stiffness and damping under cyclic loading are functions primarily of shear strain amplitude, not absolute shear strain.

4.1.3   Stiffness of sands and clays

Figure 4.2 shows typical relationships between shear strain amplitude and shear stiffness. Note the very large reduction in stiffness for shear strains exceeding 0.01%. The values for clays are for overconsolidation ratios (OCR) of 1–15. It can be seen that the stiffness of clays becomes similar to that of sands as the plasticity index (PI) approaches zero.

In Figure 4.2, the stiffness is expressed as a ratio of secant shear stiffness at the shear strain of interest, G_s to the small strain stiffness, G_0. G_0 can be measured directly on site from measurements of shear wave velocity (see Kramer, 1996) or from more conventional measurements, using empirical relationships. For sands, these relate G_0 to the blow count N

Figure 4.2 Relationship between normalised shear stiffness G_s/G_0 and cyclic shear strain

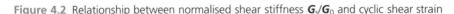

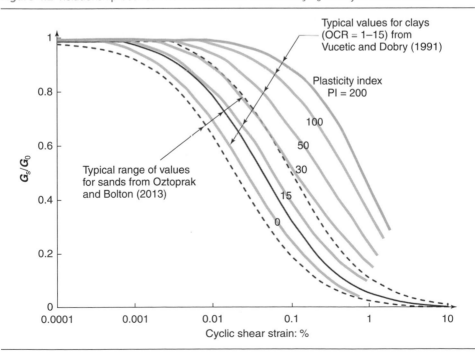

Table 4.1 G_0/c_u values (data taken from Weiler, 1988)

Plasticity index: %	Overconsolidation ratio		
	1	2	3
	G_0/c_u		
15–20	1100	900	600
20–25	700	600	500
34–35	450	380	300

for 300 mm penetration in the standard penetration test (SPT); a typical correlation between G_0 (in MPa) and blow count used in Japanese practice (Imai and Tonouchi, 1982) is $G_0 = 14.4 N^{0.68}$, but there is considerable scatter in the data. For clays, G_0 can be determined as a ratio of the undrained shear strength, c_u, as shown in Table 4.1.

4.1.4 Material damping of sands and clays

Figure 4.3 shows typical values of damping ratio; once again, the values for clay approach those for sand as the PI reduces. Note the marked increase in damping as shear strains rise above 0.001%, caused by the hysteretic energy dissipation discussed in Section 4.1.2. Stokoe *et al.* (1986) advise that the lower bound of the damping values shown for sands on the figure may be generally appropriate.

Figure 4.3 Relationship between material damping ratio and cyclic shear strain

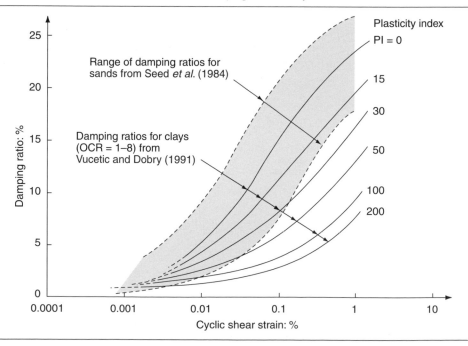

4.1.5   Stiffness and damping properties of silts

Silts have properties equivalent to clays with a PI of about 15% (Khilnani *et al.*, 1982).

4.1.6   Strength of granular soils

The cyclic loading imposed on soils during an earthquake may seriously affect soil strength. Granular materials, such as sands and gravels, rely for their strength on interparticulate friction. Although the angle of friction, ϕ', is not affected by cyclic loading, the effective stress between particles may be reduced in saturated soils if pore water pressures rise during an earthquake. The reduction in effective stress in turn reduces the shear strength. A rise in pore water pressure will occur if a loose granular material tries to densify under the action of earthquake shaking and the pressure has not had time to dissipate. In time, the pore water will find drainage paths, the pressure will release and the strength will be restored. This may, however, take a few minutes to occur, and dramatic failures can arise in the meantime (Figure 4.4). This is the phenomenon of liquefaction, which is discussed more fully in Section 4.2. The strength of granular soils is scarcely affected by the rate of loading.

4.1.7   Strength of cohesive soils

Clay particles are weakly bonded and are not subject to densification under cyclic loading. Therefore, they are unlikely to liquefy. The short-term undrained shear strength c_u, however, is affected both by the rate of loading and by the number of cycles of loading. These are now discussed in turn.

Rate effects may give rise to strength increases of up to 25% in soft clays under seismic loading conditions, compared with static strength, although the increase is less for firm clays and very stiff clays are insensitive to rate effects (Pappin, 1991).

Strength reduction under cyclic loading is progressive with the number of cycles. It is highly dependent on the OCR. Clays with high OCR are much more sensitive to cyclic loading, and their strengths revert to normally consolidated values with increasing numbers of load cycles. The strength loss is permanent, unlike that due to pore water pressure increase in sands. A normally consolidated clay (OCR = 1) can sustain 10 cycles of 90% of the undrained static shear strength c_u; this drops to 10 cycles at about 75% c_u for a clay with OCR of 4 and to 10 cycles at about 60% c_u for OCR of 10 (see Andersen *et al.*, 1982). Ten cycles of extreme loading would be unusually large except in very large-magnitude earthquakes. Further information on cyclic softening in clays and plastic silts is given by Idriss and Boulanger (2008).

4.2.   Liquefaction

4.2.1   Assessing the liquefaction potential of soils

Liquefaction is a phenomenon that occurs in loose, saturated granular soils under cyclic loading. Under such loading, pore water pressures between the soil particles build up as the soil tries to densify, until the pore water pressure overcomes the forces between soil particles (Figure 4.5) (i.e. the effective stress drops to zero). At this point, uncemented granular soils lose their shear strength, because this relies on friction forces between the soil particles. Only certain types of soils are susceptible to liquefaction, and in order for it to occur, all four of the following conditions must be present.

Figure 4.4 Foundation failure due to liquefaction

1   A soil that tends to densify under cyclic shearing.
2   The presence of water between the soil particles.
3   A soil that derives at least some of its shear strength from friction between the soil
     particles.
4   Restrictions on the drainage of water from the soil.

Condition (1) implies a loose soil; common examples are naturally deposited soils that are
geologically young (Holocene deposits younger than 12 000 years) or manmade hydraulically
placed fills. Densification and also cementation between particles (see condition (3)) tend to
increase with age, and so older deposits are less susceptible to liquefaction. Conversely,

Figure 4.5 Shearing of a loose, water-saturated granular soil in the process of liquefying (modified, with kind permission, from EERI (1994))

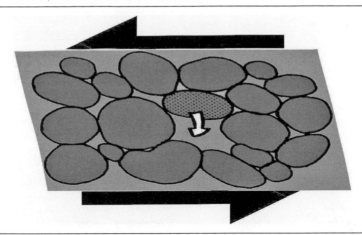

land reclaimed by pumped dredged material is highly susceptible, unless suitable measures are undertaken. Table 4.2 provides a more detailed list of the susceptibility of soils.

Condition (2) necessitates that the soil is below the water table, although liquefaction is unlikely where the water table depth is deeper than 15 m (Youd, 1998).

Condition (3) means that granular soils are the most likely to liquefy, although silts still have some potential for liquefaction. Clays do not liquefy but suffer the potential for cyclic strength degradation referred to in Section 4.1.7; the transition between these behaviours is discussed by Idriss and Boulanger (2008).

Condition (4) means that large-grained soils such as gravels are unlikely to liquefy, because any potential build-up of pore water pressure is usually dissipated rapidly by the free drainage available. As grain size decreases, the resistance to pore water drainage increases, but offsetting this is an increase in cementation between particles. The main risk of liquefaction therefore occurs in sands. However, silts may still liquefy, while coarse sands can liquefy if they are contained as lenses in larger areas of clay that inhibit dissipation of excess pore water pressures.

4.2.2 Simplified procedure for assessing liquefaction

Having established that a soil poses a potential liquefaction risk, the overall risk of it actually occurring must be related to the seismic hazard at the site; clearly the more intense the motions, the greater the risk. A simplified assessment procedure, but still the one that is most commonly used, involves the following steps, which are based on Section 3.3 of an Earthquake Engineering Research Institute (EERI) monograph by Idriss and Boulanger (2008). Idriss and Boulanger (2010) provide more comprehensive data on the assessment of

Table 4.2 Susceptibility of sedimentary deposits to liquefaction during strong shaking for preliminary design purposes (reproduced from Youd and Perkins, 1978, with permission from ASCE)

Type of deposit	Age of deposit			
	<500 Years	Holocene	Pleistocene	Pre-Pleistocene
	Likelihood that cohesionless sediments, when saturated, would be susceptible to liquefaction			
(a) Continental deposits				
River channel	Very high	High	Low	Very low
Flood plain	High	Moderate	Low	Very low
Alluvial fan and plain	Moderate	Low	Low	Very low
Marine terraces and plains	–	Low	Low	Very low
Delta and fan delta	High	Moderate	Low	Very low
Lacustrine and playa	High	Moderate	Low	Very low
Colluvium	High	Moderate	Low	Very low
Talus	Low	Low	Very low	Very low
Dunes	High	Moderate	Low	Very low
Loess	High	High	High	Unknown
Glacial till	Low	Low	Very low	Very low
Volcanic tuff	Low	Low	Very low	Very low
Volcanic tephra	High	High	?	?
Residual soils	Low	Low	Very low	Very low
Sebka	High	Moderate	Low	Very low
(b) Coastal zone: delta and estuarine				
Delta	Very high	High	Low	Very low
Estuarine	High	Moderate	Low	Very low
(c) Coastal zone: beach				
High wave energy	Moderate	Low	Very low	Very low
Low wave energy	High	Moderate	Low	Very low
Lagoonal	High	Moderate	Low	Very low
Fore shore	High	Moderate	Low	Very low
(d) Artificial fill				
Uncompacted fill	Very high	–	–	–
Compacted fill	Low	–	–	–

liquefaction potential in a University of California at Davis (UCD) report, which is freely downloadable from the UCD website; the report complements and amplifies the material on liquefaction in the EERI monograph of 2008 referred to here. Assessment of liquefaction potential, however, requires judgement and experience, and the Eurocode 8 (EC8) manual (ISE/AFPS, 2010) notes that it remains a specialist task.

1   The effective shear stress τ_e occurring in the soil at the level of interest during a design earthquake must first be calculated; a preliminary estimate of τ_e can be made from the following set of equations.

$$\tau_e = 0.65\sigma_{vo}\frac{a_g}{g}r_d \tag{4.1}$$

$$r_d = \exp(\alpha(z) + \beta(z)M) \tag{4.2}$$

$$\alpha(z) = -1.012 - 1.126\sin\left(\frac{z}{11.73} + 5.133\right) \tag{4.3}$$

$$\beta(z) = 0.106 + 0.118\sin\left(\frac{z}{11.28} + 5.142\right) \tag{4.4}$$

where a_g is the peak ground acceleration (PGA), after allowing for soil amplification effects (m/s^2), g is the acceleration due to gravity (=9.81 m/s^2), σ_{vo} is the vertical total stress at the level of interest (i.e. the total gravity overburden pressure), r_d is a reduction factor (always less than 1), accounting for dynamic effects, as explained below, z is the soil depth in metres and M is the moment magnitude of the earthquake under consideration. From simple rigid body mechanics, it can be seen that the term $(\sigma_{vo}(a_g/g))$ in Equation 4.1 is the peak shear stress that would develop at the level of interest if all the soil above it moved laterally as a rigid body. In fact, this overestimates the stress because soil accelerations tend to reduce with depth, and the acceleration at the depth of interest is less than the PGA. r_d is an empirical reduction factor depending on soil depth and earthquake magnitude, which allows for these effects. A more rigorous analysis would base the peak shear stress on a simple one-dimensional shear beam model of the soil, for example using SHAKE (1991) (see Section 4.3.1).
A further reduction is needed because the peak value of shear stress, by definition, occurs only once during an earthquake, whereas the charts for liquefaction potential are based on laboratory tests for constant amplitude loading. Therefore, the peak shear stress is multiplied by a further reduction factor of 0.65 to obtain the effective peak stress τ_e; this factor can be seen to appear in Equation 4.1. This factor is in fact rather arbitrary, but for many years the standard liquefaction potential charts have been based on this factor; other factors would have been possible but the values on the charts would then have changed accordingly.

2   τ_e is divided by the vertical effective stress σ'_{vo} at the level of interest (i.e. overburden stress less pore water pressure without allowance for liquefaction effects), to calculate the 'cyclic shear stress ratio' (CSR).

$$CSR = \tau_e/\sigma'_{vo} \tag{4.5}$$

3   The CSR (the 'demand' in the terminology of Section 3.4.1) must then be compared with a measure of liquefaction potential (the 'supply'). This is expressed as the cyclic resistance ratio, CRR, defined as the ratio of cyclic to static shear strength.

4   Figure 4.6 shows the variation of CRR with SPT value. When the CSR exceeds the CRR calculated from Equation 4.5, liquefaction is likely (Figure 4.7). Figure 4.6 is

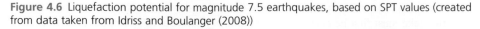

Figure 4.6 Liquefaction potential for magnitude 7.5 earthquakes, based on SPT values (created from data taken from Idriss and Boulanger (2008))

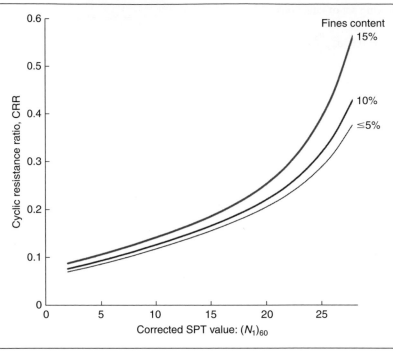

presented both for clean sands (fines content <5%) and for silty sands; the fines content is defined as the fraction of material finer than 0.075 mm. The charts for CRR are essentially empirical, being based on field observations of where liquefaction does or does not take place, and on laboratory tests of soil samples. They are presented for magnitude 7.5 earthquakes, for relatively low overburden pressure and no pre-existing static shear stress. Adjustments to allow for other conditions are now discussed.

5　Different magnitude earthquakes give rise to different numbers of cycles of loading, and liquefaction potential therefore increases with the size of the earthquake, independently of the peak shear stress. Therefore, to use Figure 4.6 for a magnitude earthquake other than 7.5, the CRR should be multiplied by a magnitude scaling factor (MSF); Table 4.3 presents values of MSF.

6　Idriss and Boulanger (2008) present correction factors to allow for the effect on CRR of both overburden pressures and static shear stresses. The effect of overburden pressures less than 2 atmospheres is quite small, but becomes significant for greater pressures, particularly when the SPT blow count exceeds 10. Idriss and Boulanger note that it is reasonable to neglect the effect of static shear in level sites or for shallow slopes, but advise that it should be included for steep slopes and for embankment dams.

Figure 4.7 Margin of safety between CSR (cyclic shear ratio) and CRR (cyclic shear ratio)

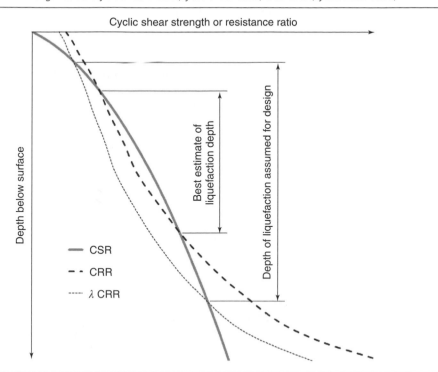

7   Figure 4.6 is based on the corrected value of SPT blow count in the soil, $(N_1)_{60}$, which is calculated as follows.

8   The SPT blowcount per 300 mm N_{SPT} is corrected to a standard value of effective vertical stress of 100 kPa by multiplying N_{SPT} by $(100/\sigma_{vo})^{1/2}$, where σ_{vo} is the effective vertical stress in kPa in the soil at the level of interest. EC8 part 5 (CEN, 2004) advises that the correction factor should lie between the values 0.5 and 2.

9   N_{SPT} is further corrected for energy ratio, by multiplying by (ER/60) where ER is the percentage of the potential energy from the hammer drop that gets delivered to driving the SPT probe (the rest being lost in friction, noise, heat, rod vibration and so on). ASTM (1986) gives a method for quantifying ER, and further discussion is provided by Abu-matar and Goble (1997).

10   $(N_1)_{60}$ in Figure 4.6 is therefore given by the following equation

$$(N_1)_{60} = N_{SPT} \, (100/\sigma'_{vo})^{1/2} \, (ER/60) \tag{4.6}$$

4.2.3   Margin between CRR and CSR

Once the CSR and CRR have been established, a check is needed that a sufficient margin exists between the two (Figure 4.7). EC8 part 5 (CEN, 2004) requires that under the 475-year return design event, CRR should be reduced by a factor λ when compared with CSR, and recommends that λ should be taken as 0.8. Cetin *et al.* (2004) provide details of a procedure to allow a probabilistic assessment to be made of liquefaction potential, rather than simply judging it to be either 'likely' or 'unlikely', and this could be used to inform the choice of margin chosen for a particular situation. A full discussion of appropriate margins is given by Idriss and Boulanger (2008).

4.2.4   Other procedures for assessing liquefaction potential

The correlations between SPT values and liquefaction potential shown in Figure 4.6 suffer from the drawbacks of all empirical relationships and are subject to very considerable uncertainty. Dynamic cone penetration test values (CPT) and shear wave velocity have also been correlated with liquefaction potential; CPT is thought to provide more reliable correlations than SPT, and shear wave velocity may be useful when CPT or SPT results are difficult to obtain. Idriss and Boulanger (2008) provide graphs for both CPT and shear wave velocity in a similar format to Figure 4.6, and discuss the relative merits of all three types of correlation.

Analytically based methods of assessing liquefaction risk have also been developed, whereby constitutive models of soil including pore water pressure generators are used in a dynamic finite element analysis. These models are still under development, and need to be supplemented by the more empirical measures described above.

4.2.5   EC8 procedure for assessing liquefaction

EC8 part 5 (CEN, 2004) provides a normative (i.e. mandatory) procedure for assessing liquefaction potential in annexe B, which has similarities to that described in Section 4.2.2, but with some important differences, as follows.

- The reduction factor r_d in Equations 4.1 and 4.2 is ignored; that is, implicitly assumed to be 1, which is a conservative assumption.
- The magnitude scaling factors recommended by EC8 are significantly different from those given in Table 4.3, and appear unconservative at earthquake magnitudes lower than 7.5 when compared with recent data.
- The recommended CRR curves at magnitude 7.5 are significantly different from those given in Figure 4.6 for silty sands, and appear unconservative when compared with recent data.

Table 4.3 Magnitude scaling factor (based on data from Idriss and Boulanger, 2008)

Surface wave magnitude M_s	Correction factor for CRR
5	(liquefaction unlikely)
5.5	1.7
6	1.5
6.5	1.3
7	1.1
7.5	1.0
8	0.9

Because of these differences, the EC8 manual (ISE/AFPS, 2010) recommends that the EC8 procedures should always be used in conjunction with specialist advice on the most recent developments in liquefaction assessment.

4.2.6 Consequences of liquefaction

Having established that the soils around a structure may liquefy, the consequences must be evaluated. The minimum consequence is that the densification associated with liquefaction gives rise to small local settlements, which may cause structural distress.

A much more serious consequence occurs when the reduction in shear strength caused by the liquefaction leads to a bearing failure (Figure 4.4). Retaining walls are particularly at risk, because they suffer not only from loss of bearing support but also from greatly increased lateral pressures, if the retained soil liquefies (Figure 4.8).

Lateral spreading can also occur, in which large surface blocks of soil move as a result of the liquefaction of underlying soil strata. The movements are usually towards a free surface such as a river bank, and are accompanied by breaking up of the displaced surface soil. Lateral spreading usually takes place on shallow slopes less than 3°. A dramatic example, which destroyed 70 houses, occurred during the Anchorage Alaska earthquake of 1964, when an area 2 km long by 300 m wide slid by up to 30 m (Figure 4.9).

The most catastrophic failure is a flow failure of soils on steep slopes (usually greater than 3°), which can give rise to displacements of large masses of soil over distances of tens of

Figure 4.8 Failure of dock wall in the port of Kobe, Japan, 1995

Figure 4.9 Liquefaction-induced lateral spreading, Alaska, 1964, showing destruction of a road and housing (courtesy of Karl V. Steinbrugge Collection, University of California, Berkeley)

metres. The flows may be composed either of completely liquefied soil, or of blocks of intact material riding on liquefied material (EERI, 1994). Movements can reach tens of kilometres, and velocities can exceed 10 km/h.

Design measures in the presence of liquefiable soils are discussed in Section 7.8.

4.3.   Site-specific seismic hazards

The next sections consider how the seismic hazard at a site may be affected by the local geology and how knowledge of the soil properties discussed in the previous sections can allow these hazards to be estimated.

4.3.1   Site amplification effects

The tendency of soft soils overlaying bedrock to amplify earthquake motions has already been discussed in Chapter 2, Section 2.6. In many cases, adequate allowance for these effects can be made by simple amplification factors provided in codes of practice. It should be noted that amplification tends to reduce with increased intensity of ground motions, because of the increase in soil damping and reduction in soil stiffness with shear strain amplitude (Figures 4.1 to 4.3). The seismic sections of the US code ASCE 7 (ASCE, 2010) allows for this effect; EC8 (CEN, 2004) does not contain a recommendation to allow for the effect, although some European countries, in their 'national annexes' to EC8, include it. For soil sites where the peak ground acceleration is less than approximately 15%, neglecting the effect may be unconservative.

In cases in which very soft materials are present, more sophisticated allowance should be made. Thus, at sites where soft clay layers are present that are deeper than 10 m and have a PI greater than 40, EC8 requires a site-specific calculation of the modification they cause in surface motions. For horizontal motions, it is usually sufficient to make this modification on the basis of simple one-dimensional shear beam models of the soil, using the soil properties discussed in Section 4.2. A range of bedrock motions appropriate to the site and to the depth of soil overlaying bedrock should be input to the base of the shear beam soil model, and the ratio of surface to bedrock motion should be calculated at a range of frequencies. These frequency-dependent amplification factors can then be used to modify design spectra appropriate for rock sites. A number of standard computer programs exist to perform this calculation; SHAKE (1991) is a well-known example. The techniques are fully discussed by Stewart et al. (2008).

Another issue that needs to be considered is the depth at which the applied seismic ground motion should be determined. Usually, buildings are designed for the surface ground motion but this may not always be appropriate. An example is when a building has a basement that is founded on rock but passes through soft soil. In this case the bedrock ground motion should be used and the basement walls designed for additional earth pressure as discussed in Section 7.7. Another case is when large diameter piles pass through soft clay over dense sand; the dense sand motion is then appropriate. Kelly (2009) contains an overview of this problem for application with ASCE/SEI 7 (ASCE, 2010).

4.3.2   Basin effects

One-dimensional shear beam models may not be adequate to describe site effects in alluvial basins where there is increasing evidence that more complex two and three-dimensional effects are at work, particularly at the basin edges (Day et al., 2008; Faccioli, 2002). These effects are not currently addressed in codes of practice, and even complex finite element modelling does not appear to yield reliable results (Adams and Jaramillo, 2002).

4.3.3   Amplification of vertical motion

The discussion so far has been on the amplification of horizontal motions. Vertical motions are much less affected; they depend mainly on the bulk rather than the shear modulus of the soil, and as the former changes less than the latter (particularly in saturated soils) when the earthquake waves pass from rock into the overlying soil, little amplification occurs.

4.3.4   Topographical effects

Damage to structures is often observed to be greater on the tops of hills or ridges than at their base. An example was seen at a housing estate in Vina del Mar after the 1985 Chilean earthquake. Celebi (1987) measured ground motions during aftershocks of this event, both at the ridge-top positions, where damage had been greatest, and at the ridge base; he found that at certain frequencies the former motions were over 10 times greater than the latter.

EC8 part 5 (CEN, 2004: annexe A) recommends that topographical effects should be accounted for in sites near long ridges and cliffs with a height greater than about 30 m; the recommended increase in acceleration is up to 40%. Faccioli (2002) provides further information.

4.3.5   Slope stability

Slope failures connected with soil liquefaction were discussed in Section 4.2.6. Even without liquefaction, the horizontal accelerations caused by an earthquake can dramatically reduce the factor of safety against movement of the slope. However, these reductions in the factor of safety are instantaneous and only lead to large soil movements if the peak forces tending to displace the slope exceed the restraining strength of the soil by a factor of at least 2; that is, when the instantaneous safety factor drops below 0.5. Relationships between the instantaneous safety factor and slope displacement were originally developed by Newmark (1965), and form the basis for many current methods both of slope design and also for checking the seismic stability of retaining structures. An excellent summary of the development of these methods is presented in Jibson (2011).

4.3.6   Fault breaks

Large earthquakes are almost always associated with rupture along fault lines. However, this rupture initiates at a depth of many kilometres and will rarely extend to the surface if the earthquake magnitude is 6 or less. Even for large earthquakes, a surface expression of the fault does not necessarily occur if large depths of soil overlay bedrock. The underlying fault movement (i.e. whether it consist of shear, tension or compression) also affects whether the fault reaches the surface.

For major active faults such as the San Andreas in California or the Northern Anatolian fault in Turkey, which have a well-recorded history of movement, the design issues are clear. Building structures should be sited away from them, and linear structures such as roads or pipelines should be designed to cope with possible fault movements. Generally, the width at risk should be taken as several hundred metres, allowing for the uncertainty as to where the fault may appear at the surface in future earthquakes. However, the potential activity of other faults may be much harder to establish and many potentially active faults have not been mapped. For very high-risk structures, extensive investigations may be needed (Mallard *et al.*, 1991).

Structural damage from fault breaks arises not only from the consequences of straddling the fault but also the high pulses of ground motion ('seismic flings') that may arise in their vicinity (Bolt, 1995). The seismic hazard maps for the USA provided in ASCE 7 (ASCE, 2010) allow for factoring of ground motion by up to 2 in the vicinity of faults. Seismic flings are further discussed in the next section.

Jonathan Bray's unpublished Joyner Memorial Lecture of 2012 was entitled 'Building near faults'; the lecture slides provide an overview of the subject in note form and can be downloaded from www.dot.ca.gov and following links.

4.3.7   Near fault ground motions ('seismic flings')

Ground motions at sites near the causative fault of a large earthquake may contain a special characteristic, known as a 'seismic fling'. This is a long period velocity pulse, with an associated large increase in displacement, caused when the fault rupture is propagating rapidly towards the site in question. It first came to wide attention after the Northridge California earthquake of 1994, when it was found to be associated with extensive structural damage. Seismic flings at sites near faults should be accounted for in design, in ways discussed by Alavi and Krawinkler (2000).

4.4.   Soil–structure interaction

Most of the previous discussion has been based on the response of soils in the 'free field' without man-made structures. The following sections discuss briefly how to account for the interaction between a structure and its supporting soil.

4.4.1   Effects of foundation flexibility

Structures founded on bedrock can be analysed assuming that their base is fixed. This assumption may be seriously in error, however, when the translational and rotational restraint offered to the structure by the soil is less than rigid. Usually, the effect of soil flexibility is to increase the fundamental period of the structure, which often takes it away from resonance with the earthquake motions. Moreover, the cyclic movement of the soils in contact with the structure's foundations causes energy to be radiated away from the structure, tending to reduce its motion. This is known as radiation damping (Figure 4.10). Generally, therefore, ignoring these soil–structure interaction (SSI) effects leads to conservative estimates of seismic forces and shear deformations in the superstructure, provided the site effects discussed in Section 4.3 have been accounted for. However, overall deformations will generally be underestimated. EC8 part 5 (CEN, 2004) lists the following instances in which soil–structure and soil–foundation interaction should be accounted for.

- Structures in which P–δ effects (Section 3.6.5) play a significant role, because of the increase in overall structural deformation.
- Tall and slender structures such as towers and chimneys, because the increased rotational flexibility may lead to significant structural effects.
- The foundations of structures with massive or deep-seated foundations, such as buildings with deep basements, may have significant interaction with the soil. Similarly, the interaction between piles and the surrounding soils during earthquakes needs to be considered when the piles pass through interfaces between very soft soils and much stiffer soils; these effects on piles are discussed further in Chapter 7.
- SSI should be more generally considered effects for structures supported on very soft soils.

Figure 4.10 Radiation damping: (a) waves radiating away from an oscillating building; (b) reduction in radiation damping with thin soil strata

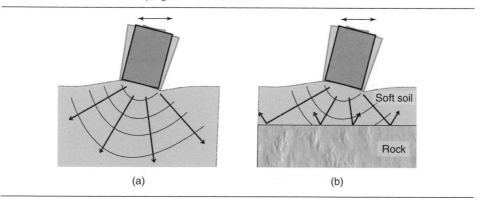

(a)   (b)

4.4.2   SSI analysis

A number of analytical techniques to investigate SSI are possible. The simplest method is to represent the soil flexibility by discrete springs connected to the foundation. For shallow foundations on deep uniform soils, the soil spring stiffness can be found from simple formulae; ASCE 4-98 (ASCE, 1998) provides standard formulae for circular and rectangular bases. These require a knowledge of the shear stiffness of the soil, which, as shown in Figure 4.2, depends on the shear strain amplitude. When linear elastic analysis is performed, a series of iterative analyses is, therefore, required to find a suitable shear stiffness consistent with the computed shear strain. Similarly, EC8 part 5 provides formulae for the effective stiffness of soil–pile systems.

The material damping associated with the soil spring is also strain-dependent (Figure 4.3); a safe value for material damping of 5% is often taken. To this may be added the radiation damping, which may be significant. ASCE 4-98 provides values of equivalent viscous damping due to radiation effects in uniform soils. These may be satisfactory when the soil depth is uniform over a depth much greater than the greatest foundation dimension. However, the presence of harder layers reflecting back radiated energy may significantly reduce radiation damping (Figure 4.10(b)), and in this case special analysis is required.

In a response spectrum analysis, the damping levels due to material and radiation damping will only affect the modes of vibration involving significant foundation movement, for which suitably reduced spectral accelerations can be assumed. The effective damping in each foundation mode will increase with the proportion of strain energy imparted to the foundation as a fraction of the total strain energy in that mode. Higher modes of vibration are unlikely to involve the foundation soils, so the damping level applied to higher modes should depend solely on the superstructure.

Modelling the foundation stiffness by conventional linear springs with properties modified as appropriate to allow for shear strain may be satisfactory in many cases (and is much used in practice). However, it is theoretically not correct. A rigorous treatment of SSI effects using soil springs requires the use of springs whose stiffness and damping properties are frequency dependent. Such an analysis can be relatively straightforward if frequency domain techniques are used. This type of analysis is discussed by Pappin *et al.* (1991) and is not treated further here.

Finite element modelling of soils is an alternative to the use of soil springs, and may be required to account for sloping or non-uniform soil strata, embedment of foundations and other complexities. The analysis is not straightforward, however, and there are special problems in treating boundaries of the portion of soil modelled in the analysis. A discussion of recent developments in these methods is provided by Radmanoviæ and Katz (2012).

REFERENCES

Abu-matar H and Goble G (1997) SPT dynamic analysis and measurements. *Journal of Geotechnical and Geoenvironmental Engineering*, American Society of Civil Engineers, Reston, VA, USA.

Adams BM and Jaramillo JD (2002) A two-dimensional study on the weak-motion seismic response of the Aburra Valley, Medellin, Colombia. *Bulletin of the New Zealand Society for Earthquake Engineering* **35/1(March)**: 17.

Alavi B and Krawinkler H (2000) *Consideration of Near-fault Motion Effects in Seismic Design.* *12WCEE*, Auckland, New Zealand, Paper no. 2665.

Andersen KH, Lacasse S, Aas P and Andanaes E (1982) *Review of foundation design principles for offshore gravity platforms*. Norwegian Geotechnical Institute Publication, No. 143.

ASCE (1998) ASCE 4-98: Seismic analysis for safety-related nuclear structures. American Society of Civil Engineers, Reston, VA, USA (A major revision is expected in 2013).

ASCE (2010) ASCE/SEI 7-10: Minimum design loads for buildings and other structures. American Society of Civil Engineers, Reston, VA, USA.

ASTM (1986) D4633-86: Standard test method for stress wave energy measurement for dynamic penetrometer testing systems. ASTM International, West Conshohocken, PA, USA.

Bolt BA (1995) *From earthquake acceleration to seismic displacement*. The Fifth Mallet–Milne Lecture. John Wiley, Chichester, UK.

Celebi M (1987) Topographical and geological amplifications determined from strong-motion and aftershock records of the 3rd March 1985 Chile earthquake. *Bulletin of the Seismological Society of America* **77(4)**: 1147–1167.

CEN (2004) EN 1998-5: 2004: Design of structures for earthquake resistance. Part 5: Foundations, retaining structures and geotechnical aspects. European Committee for Standardisation, Brussels, Belgium.

Cetin KO, Seed RB, Der Kiureghian A *et al.* (2004) Standard penetration test-based probabilistic and deterministic assessment of seismic soil liquefaction potential. *Journal of Geotechnical and Geoenvironmental Engineering, ASCE* **130(12)**: 1314–1340.

Day SM, Graves R, Bielak J *et al.* (2008) Model for basin effects on long-period response spectra in Southern California. *Earthquake Spectra* **24**: 257.

EERI (Earthquake Engineering Research Institute) (1994) *Earthquake basics: liquefaction – what it is and what to do about it*. EERI, Oakland, CA, USA.

Faccioli E (2002) 'Complex' site effects in earthquake strong motion, including topography. Keynote address, *12th European Conference on Earthquake Engineering*. Elsevier.

Idriss IM and Boulanger RW (2008) *Soil liquefaction during earthquakes*. Earthquake Engineering Research Institute, Oakland, CA, USA.

Idriss IM and Boulanger RW (2010) *SPT-based liquefaction triggering procedures*. Report no. UCD/CGM-10-02, Center for Geotechnical Modeling, Department of Civil and Environmental Engineering, University of California, Davis, CA, USA. Free download from links from nees.ucdavis.edu.

Imai T and Tonouchi K (1982) Correlations of N value with S-wave velocity and shear modulus. *Proceedings of the 2nd European Symposium on Penetration Testing*, pp. 24–27.

ISE/AFPS (Institution of Structural Engineers/Association Française du Génie Parasismique) (2010) Manual for the design of steel and concrete buildings to Eurocode 8. ISE, London, UK, and AFPS, Paris, France.

Jibson RW (2011) Methods for assessing the stability of slopes during earthquakes – a retrospective. Special Edition *Journal of Engineering Geology*, Toward the Next Generation of Research on Earthquake-induced Landslides: Current Issues and Future Challenges **122**: 43–50.

Kelly DJ (2009) Location of base for seismic design. *Structure magazine* **December**: 8–11.

Khilnani KS, Byrne PM, Yeung KK (1982) Seismic stability of the Revelstoke earthfill dam. *Canadian Geotechnical Journal* **19**: 63–75.

Kramer SL (1996) *Geotechnical Earthquake Engineering*. Prentice-Hall, USA.

Mallard DJ, Higginbottom IE, Muir Wood R and Skipp BO (1991) Recent developments in the methodology of seismic hazard assessment. In: *Civil Engineering in the Nuclear Industry*, Institution of Civil Engineers, London, pp. 75–94.

Newmark NM (1965) Effects of earthquakes on dams and embankments. *Géotechnique* **15**: 139–160.

Oztoprak S and Bolton M (2013) Stiffness of sands through a laboratory test database. *Géotechnique* **63(1)**: 54–70.

Pappin JW (1991) Design of foundation and soil structures for seismic loading. In *Cyclic Loading of Soils* (O'Reilly MP and Brown SF (eds)). Blackie, London, UK, pp. 306–366.

Radmanoviæ B and Katz C (2012) Dynamic soil–structure interaction using an efficient scaled boundary finite element method in time domain with examples. *SECED Newsletter* **23(3)**: 3–13. Society for Earthquake and Civil Engineering Dynamics, London, UK.

Seed HB, Wong RT, Idriss IM and Tokimatsu K (1984) *Moduli and Damping Factors for Dynamic Analysis of Cohesionless Soils*. Report no. UCB/EERC-84/14, College of Engineering, University of California at Berkeley, CA, USA.

SHAKE (1991) *A computer program for conducting equivalent linear seismic response analysis of horizontally layered soil deposits*. Center for Geotechnical Modeling, University of California, Davis, CA, USA.

Stewart JP, Kwok AO-L, Hashash YMA *et al.* (2008) *Benchmarking of Nonlinear Geotechnical Ground Response Analysis Procedures*. PEER Report 2008/04, Pacific Earthquake Engineering Research Center, College of Engineering, University of California, Berkeley, CA, USA.

Stokoe KH, Kim J, Sykora DW, Ladd RS and Dobry R (1986) Field and laboratory investigations of three sands subjected to the 1979 Imperial Valley earthquake. *Proceedings of the 8th European Conference on Earthquake Engineering, Lisbon*, vol. 2, pp. 5.2/57–64.

Vucetic M and Dobry R (1991) Effect of soil plasticity on cyclic response. *ASCE Journal of Geotechnical Engineering* **117(1)**: 89–109.

Weiler WA (1988) Small-strain shear modulus of clay. *Earthquake Engineering and Soil Dynamics II – Recent Advances in Ground Motion Evaluation*. ASCE Geotechnical Special Publication No. 20, pp. 331–345.

Youd TL (1998) *Screening guide for rapid assessment of liquefaction hazard at highway bridge sites*. National Center for Earthquake Engineering Research, Technical Report MCEER-98-0005.

Youd TL and Perkins SK (1978) Mapping liquefaction induced ground failure potential. *Journal of Geotechnical Engineering* **104(GT4)**: 433–446. American Society of Civil Engineers, Reston, VA, USA.

Earthquake Design Practice for Buildings
ISBN 978-0-7277-5794-4

ICE Publishing: All rights reserved
http://dx.doi.org/10.1680/edpb.57944.103

Institution of Civil Engineers

publishing

Chapter 5
Initial planning considerations

[Earthquake] safety can be expensive if you start off with the wrong system, or architectural or engineering design.

Degenkolb (1994)

This chapter covers the following topics.

- The lessons from earthquakes.
- Planning considerations and overall form.
- Site selection and the effect of foundation soils.
- Conventional framing systems.
- Special measures of earthquake resistance.
- Cost of earthquake provision.

When designing buildings in areas of high seismicity, earthquake performance is a key driver in the initial planning of a structure but many other factors – architectural, economic, functional, local availability of skills and materials, and so on – must also be considered. This chapter aims to provide an introduction to some of the factors involved in making the choices related to seismic resistance.

5.1. The lessons from earthquake damage

A guiding principle of the original edition of this book, stated on its first page, was that 'earthquake engineering is not to be learned from books only' and the study of the damage caused by past earthquakes, preferably at first hand, provides an unrivalled route to understanding what works well in creating a good earthquake-resistant building – and what does not. Many of the theoretical aspects of earthquake engineering set out in books are highly complex, and it is so easy to lose sight of the reality of the subject among these complexities. There is nothing like the experience of seeing the often disturbing consequences of a major earthquake on structures and those who live around them to rebalance one's approach and ground it in practical reality.

For this reason, field investigations of earthquakes are of great importance to structural designers. In the UK over the past 30 years, the Earthquake Engineering Field Investigation Team (EEFIT) has given over 100 engineers, coming in equal numbers from industry and universities, the opportunity to visit the sites of 29 damaging earthquakes (Booth et al., 2011). EEFIT has published reports on all these events, which, together with photos of earthquake damage, are freely available from the website of its sponsoring organisation, the Institution of Structural Engineers (www.istructe.org.uk). In the USA, the Earthquake Engineering Research Institute (www.eeri.org) similarly has been publishing excellent researched earthquake reports, and posts summaries on its website within weeks of the events occurring. The New Zealand Society for Earthquake Engineering (www.nzsee.org.nz) is another good source of English language reports.

Earthquake field reports show that common mistakes give rise to the majority of poor performance; notable failings include the following.

- Soft (weak) storeys.
- Poor detailing of reinforcement in concrete structures, and connections in steel structures.
- Inadequate design of foundations.
- Neglect of the risk of slope failure.
- Inadequate provision of lateral strength.
- Eccentricity between centres of stiffness and lateral resistance, leading to torsional response.
- Poor detailing of cladding, services and other non-structural elements.
- Inadequate account of tsunami risk.

Experience gained from past earthquakes is of course an excellent guide to decisions made during the initial planning and subsequent stages of design. For this reason, the subsequent chapters on foundation and superstructure design each start with a review of the lessons learnt from previous earthquakes.

5.2.   Design and performance objectives

Decisions made at the conceptual and planning stage usually have a crucial effect both on the cost of earthquake provision, and also on performance in an earthquake. These early decisions are often difficult to modify subsequently so that it is essential that their full consequences are understood in terms of performance and costs as early as possible.

The successful designer must be clear from the outset of the performance objectives required under earthquake loading. Table 5.1 sets out possible objectives and gives examples of the sorts of building they might apply to. The table may be compared to Table 2.1 showing the objectives defined in US and European codes, along with the associated return periods of the earthquake motions.

To date, the vast majority of buildings have been designed with the simple intention of preserving life safety, and this may be the cheapest option as far as initial costs are concerned, although this option is likely to involve costly repair or more likely rebuilding in the event that strong shaking does occur. However, society outside the engineering profession has reacted with disbelief to the notion that a building damaged beyond repair has 'performed well' in an earthquake, even if loss of life has been avoided. Thus, Kam and Pampanin (2012), commenting on the earthquakes that devastated Christchurch, New Zealand, in 2010 and 2011, state that they 'critically highlighted the mismatch between the societal

Table 5.1 Performance objectives under earthquake loading

Performance objective	Examples of building applications
Life safety of occupants and others in the building vicinity	All occupied buildings
Minimisation of structural damage	Historic and other culturally important buildings
Minimisation of cost over building lifetime, including repair costs	Depends on attitude to financial risk and availability of insurance
Protection of contents	Buildings with costly contents, such as museums with fragile artefacts and factories with expensive plant
Preservation of building function during and/or after an earthquake	Hospitals, particularly emergency hospitals Manufacturing facilities Civic control centres dealing with a post-earthquake emergency Facilities providing safety-related control in petrochemical and nuclear power installations Buildings intended as post-earthquake emergency housing
Containment of dangerous contents	Facilities housing radioactive, flammable or noxious materials

expectations over the reality of engineered buildings' seismic performance'. The trend in developed societies has therefore increasingly been towards the view that a single life safety performance objective is insufficient on its own, although of course it remains absolutely essential. Codes of practice are beginning to reflect this view, as Table 2.1 showed, although life safety currently remains the main objective dealt with in any detail by codes.

The remainder of the chapter discusses how the initial planning of a building affects its subsequent seismic performance. The various aspects considered are functional layout, site selection and choice of material and structural system. The emphasis is on conventional structural forms, but an additional section introduces some of the special measures developed to provide a performance well beyond that of simple life safety. Finally, a note is added on the cost of seismic design.

5.3.　Anatomy of a building

The functioning parts of a building affect the way in which it can accommodate its structural skeleton. For this reason, it is useful to consider the principal division of functions and how they affect the structure.

Table 5.2 provides a simple functional classification. Vertical divisions of function within the building may be a source of problems, making it difficult to avoid irregularities in mass or

Table 5.2 Functional classification of building parts

Building element	Function
Basement	Car parking, storage, mechanical and electrical plant
Street level	May be used quite differently from the rest of the building, commonly leading to a greater than typical storey height and a need for unobstructed floor space. For example, in hotels the street level may be used for reception, conference and restaurant areas in contrast with the regular pattern of rooms on the typical floors. In office buildings, the street level may include shops, banks, restaurants, etc.
Typical floors	Repetitive standard levels
Roof structures	Mechanical and electrical plant, lift motor room, water tanks, etc.
Service and access cores	Stairs, lifts, toilets and pipe ducts, which are frequently grouped together and provide potential lateral resisting elements
Usable floor	Clear spaces, usually modular. Floor diaphragms provide the vital function of distributing seismic loads back to the lateral resisting elements and for tying the structure together

stiffness. For example, the ground floor of many commercial buildings is often taller and more open than higher floors, creating a potential weak storey. However, the service cores and exterior cladding provide an opportunity to incorporate shear walls or braced panels to overcome resulting problems. An important objective of early planning is to establish the optimum locations for service cores and other structural elements providing lateral resistance that will be continuous to the foundation.

It is not unusual to find that structural and architectural requirements are in conflict at the concept planning stage but it is essential that a satisfactory compromise is reached at this time.

5.4. Planning considerations
5.4.1 The influence of site conditions

It is essential to obtain data at an early stage on the soil conditions and groundwater level at the site, because these can have a major influence on seismic design. The principal aspects to determine are the period range over which the soils may amplify seismic motions, the lique-faction potential of the soil and the stability of slopes at or near the site. Initially at least, standard tests suffice, comprising in-situ tests (standard penetration test or cone penetration test values and groundwater level measurements) and laboratory tests (soil description and standard strength tests). Additional specialist techniques such as in-situ shear wave velocity measurements and cyclic triaxial or resonant column laboratory tests may be needed in special circumstances (e.g. soil profiles S_1 and S_2 in Table 5.3). Unless the soils at the site are well understood from previous investigations, borehole data to at least 30 m (or bedrock depth if less) are required.

Table 5.3 Soil classification (derived from BS EN 1998-1:2004+A1:2013)

Description	Characteristic parameters in top 30 m		Period range T_B to T_C for peak ground motion amplification			
	Non-cohesive soils	Cohesive soils	Large earthquakes govern		Small earthquakes govern	
	N_{SPT} Blow count/ 300 mm	c_u (kPa) Undrained shear strength	T_B: s	T_C: s	T_B: s	T_C: s
A Rock or other rock-like geological formation, including at most 5 m of weaker material at the surface			0.15	0.4	0.05	0.25
B Deposits of very dense sand, gravel, or very stiff clay, at least several tens of metres in thickness, characterised by a gradual increase of mechanical properties with depth	>50	>250	0.15	0.5	0.05	0.25
C Deep deposits of dense or medium dense sand, gravel or stiff clay with thickness from several tens to many hundreds of metres	15–50	70–250	0.20	0.6	0.1	0.25
D Deposits of loose-to-medium cohesionless soil (with or without some soft cohesive layers), or of predominantly soft to firm cohesive soil	<15	<70	0.20	0.8	0.1	0.3
E A soil profile consisting of a surface alluvium layer similar to type C or D and thickness varying between about 5 m and 20 m, underlain by stiffer material with a shear wave velocity >800 m/s			0.15	0.5	0.05	0.25
S_1 Deposits consisting – or containing a layer at least 10 m thick – of soft clays/silts with high plasticity index (PI >40) and high water content		10–20	Special investigations needed			
S_2 Deposits of liquefiable soils, of sensitive clays, or any other soil profile not included in types A–E or S_1			Special investigations needed			

For other than minor projects, the soil data need to be sufficient to classify the site into one of the standard profiles described in codes of practice. Table 5.3 shows the Eurocode 8 (EC8) classification system, together with the period range T_B to T_C for peak amplification of ground motions. Structures falling into this period range may resonate with the ground motion. As a rough initial guide, the fundamental period of a building is $N/10$, where N is the number of storeys above ground level. Consequently, deep, soft soil deposits can be damaging to tall buildings, but also shallow, stiff deposits can prove troublesome for low-rise structures. If the site period is similar to that of the proposed structure, large amplification of seismic response will result and it may be worth considering ways of modifying the structural period to detune it from the earthquake motions. Increasing the stiffness (e.g. the addition of bracing or shear walls) or reducing the mass (e.g. lightweight floors, lightweight concrete) both reduce the structural period, and of course the reverse is also true. However, period depends on the square root of mass divided by stiffnesses, so large changes in mass and/or stiffness are needed for a significant change in period. Mounting the building on flexible bearings can dramatically increase the period – see Section 5.5.8.2.

Liquefaction or slope stability problems could lead to the conclusion that the site is unsuitable for development without expensive soil improvement measures or foundation solutions. An initial indication of the potential for soil liquefaction can be obtained from Table 4.2 in Chapter 4.

5.4.2   Structural layout

The experience of past earthquakes has confirmed the common sense expectation that buildings that are well tied together and have well-defined, continuous load paths to the foundation perform much better in earthquakes than structures lacking such features.

The degree of symmetry also has a significant influence on earthquake resistance. Earthquake damage is found to be five to 10 times worse in buildings with significant irregularity, compared with those with essentially regular structures. The reason is that sudden changes in section cause stress concentrations and potential failure points. The most common example is the 'weak storey', often caused by architectural requirements for openness at ground-floor level. The result is that deformations are concentrated at this level during the earthquake, giving rise to very severe structural demands. Weak storeys have caused perhaps more collapses in earthquakes than any other feature. Another important example of irregularity occurs when the centres of stiffness and mass of a structure do not coincide, giving rise to damaging coupled lateral/torsional response. Compact plan shapes are also favoured, as flexible extensions from a structure are prone to vibrate separately from the rest of the structure.

The layout of the lateral load-resisting vertical elements should therefore aim for the greatest possible regularity, compactness and torsional resistance. Irregular plan shapes can be divided into compact shapes by providing separation joints; these must be sufficiently wide (up to 50 mm for each storey height above ground in flexible structures) to prevent damaging contact from occurring as the separated parts of the building sway in an earthquake.

Figure 5.1 Buffeting damage, Mexico City, 1985

Adequate separation is even more important between adjacent buildings, because the storey heights are unlikely to coincide, and a stiff floor diaphragm of one building may impact the other at the vulnerable mid height position of columns (Figure 5.1).

The mass distribution within a building should also be considered. The characteristic swaying mode of a building during an earthquake implies that masses placed high in the building produce considerably more unfavourable effects than masses placed lower down. Massive roofs and heavy plant rooms at high level are therefore preferably avoided.

Finally, undue reliance on a few elements to provide lateral resistance should be avoided, because there is no back-up if they fail. The combination of shear walls with moment frames is an example of obtaining such redundancy.

A classic work (Arnold and Reitherman, 1982) on the need for symmetry in seismic conditions, written by an architect and a structural engineer over 30 years ago, is still worth reading and sharing with architectural colleagues.

5.4.3   Provision of adequate ductility

All of these considerations contribute towards obtaining a good earthquake-resistant design, but do not necessarily ensure that there is adequate reserve to meet an extreme earthquake attack without collapse. The strategy commonly adopted is therefore to provide sufficient strength to minimise damage in an earthquake with a high probability of occurrence, but to accept that the structure may yield in a low probability event with the accompanying risk of damage, while ensuring that the post-yield response is ductile rather than brittle. A ductile structure is one that can maintain its stability under repeated cyclical deflections considerably greater than its yield deflection. The ductile structure therefore resists the extreme earthquake not by brute force, but by allowing plastic deformations to absorb the kinetic energy induced by the ground shaking. The plastic yielding not only absorbs energy, but also softens the structure and increases its natural period, which will usually further reduce demand.

This strategy implies that considerable structural damage may occur in an extreme event, probably to the extent that the structure is not repairable. Provided this has been assessed as a low probability occurrence, and provided life safety is not impaired, this may be acceptable. Given the huge uncertainties both in predicting earthquake motions and calculating response, the provision of ductility is the surest insurance policy against destruction of human lives.

Modern earthquake codes take advantage of ductile yielding to reduce the level of seismic design force, typically to a level two to eight times lower than the strength required for the structure to remain elastic. Lower ductility demands are typically adopted in Japanese practice, but reductions of twofold or more on elastic demands during a major earthquake are still permitted. This emphasises the point that provision of adequate strength is not in itself sufficient; measures to ensure ductility are also essential.

There are two principal means of ensuring ductility. First, the capacity design procedures, described in Section 3.5, should be used to ensure that yielding takes place in ductile rather than brittle modes. At the planning stage, it is important to avoid choosing structural systems that would make this difficult to achieve – for example, frames in which the beams are larger than the columns. Second, special detailing is needed to ensure that parts of the structure designed to yield can achieve large post-yield strains. An example is the provision of horizontal confinement steel in columns. Seismic codes of practice are much concerned with such details. Detailing is of course dealt with at later stages of design than the planning stage, but the yielding modes need to be identified, and confirmed as achievable within the structural scheme adopted. For example, concrete column sizes must be large enough to accommodate the reinforcement needed to achieve sufficient ductility.

A ductile structure is much more likely to be able to be capable of satisfying the essential life safety performance objective identified in Table 5.1, but ductility by itself does not address the other objectives listed there. Many are connected with limiting the maximum or permanent deformations caused by an earthquake, and so stiffness as well as ductility is an important consideration at the planning stage, as discussed in the next paragraphs.

5.4.4   Provision of adequate stiffness

Deflections must be limited during earthquakes for a number of reasons, and hence provision of adequate stiffness is important. Relative horizontal deflections within the building (e.g. between one storey and the next, known as storey drift) must be limited. This is because non-structural elements such as cladding, partitions and pipework, must be able to accept the deflections imposed on them during an earthquake without failure. Failure of external cladding, blockage of escape routes by fallen partitions and ruptured firewater pipework all have serious safety implications. Moreover, some of the columns in a building may be only designed to resist gravity loads, with the seismic loads taken by other elements, but if deflections are too great they will fail through 'P–delta' effects (Section 3.7.5) however ductile they are. Overall deflections must also be limited to prevent impact, both across separation joints within a building, and (usually more seriously) between buildings.

For all these reasons, it is therefore essential to check that the basic structural form enables deformations to be limited to code requirements; this criterion, rather than strength, often governs section sizes, and even structural form, in tall buildings.

The discussion above has been mainly connected with life safety considerations, but other performance objectives are also involved. Peak deformations of the structure during an earthquake may cause damage to the building contents, and hence economic loss and loss of functionality, even if no life safety threat is involved. Residual deformations at the end of an earthquake due to permanent structural yielding are unlikely to be a life safety threat, but may lead to a building being viewed as uninhabitable, and hence an economic write-off. Codes of practice currently have little to say on these issues, but there has been an extensive research effort into them in the past 15 years.

5.4.5 Provision of adequate lateral strength

Ductility and stiffness are key parameters for consideration by the earthquake engineer at the planning stage, but lateral strength is also important. Adequate strength must be provided to limit ductility demands on the structure to within safe limits; structural forms with inherently low ductility, such as masonry structures, need relatively higher strength than very ductile structures to achieve life safety performance objectives. Moreover, ductile structures need sufficient strength to limit ductile demands – and hence damage – in frequent, moderate earthquakes in order to meet damage limitation performance objectives.

5.4.6 Interaction between structure and non-structure

As noted previously, structural deformations may damage building contents, such as external cladding, internal partitions and service pipework, depending on how they are attached to the building structure. Accelerations may also cause damage; the swaying of a building in an earthquake gives rise to accelerations and hence inertial forces in a building's contents, just as much as in its structure. Failure to provide simple lateral restraint – for example, to mechanical and electrical plant or to shelving in warehouses, has caused considerable damage in the past.

Design of non-structural elements and their attachment is often dealt with at the detailing stage. However, interaction of the structure with stiff non-structural elements such as infill blockwork partitions or cladding elements can result in significant and often deleterious changes to structural response. At an early stage, it should be decided whether cladding, partitions, staircases and so on are to remain separate from the main structure or to be designed to work with it in resisting seismic loads. Liaison between the structural engineer and other members of the design team, such as architects and mechanical engineers, is essential to ensure safe seismic interaction between structure and non-structure.

5.5. Structural systems
5.5.1 Foundations

General guidance on the choice of foundation system is difficult, because the relative cost and efficiency of different types depend critically on the soil conditions and type of superstructure. Some factors that should be considered in connection with seismic resistance are as follows.

5.5.1.1 General

■ When the superstructure is designed to achieve a high level of ductility, the foundation must be able to develop the superstructure's yield capacity. It is no use having a perfectly detailed ductile superstructure supported by a foundation that experiences unacceptable deformations before that ductility is achieved. EC8 (CEN, 2004) includes this requirement, but US codes do not.
■ Superstructure systems that involve large uplift forces (e.g. shear walls with a large height to width ratio) are only suitable if foundations can be built economically to resist these tension forces.

5.5.1.2 Shallow foundations

■ EC8 recommends that a rigid cellular foundation should usually be provided when the superstructure consists of discrete shear walls of different stiffnesses.
■ Individual pad foundations should be connected by tie beams to prevent relative movement in an earthquake, except on rock sites.
■ Pad foundations should be founded onto soil conditions that are similar; founding some on to soft soil and others onto hard soil or rock creates non-uniformity, which may lead to damage.

5.5.1.3 Piled foundations

■ Piles have loads imposed on them due to lateral deflection of the upper layers of softer soil during earthquakes. Small driven piles of less than 0.5 m diameter are generally sufficiently flexible to accept this movement without suffering large bending stresses. Large-diameter piles, however, may experience significant lateral forces as they are relatively stiff compared with the soil.
■ Raking piles are generally to be avoided, because they add greatly to the lateral stiffness of the pile group. Their stiffness means that they will not be able to conform to the deformations of the soft soil strata, but will receive very large lateral loads, arising from the mass of the soft soils attempting to move past the stiffened pile group. Raking piles have been found to be prone to failure during earthquakes.

5.5.1.4 Foundations on liquefiable soils

■ Piling through potentially liquefiable layers needs careful consideration, because the piles would have to transmit the lateral forces from both the superstructure and adjacent non-liquefied soil through the liquefied strata. The piles would be effectively unsupported laterally in this region and so may be subject to large bending and shear stresses that would be difficult to resist.

■ Raft foundation support by means of a basement may be an alternative solution when founding on potentially liquefiable layers, as discussed in Section 7.8.

5.5.2 Choice of structural material

The most appropriate structural material to use is influenced by a host of different factors, including relative costs, locally available skills, environmental impact, durability, architectural considerations, and so on. Some of the seismic aspects are as follows.

Steel has high strength to mass ratio, a clear advantage over concrete because seismic forces are generated through inertia. It is also easy to make steel members ductile in both flexure and shear. However, providing adequate seismic resistance of connections can be difficult, and buckling modes of failure lack ductility.

Concrete has an unfavourably low strength to mass ratio, and it is easy to produce beams and columns that are brittle in shear, and columns that are brittle in compression. However, with proper design and detailing, ductility in flexure can be excellent, ductility in compression can be greatly improved by the provision of adequate confinement steel, and failure in shear can be avoided by 'capacity design' measures. Moreover, buckling modes of failure are much less likely than in steel. Although poorly built concrete frames have an appalling record of collapse in earthquakes, concrete shear wall buildings have a much better record, even when design and construction standards are less than perfect.

Masonry, too, suffers from a high strength to mass ratio, and lacks ductility. However, good quality stone is very strong in compression. When this compressive strength is harnessed to resist earthquake forces – for example, through the use of arches and domes – the performance can be good. Unlike the traditional engineering materials of steel and concrete, unreinforced masonry structures must be designed clastically to have a large reserve against design earthquake forces, without reliance on ductility.

Timber (favourably) is strong and light, and its connections usually provide good levels of damping without suffering from the low-cycle fatigue problems that beset steel. However, timber can lose strength through fungal or insect attack, and it also burns easily.

5.5.3 Moment-resisting frames

5.5.3.1 General characteristics of moment-resisting frames

Moment-resisting (i.e. unbraced) frames derive their lateral strength not from diagonal bracing members, but from the rigidity of the beam–column connection. They consist solely of horizontal beams and vertical columns. They are in common use for both steel and concrete construction.

The advantages of using moment-resisting frames to provide seismic resistance are as follows.

■ Properly designed, they provide a potentially highly ductile system with a good degree of redundancy, which can allow freedom in the architectural planning of internal spaces and external cladding, without obstruction from bracing elements.

- Their flexibility and associated long period may serve to detune the structure from the forcing motions on stiff soil or rock sites.

The potential problems associated with moment-resisting frames are

- Poorly designed reinforced concrete moment-resisting frames have been observed to fail catastrophically in earthquakes, mainly by the formation of weak storeys and failures around beam–column junctions. Steel moment-resisting frames have performed better, but have still proved vulnerable at welded connections.
- The beam–column joint region represents an area of high stress concentration, which needs considerable skill to design and build successfully. In concrete, this often involves congested reinforcement, which needs good steel-fixing skills and good concreting to ensure proper compaction around the reinforcement. In steel, careful detailing of the connections and panel zone is needed. Ductile moment-resisting frames should not be used when these skills are not available.
- The low stiffness of moment-resisting frames tends to cause high storey drifts (inter-storey deflections), which may lead to a number of problems. These include unacceptable damage to cladding and other non-structural elements, and other serious structural problems. Moreover, the width of separation joints within the structure may need to be large to prevent buffeting during an earthquake, and this can lead to problems in detailing an acoustic, thermal and weather-tight bridge to cross the joints. A more general problem with the flexibility of moment frames, particularly in tall buildings, is that design may be governed by deflection rather than strength, leading to an inefficient use of material.

Frames with overall height to base width ratios of up to 4 are in common use. When used as the sole seismic-resistant system, higher ratios may result in uplift problems, particularly at corner columns, which tend to carry the lowest gravity load and attract the highest tensions due to lateral loads. Very slender structures are prone to deflection problems, both in excessive storey drifts and overall movement. The maximum practicable ratio depends, however, on the seismicity of the area and hence the magnitude of lateral forces that must be resisted. Moreover, wind loads as well as seismic loads need to be considered when choosing the overall slenderness.

The ratio of beam span to column height depends on a number of considerations. Internal frame geometries may be governed by the need for unrestricted internal spaces. Optimisation of the structure supporting gravity loads may also result in a larger span than would be chosen for purely seismic resistance. External frames may be less restricted in this way; the optimum beam span is likely to be 1 to 1.5 times the storey height, although a wide range of ratios is found in practice.

5.5.3.2 Grid frames and perimeter frames

Moment frames can be classified into two different types. The first, grid frames (Figure 5.2(a)), comprise a uniform grid of frames in both directions. They are highly redundant (a favourable feature), and achieve a good spread of resistance to seismic forces both within the superstructure and to the foundations. They have very good torsional resistance

Figure 5.2 Types of moment-resisting frame: (a) grid frame; (b) perimeter frame

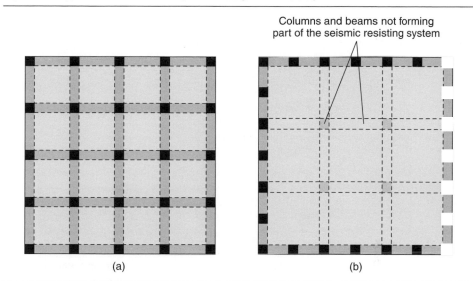

Columns and beams not forming
part of the seismic resisting system

(a)   (b)

and coupled lateral/torsional response is unlikely to be a problem, even with irregular plan shapes.

The major disadvantages of grid frames are as follows. All the columns have to be designed for biaxial loading (i.e. earthquake attack in two orthogonal directions simultaneously) and all beams and columns have to be designed and detailed for ductility. External columns (especially corner columns) carry the lowest gravity loads, but are subject to the largest seismic axial forces and so there may be uplift problems. A grid frame may restrict to some extent the freedom of architectural planning of the internal space of a building.

Grid frames generally find their application in low to medium-rise buildings, with any plan shape.

In perimeter frames (Figure 5.2(b)), the seismic-resisting frames are restricted to the outside of the building. The interior space only needs structure capable of supporting gravity loads; consequently, internal column spacing can be increased, allowing greater architectural freedom and, probably, economy. The good torsional stiffness of grid frames is retained, as is some of their redundancy.

The corner columns of perimeter frames in rectangular buildings suffer from the problems of biaxial loading and possible uplift referred to above for grid frames. Circular plan shapes are less affected by this problem.

Perimeter frames find their application in medium to high-rise structures with compact plan shapes. High-rise frames are only likely to be economic in steel.

Figure 5.3 Collapsed precast concrete frame buildings (foreground) with precast wall building behind, Spitak Armenia, 1988 (photograph source unknown)

5.5.3.3   Precast concrete frames

Precasting offers the general advantages of speed of erection, minimisation of costly form-work and falsework and the improvement in quality control possible under shop fabrication conditions. The potential seismic problems (displayed so dramatically during the Armenian earthquake of 1988, Figure 5.3) are the difficulties in ensuring ductility and continuity at the connections between precast units; the elements must be joined together so that they do not shake apart during a major earthquake. Extensive research and development in New Zealand has made the industry there confident that, properly designed and built (which the Armenian precast frame buildings certainly were not), precast frames can be safe in earthquake-prone regions, and they are commonly used in New Zealand for buildings of up to 20 storeys. EC8 (CEN, 2004) provides rules for the design of conventional precast concrete frames with fixed rigid connections, when the frame is intended to emulate the behaviour of a cast in situ concrete frame.

A recent development has been steel or precast concrete frames connected by unbonded post-tensioning cables, described in Section 5.5.8.5.

5.5.3.4   Blockwork infill in moment-resisting frames

Rigid blockwork infill of moment-resisting frames provides a good solution for providing thermal and acoustic insulation and weatherproofing. The blockwork infill causes a large increase in strength and stiffness, at the expense of a reduction in ductility, and there is evidence from recent earthquakes that such infill has protected poorly designed frames from collapse. However, particularly if the infill is not uniform across the building, unsafe conditions can result, such as the creation of a weak storey or a torsionally eccentric structure, as well as creating a hazard from falling masonry.

Unreinforced blockwork infill is not permitted in seismic areas of the USA. However, it is permitted in EC8 (CEN, 2004), and design rules are presented, based on the long experience of its use in seismic parts of southern Europe – an example of the influence of codes on an important aspect of conceptual design.

Instead of allowing the blockwork and structural frame to work together in resisting earthquake loads, they can be separated from each other so that the in-plane deformations of the frame are not imposed on the blockwork. The frame still needs to provide out-of-plane restraint to the blockwork to prevent it from falling out during strong ground motion. This is a tricky detailing problem, but potentially results in a more predictable and ductile system.

5.5.4 Concentrically braced frames

Concentrically braced frames (CBF) are conventionally designed braced frames in which the centre lines of the bracing members cross at the main joints in the structure, thus minimising residual moments in the frame (Figure 5.4). The pros and cons of braced frames are essentially the opposite of moment frames; they provide strength and stiffness at low cost but ductility is likely to be limited because buckling modes dominate failure. Moreover, the bracing may restrict architectural planning and service layout, although V bracing (Figure 5.4(e and f)) overcomes this to some extent.

Figure 5.4 distinguishes between various types of braced frame, the seismic resistance of which can be markedly different. Because of the cyclic nature of seismic loading, their behaviour under extreme lateral loads in alternating directions must be considered.

An X-braced frame (Figure 5.4(a)) has bracing members in tension for both directions of loading, and if these are sized to yield before the columns or beams fail, ductility can be developed. However, after a brace has yielded in tension due to loading in one direction, it is liable to buckle rather than yield in compression on the reverse cycle. Plastic tensile strains therefore tend to accumulate in the braces, limiting the ductility that is achievable. Moreover, this accumulated tensile strain creates a slack in the system, because after one complete cycle of loading, the deflection needs to exceed the plastic excursion achieved in the previous load cycle before the tensile brace is loaded (Figure 5.5). These effects become more marked (and unfavourable) with increasing slenderness of brace; the US code AISC 340-10 (AISC, 2010) therefore distinguishes between 'special' and 'ordinary' CBF, the former having limits on slenderness ratio, but with reduced strength requirements reflecting the increased ductility available. Ordinary frames have height restrictions imposed in areas of high seismicity and are not permitted at all in some cases; such restrictions do not appear in EC8 (CEN, 2004), another example of the applicable code affecting key decisions to be made at the planning stage.

Single bays of diagonal braces (Figure 5.4(b and c)) respond differently according to the direction of loading. Configuration (b) is much weaker and more flexible in the direction causing compression in the braces, while configuration (c) will be weaker and more flexible in the storeys with compression braces, leading to the possibility of soft storey formation. This is clearly not satisfactory, and is not permitted. However, with more than one diagonally braced bay, the performance can revert to that of X bracing if the lateral strength at each level is equal in both directions of loading (Figure 5.4(d)).

Figure 5.4 Examples of bracing schemes for concentrically braced frames

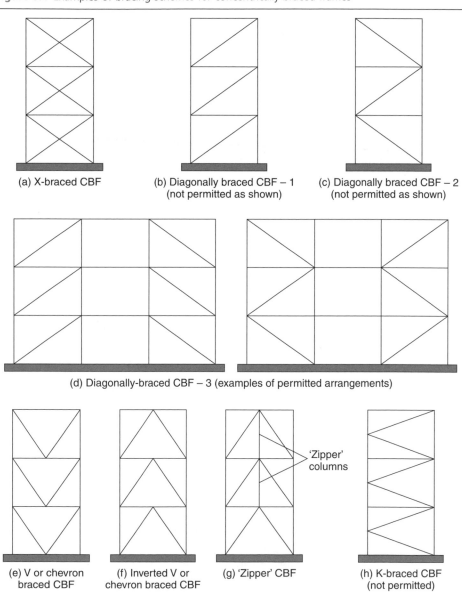

(a) X-braced CBF

(b) Diagonally braced CBF – 1 (not permitted as shown)

(c) Diagonally braced CBF – 2 (not permitted as shown)

(d) Diagonally-braced CBF – 3 (examples of permitted arrangements)

(e) V or chevron braced CBF

(f) Inverted V or chevron braced CBF

(g) 'Zipper' CBF

'Zipper' columns

(h) K-braced CBF (not permitted)

The V-braced arrangements of Figure 5.4(e and f) suffer from the fact that the buckling capacity of the compression brace is likely to be significantly less than the tension yield capacity of the tension brace. There is thus inevitably an out of balance load on the horizontal beam when the braces reach their capacity, which must be resisted in bending of the horizontal member. This restricts the amount of yielding that the braces can develop, and hence the overall ductility. When the horizontal brace has a large bending strength that can resist the out-of-balance load, the hysteretic performance of V-braced systems is improved. A more recent development has been the 'zipper' frame of Figure 5.4(g); here the large out-of-balance forces at lower levels are transferred back to the more lightly stressed upper levels by the zipper columns. Note that the zipper column restricts the relatively open arrangement of V bracing, except at the ground floor.

Figure 5.5 Failure of X-braced steel frame, Kobe Japan, 1995 (photograph courtesy of David G.E. Smith)

Figure 5.6 Examples of eccentrically braced frames

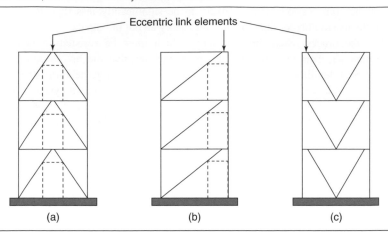

The same out-of-balance force applies to K braces (Figure 5.4(h)) when the braces reach their capacity, but this time it is a much more dangerous horizontal force applied to a column – dangerous because column failure can trigger a general collapse. For this reason, K braces are not permitted in seismic regions.

5.5.5 Eccentrically braced frames

In eccentrically braced frames (EBF), some of the bracing members are arranged so that their ends do not meet concentrically on a main member, but are separated to meet eccentrically (Figure 5.6).

The eccentric link element between the ends of the braces is designed as a weak but ductile link, which yields before any of the other frame members. It therefore provides a dependable source of ductility and, by using capacity design principles, it can prevent the lateral load in the structure from reaching the level at which buckling occurs in any of the members. The link element is relatively short and so the elastic response of the frame is similar to that of the equivalent CBF. The arrangement thus combines the advantageous stiffness of CBF in its elastic response, while providing much greater ductility and avoiding problems of buckling and irreversible yielding, which affect CBF in their post-yield phase. Arrangements such as (a) and (b) in Figure 5.6 also have architectural advantages in allowing more space for circulation between bracing members than their concentrically braced equivalent.

EBFs have been under development for 40 years, and there are extensive design rules in seismic codes, including EC8 and the US code AISC 340-10 (AISC, 2010). They are now less favoured in the USA than the buckling restrained braced frames (BRBF) described in the next section.

5.5.6 Buckling restrained braced frames

BRBFs are a more recent solution to the problem of retaining the stiffness of CBFs while greatly increasing their ductility. They are widely used in the USA and Japan. The relatively

Figure 5.7 Components of an unbonded brace (reproduced from Christopoulos and Filiatrault (2006), courtesy of IUSS Press)

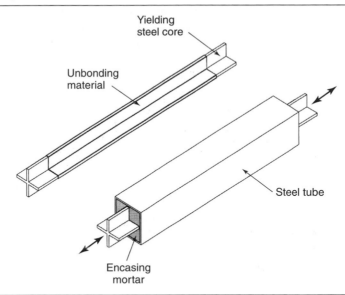

low ductility of CBFs arises from the buckling failure associated with practical sizes of compression strut. BRBFs have compression members that yielding plastically without buckling in compression as well as tension. This is achieved by placing the brace member inside a mortar-filled steel outer tube, which is not connected to the main structural frame (Figure 5.7). The brace member is debonded from the mortar, so that the brace can undergo compressive or tensile strains without stressing the mortar or outer tube, but is still restrained from buckling by the stiffness of the outer tube (see Figure 5.8).

Figure 5.8 Buckling restrained braced (BRB) frame: (a) BRB member, showing end connection; (b) frame with BRB (photographs courtesy of Star Seismic, Park City, UT, USA)

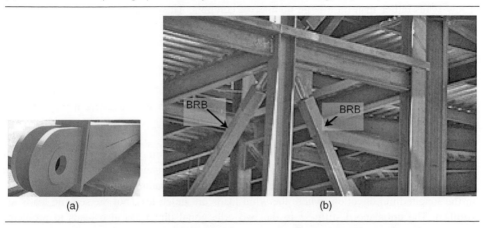

This proprietary system, which is sold under licence, has been shown to be capable of undergoing many reversing cycles of tensile and compressive plastic strain without strength or stiffness degradation, and hence should survive earthquake loading without the need for repair. Should post-earthquake replacement be necessary, bolted end connections would make removal and replacement relatively straightforward. The yielding portions of EBFs, being parts of beams, are much harder to replace; moreover, being subject to much more complex stress states, they are more likely to require replacement.

Any of the permitted arrangements for CBFs shown in Figure 5.4 would be possible, with the diagonal braces being formed of BRBs. AISC 349-10 (AISC, 2010) gives design rules for BRBFs, which are assigned the same force reduction (ductility) factor as EBFs. EC8 does not treat BRBFs.

5.5.7 Shear walls
5.5.7.1 General

Shear walls are more rationally known as 'structural walls' in New Zealand, because their flexural behaviour is usually more important than their shear behaviour. Their favourable features are the provision of strength and stiffness at low cost. The discussion below concentrates on reinforced concrete shear walls, although plywood shear walls are widely used as an efficient bracing system in low-rise timber frame housing, particularly in California, and steel shear walls have also been used (Section 5.5.7.6).

The behaviour of concrete shear walls in earthquakes has generally been good; they are not prone to the 'pancake' collapses that can flatten frames, and prove so lethal to their occupants. The shear wall can be thought of as the ultimate 'strong column', which prevents the formation of a soft storey. Moreover, shear walls avoid the stress concentrations found at the beam–column joint regions of reinforced concrete frames, and avoid some of the dependence on good formwork and steel-fixing skills associated with frames. Considerable ductility is possible in slender shear walls (those with a high ratio of height to width), which reach their ultimate strength in flexure before shear. Stocky shear walls may be harder to make ductile, but their large potential strength reduces the need for ductility. Offsetting these advantages to some extent, lateral load resistance in shear wall buildings is usually concentrated on a few walls rather than on a large number of columns. This implies lower redundancy and possible foundation problems, including those of uplift. The openings that are inevitably needed in the walls also create stress concentrations and points of weakness.

Concrete shear walls often form the access cores of a building, carrying lifts, staircases and service ducts. These can be readily employed as seismic-resisting elements, but the stiffness needs to be balanced on plan to prevent torsional problems arising from eccentricity between centres of stiffness and mass.

Regular cross-walls are also often found in rectangular buildings between office spaces or hotel bedrooms (Figure 5.9). Often this provides adequate seismic strength in the transverse direction of the building (where it is needed most for wind loadings) but inadequate strength in the longitudinal direction (where the wind loads are much less, but the seismic loads are similar). The unbalanced stiffness in the two orthogonal directions also leads to problems

Figure 5.9 Cross-wall construction

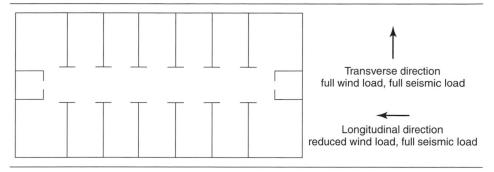

Transverse direction
full wind load, full seismic load

Longitudinal direction
reduced wind load, full seismic load

of torsional instability. Another potential danger of the arrangement in Figure 5.9 is that the partition walls are needed on upper storeys, but are discontinued at ground floor for architectural reasons, creating a potentially lethal weak storey. Shear walls at other than service cores and partition walls present barriers that may interfere with architectural and services requirements.

Non-symmetrical shear walls, for example, forming T or L shapes, must be used with care to ensure that seismic response in one direction is the same as in the opposite direction. The reasons are the same as those for not permitting the CBF configurations shown in Figure 5.4(b and c); yielding deflections in non-symmetrical cases will be greater in the weaker than in the stronger directions, and seismic deflections will therefore accumulate in the weak direction.

Shear walls on their own are a highly suitable solution for medium-rise buildings up to about 20 storeys. In taller buildings, it is likely that they need to be combined with frames to provide sufficient overall stability and stiffness.

5.5.7.2 Single or isolated shear walls

The aspect ratio of a shear wall (the ratio of its height to width in the plane of loading) should normally be restricted to about 7. Higher ratios may result in inadequate stiffness, problems in anchoring the tension side of the shear wall base, and possibly significant amplifications due to 'P–delta effects'.

Aspect ratios below about 2 mark the transition from 'slender' to 'stocky' behaviour, and walls with such dimensions require considerable care in design if a ductile failure mode is required. Without this care, stocky shear walls are likely to fail in brittle failure modes such as diagonal tension or sliding shear, rather than undergoing the more ductile flexural failure possible in slender walls. Stocky shear walls may need increased strength or special detailing, including diagonal steel, to overcome these problems.

5.5.7.3 Large panel precast wall systems

Large panel systems have been extensively used to provide rapid construction of medium-rise housing in seismic areas, particularly in the Balkan region and the former Soviet Union. In contrast with the disastrous performance of precast frames in the 1988 Armenian earthquake, panel housing performed quite well both in that event and the 1978 Bucharest earthquake. It appears that deficiencies in construction quality and lack of ductility were more than compensated for by high strength.

'Tilt-up' construction is a form of precasting extensively used in seismic areas of the USA and elsewhere for low-cost one and two-storey industrial sheds. The wall panels are cast horizontally on the ground at site and then 'tilted up' when they have achieved sufficient strength. They have been prone to fail in earthquakes at their connections with the roof; provided adequate strength is supplied at this connection, the system can perform satisfactorily.

5.5.7.4 Frame–wall or dual systems

Combinations of moment-resisting frames with shear walls are known as frame–wall or dual systems. This combination can be structurally efficient and is favoured in both US and Japanese practice as providing good redundancy. One advantage of frame–wall systems is that the shear wall can be used to prevent a 'weak storey' forming in the moment-resisting frame. This means that the relative strength requirements to ensure a 'strong column/weak beam' frame may theoretically be relaxed. This gives more freedom in selecting beam and column sizes and there is less concern about the strengthening effect that floor slabs have on beams. EC8 and New Zealand codes, but not US codes, allow for this.

A common application of frame–wall systems is in medium to high-rise buildings, where perimeter frames are used in conjunction with central shear wall cores. In buildings of over 50 storeys in which wind-induced motions must be controlled, 'outriggers' between the core and perimeter frame are often used (Figure 5.10) to increase stiffness. In structures that require earthquake resistance, careful consideration must be given to the capacity design implications of using outriggers. Good ductility requires that yield occurs first in ductile modes, and it must be ensured that the outriggers do not force a brittle mode, such as crushing or budding in the perimeter columns connected to the outriggers. A possible solution would be to design the outriggers to yield in a ductile manner at a load less than that corresponding to brittle failure of the columns.

5.5.7.5 Coupled shear walls

Coupled shear walls consist of two or more walls linked by horizontal coupling beams (Figure 5.11). The beams are often formed as a result of openings required through the wall at each floor level; the resulting structure becomes effectively a frame with very strong columns and weak beams. Most of the yielding is therefore confined to the coupling beams and the bases of the walls. Provided they are adequately designed, which for the coupling beams often involves the use of diagonal steel, excellent ductility can be obtained, accompanied by good stiffness. Redundancy is also good, in that plastic energy dissipation (with the attendant risk of failure) is distributed between a number of coupling beams. It should be noted that EC8 (CEN, 2004) observes that slabs are ineffective as coupling elements between pairs of shear walls, and should not be used as such.

Limiting overall aspect ratios of coupled shear walls are similar to those for a similar unperforated single shear wall. Satisfactory efficiency of coupling beams is defined by EC8 to

Figure 5.10 Outriggers in a shear core/perimeter frame building

Outrigger beams   Perimeter frame

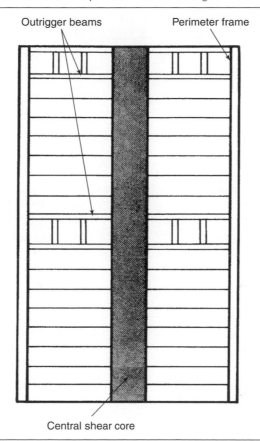

Central shear core

occur when the proportion of base moment resisted by push–pull axial forces in the shear walls is at least three-quarters of the total base overturning moment.

5.5.7.6 Steel plate shear walls

Steel plate shear walls (Figure 5.12) consist of thin steel web plates surrounded by stiffer steel column and beam elements that act as restraining elements. Usually, the web plates are unstiffened and buckle under relatively low seismic shears, but the restraining elements and their vertical and horizontal boundaries enable them to develop considerable post-buckling deformation capacity through the formation of a 'tension field' within the plate (Figure 5.12(b)). They therefore provide lateral strength and stiffness in the pre-buckling phase, and ductility post-buckling and offer some of the advantages of concrete shear walls at much lower mass and weight. US standards provide advice on their design.

5.5.8 Special methods of improving earthquake resistance

5.5.8.1 Overview

In the past 25 years, special systems have been developed to improve the earthquake-resisting characteristics of conventional structures by modifying their dynamic characteristics, with the

Figure 5.11 Coupled shear walls

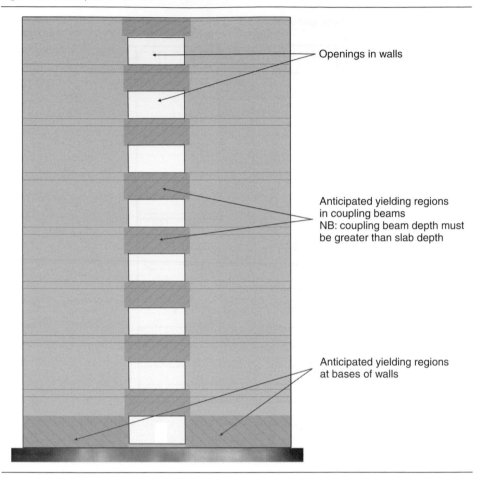

Openings in walls

Anticipated yielding regions
in coupling beams
NB: coupling beam depth must
be greater than slab depth

Anticipated yielding regions
at bases of walls

principal aim of extending the range of performance targets beyond simply 'life safety' to others shown in Table 5.1. This has generally been done by either 'passive' or 'active' devices. The passive devices either change the period of the structure or increase its damping or, more usually, do both in combination. According to Martelli *et al.* (2012), approximately 20 000 buildings have been built in the past 25 years that employ passive devices, including 6600 in Japan and nearly 2500 in China; however, they still form a small minority of buildings in seismically active regions. With very few exceptions, the many buildings with passive response control which have been subjected to damaging earthquakes have performed extremely well.

Active devices modify the structure's dynamic characteristics continuously during the course of an earthquake to provide an optimal configuration; however, they require an external power source to effect these modifications and raise serious questions about whether they

Figure 5.12 Steel plate shear walls: (a) three-storey steel plate shear wall; (b) tension field developing in buckled steel plate shear wall under lateral loading (part (b) reproduced from Berman and Bruneau, 2003; courtesy of Michel Bruneau, Dept. of Civil, Structural, and Environmental Engineering, University at Buffalo)

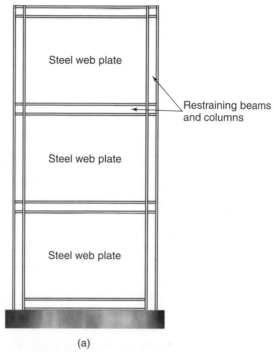

(a)

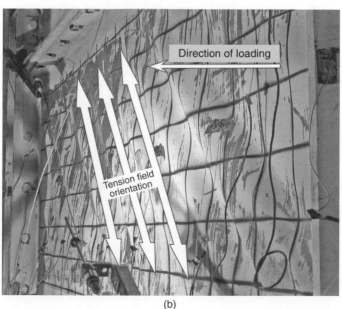

(b)

can be made reliable enough to function dependably during the chaotic conditions of an earthquake. As a result, active devices have not been used in practice for earthquake resistance, although 'semi-active' devices, which overcome some of these reliability concerns and are described in Section 5.5.8.4 below, have been built in limited numbers.

Chapter 13 provides a more detailed introduction to one of these systems, namely seismic isolation. A comprehensive treatment is given by Christopoulos and Filiatrault (2006). Mayes *et al.* (2013) compares the seismic performance and cost of some of the passive systems.

5.5.8.2 Seismic isolation in buildings

Seismic isolation involves mounting a building on bearings of low lateral stiffness. Laminated natural or synthetic rubber bearings are the most commonly used forms, with a typical plan dimension of 600 mm and thickness of 150–250 mm, although sliding bearings are also widely used. A typical vertical load capacity per bearing is 60 t. The intention is to increase the natural period of the building to take it away from resonance with the forcing motions of the earthquake. The bearings, because they experience high cyclic strains, also provide suitable locations for introducing hysteretic, viscous or frictional damping elements, to reduce response still further.

Seismic isolation is therefore most suitable for low to medium-rise buildings on relatively stiff soil sites. A building period of around 1 s is usually recommended as the upper bound, implying a maximum height of 12–15 storeys in shear wall structures and about 10 storeys in frame buildings. Taller buildings are less suitable, partly because their period is already likely to be well away from resonance and partly for the practical reason that overturning forces would result in large uplifts on the bearings, which may be difficult to sustain. However, a number of buildings over 20 storeys high have been built in Japan, and successfully survived the great Tohoku Japan earthquake of 2011.

Geological as well as structural characteristics are also important. Sites with deep soft soil deposits, with a site period exceeding 1.5 s, are unlikely to be suitable, because the long-period earthquake motions associated with them mean that a shortening, not a lengthening, of the building period is usually needed. Sites near active faults may also be unsuitable, because of the special nature of near-fault motions (Kelly *et al.*, 2010). When wind loads exceed 10% of the building weight, the advantages of isolation diminish considerably, although such a large percentage is very unlikely in concrete buildings.

Seismic isolation can reduce design forces on the superstructure by a factor of up to two or three. Just as importantly, by filtering out high-frequency accelerations and limiting the storey drifts (inter-storey deflections), it can increase protection to non-structural elements very significantly and result in 'immediate occupancy' performance even after a strong earthquake. Moreover, occupants of seismically isolated buildings are less aware of the motion caused by moderate events; this factor is valued in Japan, where perceptible earthquakes are common. By limiting or eliminating structural and non-structural damage, the technique is attractive for retrofitting historic buildings, in which the structure and contents need to be preserved intact for future generations.

Figure 5.13 Crawl space under a base-isolated building (photograph courtesy of Kajima Corporation, Tokyo)

Large horizontal deflections occur between the top and bottom of the bearing during a large earthquake, and these must be allowed for by the provision of a sufficiently wide gap. It is important that the gap does not become bridged or filled during the lifetime of the building. Services entering the building and finishing at ground level may have to accommodate deflections of the order of 100–200 mm. Flexible loops in services and suitable detailing of finishes have enabled these problems to be successfully overcome in practice.

Isolation bearings and associated energy absorbers are usually located so that they can be inspected during the lifetime of the building. This is often achieved by providing a crawl space under the building in which the bearings are located (Figure 5.13), which adds to the cost of seismic isolation.

Seismic isolation of buildings is a relatively mature technology, which has been subject to intensive theoretical and laboratory investigation (Kelly *et al.*, 2010). Many base-isolated buildings have been tested by large earthquakes in Japan, the USA, New Zealand and Chile, some to their design limits or above, and with few exceptions the response has been excellent. However, European and US codes of practice, although they cover seismic isolation, require relatively conservative design procedures, and this may have prevented more widespread use of the technology.

5.5.8.3 Supplemental damping
Damping devices are almost always used in conjunction with seismic isolators to limit deflections. However, they have also been used independently, and equivalent viscous damping values of up to 20% may be achieved. Two possible configurations are now discussed. In Figure 5.14(a), the damping devices remain very stiff or rigid up to a threshold shear force value, which should be less than the design wind load, and the frame responds as a conventional CBF. Above the threshold, the damping device loses stiffness but provides energy

Figure 5.14 Supplemental damping systems

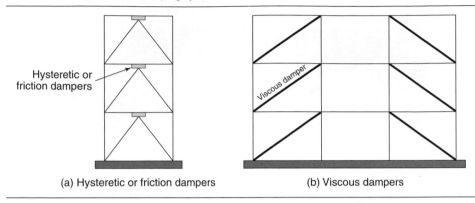

Hysteretic or friction dampers

Viscous damper

(a) Hysteretic or friction dampers   (b) Viscous dampers

dissipation – and hence damping. This can be achieved by a metal damper that yields at the threshold shear, providing hysteretic damping, or friction devices that slip at the threshold, providing friction damping. There are other more sophisticated possibilities, discussed by Christopoulos and Filiatrault (2006).

Figure 5.14(b) shows the damper effectively replacing the conventional bracing. Here, the damper is a viscous device (similar in principle to the shock absorber of a car) and the damping force (unlike the case for hysteretic dampers) depends on the relative velocity between the two ends of the damper, rather than the relative displacement. For low-level earthquakes or for wind loading, the relative velocity is low, and the frame acts as an unbraced (moment-resisting) structure. As the seismic excitation increases, the damping force and energy dissipation also increases. The same result would be obtained by using a viscous instead of hysteretic or friction damper in Figure 5.14(a). Viscous dampers have a number of different, and mainly advantageous, characteristics compared with friction or hysteretic dampers, as follows.

- The maximum force imposed by a viscous damper on the rest of the structure is at a maximum when the structure is passing through its undeflected and hence least stressed position, because this is when the velocity is greatest. For a hysteretic or friction damper, the opposite is true; the highest forces are imposed on the structure at its most deflected and stressed state; this may be a particularly significant consideration when retrofitting a weak existing building.
- If the viscous dampers protect the structure from yield, it will return to its undisturbed position at the end of the earthquake, which is much less likely for hysteretic or friction dampers, which are likely to impose permanent displacements.
- The low initial stiffness of a viscously damped system means that high frequency ground motions, which can be damaging for building contents, are filtered out.
- However, viscous dampers are likely to be considerably more expensive than their hysteretic or friction equivalents.
- They are also less efficient than hysteretic or friction dampers at responding to the sudden impulsive ground motions ('flings' – see Section 4.3.7) that may occur near to the causative fault of an earthquake.

Supplemental damping of the kind shown in Figure 5.14 is suitable as a retrofit measure for improving the seismic performance of inadequate frame buildings, as well as for new buildings. Although not as extensively used as base isolation, many such systems have been incorporated in buildings in Japan, the USA and elsewhere. Their record of performance in earthquakes to date has been good.

Codes of practice do not currently provide design rules for passive supplemental damping, but extensive guidance exists – see Chapter 13.

5.5.8.4 Semi-active damping systems

The first building with 'semi-active' seismic control was built in Japan in 1997. In this system, the level of damping is switched between high and low during an earthquake depending on the response of the building measured in real time, with the aim of optimising stiffness and energy dissipation. Failure of the switching control is designed to reset the damping level to an intermediate level of passive damping, which could still control response adequately, although not optimally. This failsafe feature implies that there is much less reliance on the control system to work in the rare and chaotic conditions of a damaging earthquake than is the case for fully active systems. Figure 5.15 gives details of a 26-storey building in Tokyo, which in fact functioned as intended during the 2011 Tohoku Japan earthquake. The building was base isolated and the viscous dampers at isolation level were semi-actively controlled. The building site experienced a peak ground acceleration of 6% g in the longitudinal direction (shown in the figure) and 10% g in the transverse direction. Based on calculations, the semi-control system reduced building response by 10–20% compared with purely passive dampers. A reduction of up to 40% is claimed as possible if semi-active dampers had been deployed up the height of the building, instead of only at the isolation level. Full details of the building are given by Nagashima *et al.* (2012).

5.5.8.5 Self-centering systems

Christopoulos and Filiatrault (2006) describe four types of system intended to ensure that the building returns to its undisturbed position at the end of an earthquake, and hence potentially achieve an 'immediate occupancy' performance. Three of these systems effectively replace the hysteretic or friction damper of Figure 5.14(a) with (respectively) shape memory alloy devices, special energy dissipating restraints or ring springs; for details see Christopoulos and Filiatrault (2006).

A fourth self-centering device is the post-tensioned frame (Figure 5.16). This comprises a precast concrete frame or wall structure that is connected by post-tensioned cables passing unbonded through their ducts. At a critical displacement, designed to be below the yielding capacity of the frame, the joints between the beams and columns open up, limiting further stressing of the frame, and concentrating further deformation at the opening beam–column joints. Mild non-prestressed reinforcement is placed across these joints to provide hysteretic energy dissipation (and hence damping) when the joints open. As the frame does not yield during the earthquake, the post-tensioned cables can pull the structure back to its initial undisturbed position at the end of an earthquake, and the structure at least returns to an 'immediate occupancy' performance state. Similar systems are possible with steel frames and also with precast concrete walls (Figure 5.17). A number of buildings up to 39 storeys high, which

Figure 5.15 Performance of Tokyo building with semi-active damping during the 2011 Tohoku Japan earthquake (reproduced by kind permission of Dr I. Nagashima, Taisei Corporation, Yokohama)

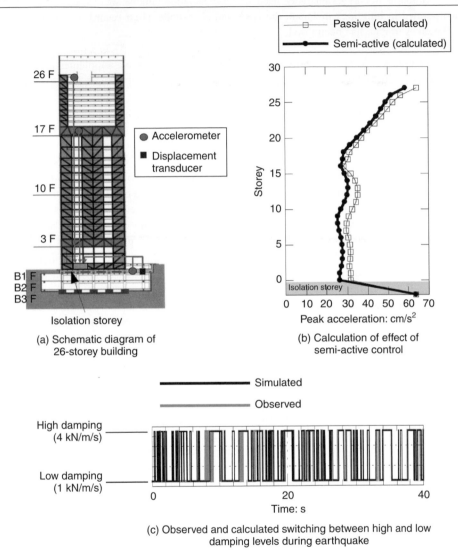

(a) Schematic diagram of
26-storey building

(b) Calculation of effect of
semi-active control

(c) Observed and calculated switching between high and low
damping levels during earthquake

incorporate rocking precast walls, have been built in the USA, New Zealand, Chile and elsewhere, and one performed very well in the 2011 Christchurch New Zealand earthquake (Pampanin, 2012).

5.6.   Cost of providing seismic resistance

The additional cost of providing seismic resistance is hard to establish, because buildings tend to be unique projects, and it is difficult to compare sufficiently similar buildings that differ only in their need for seismic resistance. Indicative figures for areas of high seismicity are

Figure 5.16 Post-tensioned precast frame (reproduced from Stanton and Nakaki, 2002)

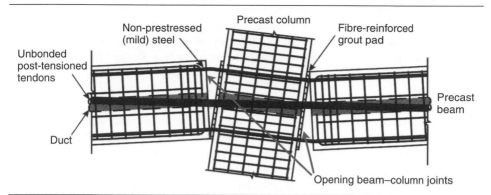

20% on structural design costs, 10% on structural construction costs and significantly less on overall project costs, once building contents, services, land cost, etc. are taken into account. However, for special projects such as casualty hospitals or nuclear power-related projects, the costs could be very considerably greater.

On the cost of providing conventional means of seismic resistance, the Royal Commission notes

'Another interesting finding from examining Rawlinson [New Zealand construction handbook, 25th edition] is that the cost of an office building does not necessarily correlate to the seismicity of the region. Wellington and Auckland have similar building cost ranges, despite the difference in their hazard factors (0.4 and 0.13 respectively). This shows that there are significant other considerations driving the overall cost of buildings.'

The economic implications of seismic isolation are also difficult to establish, partly because buildings are usually prototypes and so sufficiently comparable isolated and non-isolated

Figure 5.17 Post-tensioned rocking wall system (reproduced from Stanton and Nakaki, 2002)

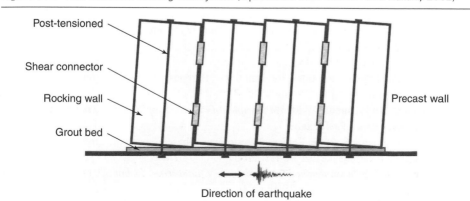

buildings are hard to find, and partly because a conservative view has generally been taken of the reduction in superstructure forces possible because of the seismic isolation. The cost of the bearings is typically about 10% of the total structural cost, and a lower proportion of the total building cost. A discussion of the costs of base isolation and other means of improving seismic performance is given by the report of the New Zealand Royal Commission on the Canterbury earthquakes (Canterbury Earthquakes Royal Commission, 2012).

An overall cost increase of less than 10% appears small, but it underestimates the problem of providing satisfactory seismic resistance, because cost is not the only issue. Catastrophic destruction of buildings, such as occurred in the Kocaeli, Turkey earthquake of 1999, has not occurred primarily as a result of pennypinching by developers or contractors, but from the absence of engineers qualified in the design and construction skill required, and the absence of checking and enforcement procedures.

These relatively low values of cost increase apply when seismic resistance is taken into account at the beginning of a project. The cost of trying to add in seismic resistance at a late stage in design is likely to be much greater. Providing seismic resistance to inadequate existing buildings is even more expensive, and can often exceed 60% of the replacement cost. Indicative cost figures for retrofit are given by Calvi, 2012.

REFERENCES

AISC (2010) ANSI/AISC 340-10: Seismic provisions for structural steel buildings. American Institute of Steel Construction, Chicago, IL, USA.

Arnold C and Reitherman R (1982) *Building Configuration and Seismic Design*. Wiley, New York, USA.

Berman JW and Bruneau MR (2003) *Experimental investigation of light gauge steel plate shear walls for the seismic retrofit of buildings*. Report MCEER-03-0001, MCEER, Buffalo, NY, USA.

Booth E, Wilkinson S, Spence R, Free M and Rosetto T (2011) EEFIT: the UK Earthquake Field Investigation Team. *Proceedings of the Institution of Civil Engineers – Forensic Engineering* **164(3)**: 117–123.

Calvi GM (2012) Alternative choices and criteria for seismic strengthening. *Proceedings of the 15th World Conference on Earthquake Engineering, Lisbon*.

Canterbury Earthquakes Royal Commission (2012) *Final Report Volume 2: The performance of Christchurch CBD buildings*. See http://canterbury.royalcommission.govt.nz/Final-Report – Volumes 1–2 and 3.

CEN (2004) EN 1998-1: 2004: Design of structures for earthquake resistance. Part 1: General rules, seismic actions and rules for buildings. European Committee for Standardisation, Brussels, Belgium.

Christopoulos C and Filiatrault A (2006) *Principles of Passive Supplemental Damping and Seismic Isolation*. IUSS Press, Pavia, Italy.

Kam WK and Pampanin S (2012) Revisiting performance-based seismic design in the aftermath of the Christchurch earthquake 2010–2011: raising the bar to meet societal expectations. *Proceedings of the 15th World Conference on Earthquake Engineering, Lisbon*.

Kelly T, Skinner RI and Robinson WH (2010) *Seismic isolation for designers and structural engineers*. National Information Centre of Earthquake Engineering, IIT Kanpur, India.

Martelli A, Forni M and Clemente P (2012) Recent worldwide application of seismic isolation and energy dissipation and conditions for their correct use. *Proceedings of the 15th World Conference on Earthquake Engineering, Lisbon.*

Mayes R, Wetzel N, Weaver B *et al.* (2013) Performance based design of buildings to assess damage and downtime and implement a rating system. *Bulletin of the New Zealand Society for Earthquake Engineering* **46(1) March**: 40–55.

Nagashima I, Maseki R, Shinozaki Y, Toyama J and Kohiyama M (2012) Study on performance of semi-active base-isolation system using earthquake observation records. *Proceedings of the International Symposium on Reliability and Risk Management (ISRERM), Kanagawa University, Yokohama.*

Pampanin S (2012) Reality-check and renewed challenges in earthquake engineering: implementing low-damage structural systems – from theory to practice. *Proceedings of the 15th World Conference on Earthquake Engineering, Lisbon.*

Stanton J and Nakaki SD (2002) Design guidelines for precast concrete seismic structural systems. Press report No. 01/03-09 (available from Precast/prestressed Concrete Institute, www.pci.org).

Earthquake Design Practice for Buildings
ISBN 978-0-7277-5794-4

ICE Publishing: All rights reserved
http://dx.doi.org/10.1680/edpb.57944.137

Chapter 6
Seismic codes of practice

I consider that codes of practice have stultified the engineering
profession, and I wonder whether an engineer can now act professionally,
i.e. use his judgement.

Francis Walley, *The Structural Engineer*, February 2001

This chapter covers the following topics.

- The development and philosophy of codes.
- Overview of code requirements for analysis, strength, deflection, detailing, foundation design and non-structural elements.
- Sources of code guidance and advice.

The chapter provides a broad description of the principal features of European and US codes, highlighting the differences between them. More details are given about rules for concrete, steel and other materials in the following chapters.

6.1.   Role of seismic codes in design

In most earthquake-prone areas, building construction is subject to a code of practice or standard, often legally enforceable, which establishes minimum requirements. Even where this is not so, common practice or contractual requirements will require compliance with a code; for example, US seismic codes have very often been used in seismic areas outside the USA in the past, and the same applies to the use of Eurocode 8 (EC8) outside Europe. In consequence, a normal part of design is to ensure that a set of minimum code-based acceptance criteria have been met. As with any other part of the design process, however, the use of codes should not be a substitute for the use of sound engineering judgement. Codes describe minimum rules for standard conditions and cannot cover every eventuality. Buildings respond to ground shaking in strict accordance with the laws of physics, not in accordance with rules laid down, perhaps many years ago, by a (sometimes fallible) code drafting committee.

It should further be remembered that seismic safety results not only from the use of appropriate codes being applied with understanding in the design office, but also from the resulting designs being implemented correctly on site. Disasters can happen even when reliable seismic codes of practice have statutory force; the great destruction in western Turkey caused by earthquakes in 1999 is an often quoted example.

Seismic codes are essential tools for seismic designers; at best, they are repositories of the current state of practice based on decades of experience and research. They can, however, restrict the designer in ways that do not necessarily improve seismic safety; several examples were quoted in the previous chapter, and they form only one part of the design process.

6.2.   Development of codes

Early codes were based directly on the practical lessons learned from earthquakes, relating primarily to types of construction. In 1909, following the Messina earthquake, which caused 160 000 deaths, an Italian commission recommended the use of lateral forces equal to 1/12th of the weight supported. This was later increased to 1/8th for the ground storey. The concept of lateral forces also became accepted in Japan, although there was a division of opinion on the merits of rigidity as opposed to flexibility. After the 1923 Tokyo earthquake a lateral force factor of 1/10th was recommended and a 33 m height limit imposed. In California lateral force requirements were not adopted by statute until after the 1933 Long Beach earthquake.

The use of lateral forces in design became widely used, the value of the coefficients being based almost entirely on experience of earthquake damage. In 1943, the city of Los Angeles related lateral forces to the principal vibration period of the building and varied the coefficient through the height of multi-storey buildings. By 1948, information on strong motion and its frequency distribution was available and the Structural Engineers Association of California recommended the use of a base shear related to the fundamental period of the building. Once information was available on the response spectra of earthquake ground motion, the arguments over flexibility against rigidity could be resolved. The flexible structure was subjected to lower dynamic forces but was usually weaker and suffered larger displacements.

The next important step grew out of advances in the study of the dynamic response of structures. This led to the base shear being distributed through the height of the building according to the mode shape of the fundamental mode, as originally proposed in the 1960 Structural Engineers Association of California code.

At this stage lateral forces had undergone a quiet revolution from an arbitrary set of forces based on earthquake damage studies to a set of forces which, applied as static loads, would reproduce approximately the peak dynamic response of the structure to the design earthquake. This, however, is not quite the end of the matter for lateral loads, because structural response to strong earthquakes involves yielding of the structure so that the response is inelastic.

As discussed in Section 3.4, much larger design forces are required for an elastic structure without ductility than for one that can tolerate substantial plastic deformations. Because it is found in practice that the increased cost of elastic design requirements is unacceptably large, it is almost universally accepted that ductile design should apply for major earthquakes. Exceptions to this are made for structures of special importance, or when the consequences of damage are unacceptable. Although modern codes contain much useful guidance on other matters, it is the calculation of lateral design forces and the means of providing sufficient ductility that constitute, in practice, the two most vital elements for the structural engineer.

6.3.   International seismic codes

The discussion that follows concentrates on European and US practice, although other international seismic codes are touched on more briefly. Free downloads of seismic loading provisions from 64 codes worldwide (not necessarily in the current versions) are available at http://iisee.kenken.go.jp/worldlist/Web/WorldList_TOP.htm.

European seismic practice is taken here as represented by EC8 (CEN, 2004), which has been adopted by most (although not all) European countries that are members of the European Committee for Standardisation (CEN) (www.cen.eu), and has been used as a model by a few countries outside Europe, such as Egypt and Vietnam. The basic version of EC8 must be used in conjunction with a 'national annexe' for the country where it is being applied, which will provide data for the seismic hazard values to be taken in that country, and on other aspects (usually connected with the margin of safety to be provided) when the basic EC8 allows national choice. A freely downloadable introduction to all of the structural Eurocodes is provided by the British Standards Institution (BSI, 2009a).

The closest US equivalent to the Eurocodes is the international building code (ICC, 2012), formerly called the uniform building code. The provisions of the international building code refer to other US codes – ASCE 7 for loading, ACI 318 for concrete, ANSI/AISC 341 for steel and so on – and it is generally these referenced codes that are referred to in what follows.

6.4.   Design and performance objectives

As discussed in section 5.2, EC8 (CEN, 2004) and the US codes for new buildings currently concentrate on the objective of ensuring life safety in earthquakes. Limiting damage (as opposed to saving lives) is stated as an additional objective, but as at the time of writing (2013) both US and European codes essentially give provisions for only one limit state, it is hardly surprising that only one performance objective (i.e. life safety) is comprehensively addressed. Since the mid-1990s, a large research effort has been conducted into developing 'performance-based' rules, suitable for adoption in codes of practice (Fajfar and Krawinkler, 2004), which extend the objectives to others (see Table 5.1 in the last chapter), and it is likely that this will be reflected in future code revisions. The lack of code provisions has not prevented major projects and those incorporating specially engineered features such as seismic isolation or supplemental damping from explicitly considering multiple performance objectives; see for example Pampanin (2012).

The situation is different for rules for strengthening existing buildings, in which both EC8 part 3 (CEN, 2005) and ASCE 41-06 (ASCE, 2006) consider collapse, life safety and damage limitation performance objectives.

By contrast, in Japan a two-stage check has been required for new buildings since the early 1980s. The structure is designed to survive a 'first-phase' event, which has a low but non-negligible probability of occurring during the building lifetime; yield strains may approach but not exceed their elastic limit based on an elastic analysis. The structure is then checked for its ability to survive the 'second-phase event', which is roughly equivalent to the maximum recorded earthquake. For low-rise buildings, this involves checking a set of 'deemed to satisfy' rules (i.e. simple rules not requiring a detailed analysis), while for buildings taller than 61 m, non-linear dynamic analysis is required.

6.5.   Code requirements for analysis
6.5.1   Equivalent static design
Most codes specify a procedure whereby a minimum lateral strength is calculated; the corresponding base shear force is then applied to the structure as a set of equivalent static forces distributed up the height of the building. This procedure is permitted for low-rise buildings without significant structural irregularities; more complex analyses are required in other cases.

The lateral strength requirement is calculated as a function of the following parameters.

1   Building mass: This is calculated as the structural mass arising from the dead load, plus a proportion of the variable mass arising from the live load. EC8 typically specifies that 30% of the live load in office and residential loading should be included, but this might fall to 0% of the snow load in areas where snow is relatively rare and rise to up to 100% of live load for warehouses and archive buildings, and for permanent equipment.

2   Basic seismicity: In EC8, this is expressed as the design peak ground acceleration a_{gR} expected on rock sites, for the reference return period (usually 475 years). This is the equivalent of the Z factor in earlier US codes. ASCE 7-10 (ASCE, 2010) now provides maps for the USA of spectral accelerations expected on rock at 0.2 s and 1.0 s periods, for the 'maximum considered earthquake' with a return period of approximately 2500 years. In the basic version of EC8 (CEN, 2004), no maps are provided; countries adopting the code provide the design accelerations in 'national annexes' published in conjunction with the national edition of the code.

3   Earthquake magnitude and distance: In EC8, two types of site are recognised, one dominated by large magnitude earthquakes, the other by smaller magnitude but closer events. Different design response spectrum shapes apply to each, and the 'national annexe' is supposed to specify which one should be used for a particular region. ASCE 7-10 does not have this explicit distinction, but allows for the effect of different magnitudes and distances of causative earthquakes on the shape of the design spectrum in a more satisfactory way. This is done by varying the relative values of the 0.2 s and 1.0 s spectral accelerations, which are specified in the seismic hazard maps of the USA referred to above.

4   Site classification: The basic information on seismicity is presented for rock sites, but the soils overlying rock can make an enormous difference to earthquake intensity. In both EC8 and ASCE 7-10, sites must be classified into one of several categories, ranging from rock to very soft soils, although the exact descriptions of the site categories vary somewhat between the two codes. In EC8, the site classification determines a factor called S, which amplifies the rock accelerations. It also determines the periods T_B, T_C and T_D, which modify the shape of the response spectrum. In ASCE 7-10, the site classification together with the basic seismicity together determine the modifications arising from soil; unlike EC8, the tendency of amplifications to reduce with intensity of earthquake is included (see Section 4.3.1).

5   Building function: Some buildings, such as emergency hospitals, may have a need for enhanced protection during an earthquake. ASCE 7-10 allows for this with an importance factor I_E, which varies between 1 and 1.5. The structural factor R (see

below) is divided by I_E, so effectively design forces are increased directly in proportion to I_E. In EC8, a similar factor γ_I is used to multiply the design ground acceleration on rock; for a linear analysis, design forces increase by the same factor, but for a non-linear time history analysis, this is not necessarily the case. In EC8, recommended values of γ_I vary between 0.8 for agricultural buildings without permanent occupancy to 1.4 for emergency hospitals.

6   Structural factor: This allows for the inherent ductility of the structure, and also the fact that during the peak transient loading of an earthquake, it is acceptable to utilise more of the 'overstrength' inherent in most structures (i.e. the ratio between ultimate lateral strength and nominal design strength) than would be the case for permanent loads or wind loads. In ASCE 7-10, the structural factor is called R and is a straight divisor on the required strength. R factors range from 8 for specially designed and detailed ductile frames to 1.25 for highly non-ductile systems. In EC8, the structural factor is called q, and for medium to long period buildings is also a simple divisor on required lateral forces. For very short period buildings, however, the reduction due to q is limited (see Table 3.2). q Factors range from 8 for very ductile structures to 1.5 for structures without seismic detailing. Unlike R in ASCE 7-10, q in EC8 reduces when significant structural irregularity exists, and also depends explicitly on the 'overstrength' (ultimate lateral strength divided by lateral strength at first yield), which arises from a redistribution of forces after plastic yielding. In both ASCE 7-10 and EC8 the use of low ductility structural types is restricted to areas of low seismicity; in ASCE 7-10, the restrictions are more extensive than in the Eurocode, and ASCE 7-10 also places height limits on certain structural forms.

7   Building period: In practically every modern seismic code, the required lateral strength varies with the fundamental period of the building. This can either be assessed directly from the mass and stiffness of the structure, usually using a computer program, or from empirical formulae based on building height and structural form. ASCE 7-10 recognises that the empirical formulae account for the stiffening effect of non-structural elements such as cladding in lowering structural period, which will usually result in an increase in seismic load. ASCE 7-10 therefore restricts the advantage that can be gained from using a lower period based on a 'direct' analysis, which will generally ignore these stiffening effects. EC8 has no similar restriction.

The lateral strength requirement calculated from these procedures is then equal to the design shear at the base of the building or 'seismic base shear'. In order to assess strength requirements in other parts of the building, the base shear must be distributed up the height of the building. A commonly adopted formula assumes that the fundamental mode of the building is a straight line, leading to

$$F_i = F_b \frac{z_i m_i}{\sum z_j m_j} \tag{6.1}$$

where F_i is the force at level i, F_b is the seismic base shear, m_i and z_i are the mass and height at level i, and the summation $\sum m_j z_j$ is carried out for all masses from the effective base to the top of the building.

Recognising that relatively greater seismic loads may occur at the top of tall buildings due to higher mode effects, ASCE 7-10 modifies this formula slightly to

$$F_i = F_b \frac{(z_i m_i)^k}{\sum (z_j m_j)^k} \tag{6.2}$$

where k equals 1 for building periods less than 0.5 s (i.e. retain Equation 6.1), and k equals 2 for periods exceeding 2.5 s, with a linear interpolation for intermediate periods. EC8 allows Equation 6.1, where the fundamental mode shape is approximately linear, otherwise requiring z_i and z_j to be replaced by the mode shape of the fundamental building mode.

The horizontal forces must also be distributed in plan at each level. As the seismic forces arise from inertia effects, F_i is distributed in proportion to the mass at that level. However, special allowance is usually made for torsional effects (Section 3.6.4); procedures vary between codes.

This gives sufficient information to calculate not only the shears but also the bending moments at each level in the building. For foundation design, ASCE 7-10 allows a reduction of up to 25% in the overturning moment at the soil–foundation interface, but this reduction is not permitted in EC8, which requires foundations to be designed for the yield capacity of the superstructure (see Section 7.2 in the next chapter).

6.5.2 Response spectrum analysis

Both EC8 and ASCE 7-10 allow the response spectrum used as the basis for equivalent static design to be used to carry out a response spectrum analysis, and both make this mandatory for tall buildings or buildings with significant structural irregularities. There are differences, however. In EC8, the results from the response spectrum analysis can be used directly. In ASCE 7-10, the procedure is more complex. An equivalent static analysis must first be carried out to determine the base shear V_{base} corresponding to an empirically determined period T, and also a modified (and usually lower) value of base shear V'_{base} corresponding to a period $c_u T$, where c_u ranges from 1.7 for low buildings to 1.4 for tall buildings. The total base shear from response spectrum analysis must then be adjusted to equal at least $0.85 V'_{base}$.

6.5.3 Time history analysis

Linear or non-linear time history analysis is referred to in both EC8 and ASCE 7-10. A major issue is the selection of appropriate time histories. EC8 requires at least three time histories to be used, which on average match the specified design peak ground acceleration a_g, and the average 5% damped spectral values must also be within 90% of the design response spectrum for the appropriate ground conditions. Either artificially generated time histories may be used, or real time histories with appropriate seismological characteristics (i.e. magnitude, distance, soil type, etc.). ASCE 7-10 similarly permits a minimum of three sets of artificial or real time histories, each set consisting of a pair of horizontal motions in two orthogonal directions. The average spectral values must match the design spectrum between 0.2 and 1.5 s periods.

In both EC8 and ASCE 7-10, in which three time histories are used, the maximum response value from the three separate analyses conducted must be used, but with seven time histories, use of the average value is permitted, although as noted in Section 2.8.3, more time histories will normally need to be used to obtain a reliable result.

6.5.4   Non-linear static analysis

EC8 permits this type of non-linear static (pushover) analysis (Section 3.7.9) for the following purposes in buildings.

- To verify or establish the 'overstrength' ratios (ultimate lateral strength divided by lateral strength at first yield), which is used in the calculation of the structural or behaviour factor q.
- To estimate where plastic deformations will occur, and in what form.
- To assess the performance of existing or strengthened buildings, when using EC8 part 3.
- To design new buildings as an alternative to the standard procedures based on dividing the results of elastic analysis by the behaviour factor q.

A detailed procedure is provided in EC8 for carrying out a non-linear static analysis, which is permitted for new buildings. Special rules are included for torsionally eccentric buildings.

ASCE 7-10 does not currently refer to non-linear static procedures. However, ASCE 41-06 (ASCE, 2006) sets out detailed procedures for carrying out such an analysis for existing or retrofitted buildings.

6.5.5   Limiting plastic strains

When performing a non-linear analysis (either static or time history), the designer needs to have values of the maximum plastic response consistent with a specified performance objective, such as damage limitation or collapse prevention. Neither EC8 part 1 nor ASCE 7-10 provides much help here. For example, EC8 part 1 does not specify the maximum permitted rotation of a plastic hinge in a steel or concrete beam or the maximum plastic extension of a steel brace. It does, however, provide values of maximum permissible curvatures in reinforced concrete beams, and these can be converted to plastic hinge rotations if an equivalent plastic hinge length is assumed (see Section 8.4.2). However, EC8 part 1 gives no such clues for steel members in new buildings, although values are provided for both steel and concrete members in informative annexes to EC8 part 3 for existing buildings.

In US practice, ASCE 7-10 gives no information on maximum plastic demands, but ASCE 41-06 provides values of permissible plastic rotations in steel and concrete elements (including concrete shear walls) and plastic deformations in steel tension and compression braces. As noted in Section 3.7.9.5, these permissible plastic deformations are related to the performance goals of 'immediate occupancy', 'life safety' and 'collapse prevention'.

6.6.   Code requirements for strength

In general, both EC8 and US codes specify the same design strength for resisting seismic loads as for gravity, wind or other types of load. There are, however, important exceptions. First, in both EC8 and US codes, an important step is to check that non-ductile elements have sufficient strength so that their capacity is never exceeded. Second, in ACI 318-08

(ACI, 2008) the concrete contribution to shear strength is usually ignored, because this tends to degrade under cyclic loading, although this only applies to high ductility concrete structures in EC8.

6.7.   Code requirements for deflection

Storey drifts (the difference in horizontal deflection between the top and bottom of any storey) must be checked and compared with specified limits in both codes, principally to limit damage to non-structural elements. Under 475-year return events, ASCE 7-10 sets the maximum drift for normal buildings at between 0.7% and 2.5% of storey height, while EC8 specifies between 1% and 1.5%. P–delta effects and separations between structures to prevent pounding must also be checked. Specific elements such as external cladding and columns sized for vertical loads but not seismically detailed must also be checked to confirm that they can withstand the deflections imposed on them during the design earthquake.

The calculation of deflections in EC8 directly follows Equation 3.14 – that is, the elastic deflection corresponding to the design seismic forces is multiplied by the structural or behaviour factor q. Hence design deflections are independent of q factor; the lower analysis forces associated with large q result in lower deflections, but these must then be increased by q. In ASCE 7-10, there is a similar requirement, but the multiplier instead of being the structural factor R is the 'deflection amplification factor' C_d, which is generally lower than R. The independence of design deflection from q and R factor means that when deflection rather than strength governs design, high values of q and R are of little use.

6.8.   Load combinations

The seismic load combinations required by EC8 and ASCE 7-10 can be summarised as follows.

In EC8, the 'design action effect' (i.e. the ultimate load) is taken as being due to the unfactored combination of permanent plus earthquake loads, plus a reduced amount of variable loads, such as live or snow loads. Wind loads are never included with seismic loads; that is, ψ_2 is always taken as zero for wind loads. Using Eurocode notation, this is expressed as

$$
\underset{\text{Design load}}{E_d} \quad = \quad \underset{\text{Permanent}}{\sum G_{kj}} \quad + \quad \underset{\text{Earthquake}}{A_{Ed}} \quad + \quad \underset{\text{Reduced variable load}}{\sum \psi_{2i} Q_{ki}} \quad (6.3)
$$

In ASCE 7-10, essentially two load combinations must be considered, as follows (using ASCE 7-10 notation).

$$
\begin{aligned}
\text{Design load} \quad = \quad & 1.2D \quad + \quad 1.0E \quad + \quad f_1 L \quad + \quad 0.2S \quad\quad (6.4)\\
& 0.9D \quad + \quad 1.0E \quad\quad\quad\quad\quad\quad\quad\quad\quad\quad (6.5)\\
& \text{Dead} \quad\quad \text{Earthquake} \quad\quad \text{Live} \quad\quad \text{Snow}
\end{aligned}
$$

f_1, the factor on live load, is between 0.5 and 1.0. Thus, member forces due to the unfactored earthquake load are combined either with 120% of forces due to gravity loads and a reduced proportion of forces due to live and snow (but not wind) loads. In the second load combination, which for example may govern the design of columns or walls subject to uplift, the unfactored effects of earthquake load are combined with 90% of the dead load and no live load.

There is an additional requirement in ASCE 7-10, which is not found in EC8. The earthquake load E is calculated as follows

$$E = \rho Q_E \pm 0.2\, S_S D \tag{6.6}$$

where ρ is a reliability factor, Q_E is the effect of horizontal seismic forces, S_S equals design 5% damped spectral acceleration at short period and D is dead load.

The reliability factor ρ allows for the system redundancy; it lies between 1.0 for structures in which the structure is highly redundant (i.e. the lateral strength is not greatly reduced by the loss of any single member) to a maximum of 1.3. Q_E is the seismic load calculated in accordance with the analysis procedures outlined in Section 6.4.

The term $0.2\, S_{DS} D$ is stated to account for vertical seismic loads. S_{DS} corresponds to the spectral peak, which is typically 2.5 times the peak ground acceleration. Therefore, for an area of high seismicity with a peak ground acceleration of 40% g, the term $0.2\, S_{DS} D$ represents $0.2 \times 2.5 \times 0.4$ or 20% of the dead load.

6.9.　Code requirements for detailing

A large proportion of the seismic provisions of EC8 and the seismic sections of US steel or concrete codes are concerned with detailing rules to ensure adequate ductility. EC8 and a variety of US codes provide rules for steel, concrete, masonry and timber elements, as well as steel–concrete composite structures. There are detailed differences between the two procedures (see e.g. Booth *et al.*, 1998 for a discussion on the rules for concrete) but many broad similarities exist.

6.10.　Code requirements for foundations

EC8 states explicitly that capacity design considerations must apply to foundations; that is, they must be designed so that the intended plastic yielding can take place in the superstructure without substantial deformation occurring in the foundations. One way of showing this would be to design the foundation for the maximum forces derived from a pushover analysis. EC8 provides an alternative rule for ductile structures, whereby the foundation is designed for the load combination of Equation 6.1, but with the earthquake load increased by a factor 1.2Ω (reducing to 1.0Ω for $q \le 3$) where Ω is the ratio of provided strength to design strength for the superstructure element most affecting the foundation forces. There are no similar capacity design rules in ASCE 7-10.

Both codes give rules for seismic detailing of piled foundations and for site investigation requirements. The information given in EC8 part 5 is more extensive than anything appearing in current US codes. For example, it contains information on liquefaction assessment, design bearing pressures for shallow foundations, soil structure interaction analysis and retaining wall design.

6.11. Code requirements for non-structural elements and building contents

Both ASCE 7-10 and EC8 provide simplified formulae for the forces required to anchor non-structural elements back to the main structure, in terms of the ground accelerations, the height of the non-structural element within the building (to allow for the increased accelerations at higher levels) and the ratio of the natural period of the element to that of the building (to allow for resonance effects).

Rules are also given to check that items that are attached to more than one part of the structure (pipe runs, cladding elements, etc.) can withstand the relative deformations imposed on them. Extensive rules are given in EC8 for unreinforced masonry infill, which is not permitted in high seismicity areas of the USA.

6.12. Other considerations

6.12.1 Combinations of forces in two horizontal directions

In buildings without significant torsional eccentricities and in which lateral resistance is provided by walls or independent bracing systems in the two orthogonal directions, EC8 allows seismic forces in the two orthogonal directions to be considered separately, without combination. Otherwise, the forces due to each direction must be combined either by a square root sum square (SRSS) combination or by taking 100% of forces due to loading in one direction with 30% in the other – the 100:30 rule. The requirements of ASCE 7-10 are essentially the same.

Care needs to be taken when considering the effects of biaxial loading, for example, axial load and moment in a column, or shear and vertical load on a foundation. Combining each load component separately by SRSS and then considering them to act together may be very conservative, because the two maxima do not occur simultaneously. Table 7.1 in Chapter 7 shows an example calculation of how this can be dealt with for the case of the sliding of a foundation. The 100:30 rule is more straightforward to apply, but tends to be conservative.

6.12.2 Vertical seismic loads

EC8 requires vertical seismic loading to be considered in areas of high seismicity in the design of the following types of structural element.

■ Beams exceeding 20 m span.
■ Cantilever beams exceeding 5 m.
■ Prestressed concrete beams.
■ Beams supporting columns.
■ Base-isolated structures.

Rules are given for the vertical response spectrum, which is independent of the soil type. Vertical and horizontal seismic effects can be combined either using an SRSS rule or 100% + 30% + 30% rule, similar to that discussed above for the horizontal directions.

ASCE 7-10 requires that a vertical seismic load should be considered in all structures. This is calculated simply as a proportion of the dead load, the proportion increasing with the site seismicity of the site (Equation 6.6).

6.13. Guidance material

6.13.1 Eurocode 8

EC8 contains no commentary. However, guidance material written by some of the original drafters of the code is given in Fardis *et al.* (2005). The Institution of Structural Engineers' manual on EC8 (ISE/AFPS, 2010) provides a selection of the code sufficient for the seismic design of straightforward buildings, with an extensive commentary. The *Design of buildings to Eurocode 8* (Elghazouli, 2009) is a textbook providing explanation of the code, and giving worked examples. The Joint Research Council (JRC, 2011) provides a freely downloadable set of worked examples of building design to EC8.

For the application of EC8 in the UK, PD 6698 (BSI, 2009b) provides guidance, and Booth and Skipp (2004) give a general discussion.

6.13.2 US codes

Unlike EC8, US codes do contain commentaries. In addition, Naeim (2001) and Chen and Scawthorn (2003; new edition currently being prepared) provide extensive information on US methods of earthquake engineering design and analysis, which includes many references to US codes. For many years, the US National Earthquake Hazard Reduction Program has at regular intervals published recommended seismic provisions with commentaries, which have subsequently become the basis for revising US codes. The current edition is FEMA 750 (BSSC, 2009).

REFERENCES

ACI (2008) ACI 318-08: Building code requirements for structural concrete and commentary. American Concrete Institute, Farmington Hills, MI, USA.

ASCE (2006) ASCE/SEI 41-06: Seismic rehabilitation of buildings. American Society of Civil Engineers, Reston, VA, USA.

ASCE (2010) ASCE/SEI 7-10: Minimum design loads for buildings and other structures. American Society of Civil Engineers, Reston, VA, USA.

Booth E and Skipp B (2004) Eurocode 8 and its implications for UK-based structural engineers. Institution of Structural Engineers, London, UK. *The Structural Engineer* **82(3)**: 39–45.

Booth E, Kappos AJ and Park R (1998) A critical review of international practice on seismic design of reinforced concrete buildings. *The Structural Engineer* **76(11)**: 213–220.

BSI (2009a) BSI Structural Eurocodes companion. See http://shop.bsigroup.com/en/Browse-by-Subject/Eurocodes/The-BSI-Companion-to-the-Structural-Eurocodes/.

BSI (2009b) PD 6698: 2009: Background paper to the UK national annexes to BS EN 1998-1, BS EN 1998-2, BS EN 1998-4, BS EN 1998-5 and BS EN 1998-6. BSI, Chiswick, UK.

BSSC (Building Seismic Safety Council) (2009) FEMA 750: Recommended seismic provisions for new buildings and other structures. BSSC, Washington, DC, USA. See www.fema.gov/library/viewRecord.do?id=4103.

CEN (2004) EN 1998-1: 2004: Design of structures for earthquake resistance. Part 1: General rules, seismic actions and rules for buildings. European Committee for Standardisation, Brussels, Belgium.

CEN (2005) EN 1998-3: 2005: Design of structures for earthquake resistance. Part 3: Assessment and retrofitting of buildings. European Committee for Standardisation, Brussels, Belgium.

Chen W-F and Scawthorn C (2003) Earthquake engineering handbook. CRC Press, Boca Raton, FL, USA.

Elghazouli A (ed.) (2009) *Design of buildings to Eurocode 8*. Taylor & Francis, London and New York.

Fajfar P and Krawinkler H (eds) (2004) *Performance-based seismic design – concepts and implementation*. PEER Report 2004/15. Pacific Earthquake Engineering Research Center, University of California, Berkeley, CA, USA.

Fardis M, Carvalho E, Elnashai A *et al.* (2005) *Designers' guide to EN 1998-1 and EN 1998-5*. Thomas Telford, London, UK.

ICC (2012) International Building Code. International Code Council, Whittier, CA, USA.

ISE/AFPS (Institution of Structural Engineers/Association Française du Génie Parasismique) (2010) *Manual for the design of steel and concrete buildings to Eurocode 8*. Institution of Structural Engineers, London, UK.

JRC (Joint Research Council) (2011) Eurocode 8: seismic design of buildings – worked examples. See http://eurocodes.jrc.ec.europa.eu/showpage.php?id=335_2.

Naeim F (2001) The seismic design handbook. Kluwer Academic Publishers, Boston, MA, USA.

Pampanin S (2012) Reality-check and renewed challenges in earthquake engineering: implementing low-damage structural systems – from theory to practice. *Proceedings of the 15th World Conference on Earthquake Engineering, Lisbon*.

Standards New Zealand (2004) NZS 1170: Part 5: 2004 – Earthquake actions – New Zealand. Standards New Zealand, Wellington, NZ.

Earthquake Design Practice for Buildings
ISBN 978-0-7277-5794-4

http://dx.doi.org/10.1680/edpb.57944.149

Chapter 7
Foundations

There is no glory in foundations.

Professor K. Terzaghi

This chapter covers the following topics.

- Design objectives and capacity design.
- Shallow bearing foundations.
- Piled foundations.
- Deep basements.
- Retaining walls.
- Foundations in the presence of liquefiable soils.

Foundation design, particularly for cases in which there is potential for earthquake loading, is usually carried out by geotechnical specialists. However, the superstructure designer also needs to have an understanding of the issues affecting foundation design that are discussed in this chapter, as well being aware of the soil analysis issues discussed in Chapter 4.

7.1. Design objectives

Life-threatening collapse of structures due to foundation failure in earthquakes is comparatively rare even in the extreme circumstance when soil liquefaction occurs. This is because failure of the foundation limits the amount of shaking that is transmitted into the superstructure; it is a type of uncontrolled base isolation. Foundation failure accompanied by catastrophic collapse of superstructure therefore does not often occur. However, foundation failures can be extremely costly; for example, liquefaction-induced failures in the port of Kobe, Japan, in 1995 are estimated to have cost many billions of pounds of structural damage, with a roughly equal loss arising from the economic consequences of the port's closure.

The main features to consider in the seismic design of foundations are as follows.

- A primary design requirement is that the soil–foundation system must be able to maintain the overall vertical and horizontal stability of the superstructure in the event of the largest credible earthquake.
- The foundation should be able to transmit the static and dynamic forces developed between the superstructure and soils during the design earthquake without inducing excessive movement.

- The possibility of soil strength being reduced during an earthquake needs to be considered.
- It is not sensible to design a perfectly detailed ductile superstructure supported by a foundation that fails before the superstructure can develop its yield capacity (Section 7.2).
- Just as design of the superstructure should minimise irregularity, so irregular features in foundations need to be avoided. These include mixed foundation types under different parts of the structure and founding at different levels or onto strata of differing characteristics.
- Special measures are needed if liquefaction is a possibility (Section 7.8).
- Special considerations apply to piled foundations (Section 7.6).

7.2. 'Capacity design' considerations for foundations

Capacity design (Section 3.5) is accepted as a standard procedure for superstructures, but has been less widely adopted for foundations, and is uncommon in US practice. However, it is required by Eurocode 8 (EC8) part 5 (CEN, 2004), and in the author's opinion is just as valuable below ground as above. The next sections consider its application both to the substructural elements forming the foundations and to the surrounding soil. The basic principle is that the order of formation of yielding mechanisms must be determined, and the relative strength of superstructure, foundations and soil must be arranged so that the designer's intentions are realised in the event of a damaging earthquake.

7.2.1 Strength and ductility of foundation structures

The most straightforward case is when the strength of the foundation and its underlying soils is sufficient to support the actions corresponding to the ductile yielding mechanism chosen for the superstructure. In this case, the foundations can be assumed to remain elastic, even in the most severe earthquake, and special ductile detailing of the foundations is not required. EC8 part 5 provides a simplified rule for estimating the required capacity of foundations for this to occur (described previously in Section 6.10), and in this case exempts foundations (except piles) from special seismic detailing requirements. Similarly, when the strength of the foundation exceeds that required to resist the forces imposed on it by a superstructure designed to remain elastic, no special foundation detailing is needed. Piles are an exception, and require special detailing under most circumstances; these aspects are discussed further in Section 7.6. Similar capacity design rules are found in the New Zealand concrete code (NZS, 2006), but they do not appear in US codes.

When the elastic strength of the foundation structure is likely to be exceeded, some degree of ductile detailing is needed. EC8 part 5 allows for this possibility. However, foundation structures are difficult to inspect for possible damage and also to repair if found necessary; this should be borne in mind when considering whether to allow them to yield during a severe earthquake.

7.2.2 Soil response

In several earthquakes, structures in which the foundation soils have failed have been observed to be less damaged than those in which soil failure did not occur. Capacity design involving a yielding mechanism in the soil is therefore a possible strategy, with soil failure acting as a fuse to prevent damage of the superstructure. As, in most cases, soils retain their shear strength even at large deformations, ductile response is usually possible, unless accompanied by excessive rotation and associated P–delta effects. A notable recent example is the design of the Rion Antirion Bridge founded on poor soil and spanning the highly seismic Gulf of Corinth in Greece. Pecker (2004) describes how the bridge piers are designed to slide, in order to limit the superstructure response.

However, such a strategy is unusual, and requires extensive study. The difficulty arises from assessing with any certainty both the upper and lower bounds of the soil foundation capacity in seismic conditions. For the Rion Antirion Bridge, special measures were taken to prevent a slip circle failure in the soil, and a carefully laid top surface of gravel was placed immediately under the foundations to ensure that sliding was reasonably well controlled. Given current knowledge, the prudent course for occupied buildings will usually be to design to avoid soil bearing failure under code-specified seismic loads, and to keep ductile soil response in reserve to withstand more extreme events. However, a limited amount of sliding failure may be acceptable under some circumstances and is permitted by EC8; services such as gas and water pipes entering the building will need to be designed to accommodate such sliding. Permanent soil deformations under design earthquake loading are considered more generally acceptable in retaining walls, depending on the required performance of the wall after the earthquake, and are used to justify reduced design loads (Section 7.7).

7.3. Safety factors for seismic design of foundations
7.3.1 Load factors
The advantage of carrying out a capacity design to ensure that foundations remain elastic is that there should be considerable confidence that these forces cannot be substantially exceeded. Therefore, load factors of unity on seismic loads are justified. Of course, to carry out a capacity design, information is needed about the actual yielding strength of the superstructure; when foundation construction starts before the superstructure design is complete, suitably conservative assumptions are essential.

When the superstructure is designed for an essentially elastic response to the design earthquake, no reliance is placed on the formation of ductile yielding to limit response and load factors of unity are probably still appropriate. However, the design earthquake is unlikely to be the maximum conceivable event, and it would be prudent for the designer to consider what might happen if the design forces are somewhat exceeded, to ensure that a brittle or unstable response is avoided. This can be particularly important in areas of moderate seismicity, where the 475-year return period motions often considered in design may be significantly exceeded.

7.3.2 Soil design strengths and material safety factors
Soil strength parameters for seismic design of the foundation structure should normally be those used for static design. However, as discussed in Chapter 4, strength degradation under cyclic loading may occur in both granular and cohesive soils, and specialist advice must be sought if this is a possibility. Conversely, the high rate of loading during an earthquake may lead to strength increases of up to 25% in some cohesive soils.

In European practice, checks on the acceptability of soil stresses follow limit state design principles, and the soil stresses are compared with soil strength divided by a material factor γ_m, which EC8 part 5 proposes should be between 1.25 and 1.4 for seismic design. In US practice, ASCE 7-10 (ASCE, 2010) states that for the load combination including earthquake 'the dynamic properties of the soil should be included in the determination of the foundation design criteria'; further advice on soil capacity is not provided, although the structural design of the foundations is covered.

7.4. Pad and strip foundations
7.4.1 Failure modes

In addition to transferring vertical loads safely into the soil, shallow foundations in the form of pads or strips must also transfer the horizontal forces and overturning moments arising during an earthquake. The associated potential failure modes in the soil and the foundation structure illustrated in Figure 7.1 are now considered in turn.

Figure 7.1 Modes of failure in pad foundations

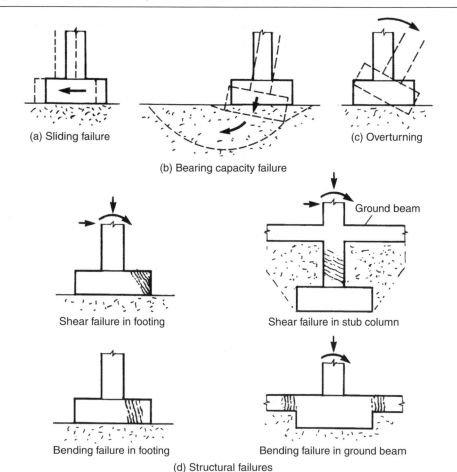

(a) Sliding failure

(b) Bearing capacity failure

(c) Overturning

Shear failure in footing

Ground beam

Shear failure in stub column

Bending failure in footing

Bending failure in ground beam

(d) Structural failures

7.4.1.1 Sliding failure

Resistance to sliding in shallow footings will usually be mobilised from the shear strength of the soil interfacing with the footing (see Figure 7.1(a)). Passive resistance, even if significant, would only be mobilised at much larger deflections, and should generally be ignored, unless associated with a deep retaining structure such as a deep basement.

In granular materials, the minimum vertical load that could occur concurrently with the maximum horizontal force must be considered, because this condition will minimise shear resistance. This creates a difficulty – the vertical load is at a minimum when the uplift due to seismic effects is at a maximum but seismic uplift results from two separate components. The first component is due to overturning, which is associated with horizontal accelerations, while the second is due to vertical seismic accelerations. As vertical and horizontal accelerations are not in general well correlated, it would be conservative to take a simple arithmetic sum of the shears due to these two components and then calculate the margin against sliding assuming that the two uplifts are also summed. Instead, the ratio of shear to vertical load on the footing can be calculated separately for horizontal and vertical accelerations and then combined by the square root sum of squares (SRSS) method. The combined result gives an estimate of the least favourable margin against sliding (see Table 7.1). Alternatively, the shears and vertical loads due to horizontal and vertical accelerations can be combined by the 100:30 rule (Section 6.12). This is a more straightforward procedure, but usually more conservative.

7.4.1.2 Bearing capacity failure

Static bearing capacity can be determined from formulae that allow for the inclination and eccentricity of the applied load (see Figure 7.1(b)). EC8 part 5 (CEN, 2004) annexe F provides expressions.

Table 7.1 Example calculation of the margin against sliding for a footing

	Loads on footing: kN			Shear load
	Gravity	Seismic uplift	Shear	Vertical load
Due to horizontal accelerations	1000	50	400	$400/(1000 - 50) = 0.42$
Due to vertical accelerations	1000	300	10	$10/(1000 - 300) = 0.014$

Combination of shear to vertical load ratio

SRSS method

Ratio $= (0.42^2 + 0.014^2)^{1/2}$ $= 0.42$

100:30 rule

Ratio $= (400 + 10 \times 0.3)/(1000(50 + 0.3 \times 300))$ $= 0.47$

or $= (0.3 \times 400 + 10)/(1000(0.3 \times 50 + 300))$ $= 0.19$

The higher ratio of 0.47 governs design

The shear to vertical load ratio should be compared to the tangent of the friction angle between footing and soil, tan ϕ', to calculate the margin against sliding

7.4.1.3   Rotational failure

When the soil is strong, the foundation may start to rotate before a bearing capacity failure occurs, particularly if the vertical load is small (see Figure 7.1(c)). In the case of pad foundations supporting a moment-resisting frame, such a rotation may be acceptable, because a frame with pinned column bases still retains lateral stability. However, the associated redistribution of moments would lead to increased moments at the top of the lower lift of columns, which would need to be designed for.

In contrast, an isolated cantilever shear wall is not statically stable with a pinned base. Rocking should, therefore, be prevented under design forces in most circumstances. Uplift can be prevented by the provision of additional weight or by piles or anchors to resist the transient vertical loads, or by a wider foundation.

7.4.1.4   Structural failure in the foundation

Sufficient strength must be provided to prevent brittle failure modes in the foundation structure, such as shear failure in footings or stub columns (see Figure 7.1(d)). Ductile flexural yielding may in unusual circumstances be permitted (Section 7.2.1), provided the ductile detailing provisions described in other chapters are present.

7.4.2   Ties between footings

Some form of connection is usually needed at ground level to link isolated footings supporting a moment-resisting frame. The ties prevent excessive lateral deflection between individual footings, caused by locally soft material or local differences in seismic motion. When the footings are founded on rock or very stiff soil, however, the tendency for relative movement is much less and the ties are generally not required.

The connection can take the form of a ground beam, which will also assist in providing additional fixity to the column bases and will help to resist overturning. Alternatively, the ground-floor slab can be specially reinforced to provide the restraint. EC8 part 5 section 5.4.1.2 gives design values for the tie force that increases with seismicity, soil flexibility and axial load in the restrained columns.

7.5.   Raft foundations

All of the soil failure modes illustrated in Figure 7.1(a–d) may apply to raft foundations, and Figure 7.2 shows a bearing capacity failure under a 13-storey building in the 1985 Mexico earthquake. In most cases, however, general soil failure is unlikely and the main consideration is the ability of the raft structure to distribute concentrated loads from columns or walls safely into the soil. At its simplest, the analysis would assume a uniform soil pressure distribution in equilibrium with the peak applied loads. Figure 7.3 shows that this may lead to an underestimate of shears and moments within the raft near its edge, because the soil, being poorly restrained, has low bearing capacity there. More complex analysis would allow for soil non-linearity and dynamic effects.

Partial uplift on one side of the raft under seismic overturning moments may be tolerable in the raft foundations supporting relatively flexible structures such as large fluid storage tanks. Once again, however, the effect of the uplift on internal forces within the raft

Figure 7.2 Bearing capacity failure in Mexico City, 1985 (photograph courtesy of J. Pappin)

foundation and superstructure must be accounted for, as discussed by the New Zealand Society for Earthquake Engineering (NZSEE, 2009).

7.6. Piled foundations
7.6.1 Vertical and horizontal effects
Vertical loading on pile groups during an earthquake arises from gravity loads, seismic overturning moments and vertical seismic accelerations. As the two latter effects are not correlated, they can be combined by the SRSS method, and added to the gravity load. The procedures are straightforward, and the design of end-bearing piles is similar to that for

Figure 7.3 Pressure distribution near the edge of a raft under seismic loading (reproduced from Pappin, 1991)

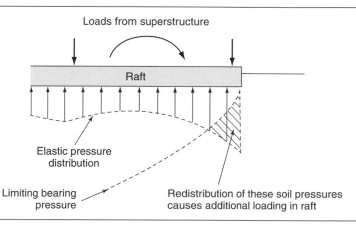

Loads from superstructure

Raft

Elastic pressure distribution

Limiting bearing pressure

Redistribution of these soil pressures causes additional loading in raft

Figure 7.4 Inertial and kinematic loading on piles (reproduced from Pappin, 1991)

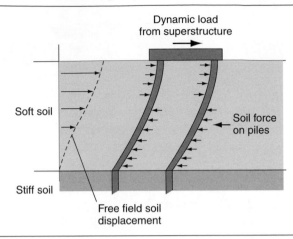

static vertical loads. Friction piles may be less effective under earthquake conditions and require special consideration.

Horizontal response is much less easy to calculate, because the inertia loads arising from the superstructure must be combined with the effects of the soil attempting to move past the piles (Figure 7.4). The severity of the latter effect (often called the kinematic effect) is related to the pile diameter and hence to its stiffness; flexible piles may be able to conform to the deflected soil profile without distress, but large-diameter piles are relatively much stiffer than the soil and large forces may be generated.

The most straightforward analysis for the kinematic effect assumes that the pile adopts the deflected soil profile, which may be assessed from a one-dimensional shear beam model of the soil (Section 4.3.1). This may be excessively conservative, because it neglects the effect of local soil failure that will tend to reduce the curvature imposed on the pile and hence the induced moments and shears. Pappin *et al.* (1998) propose modelling the soil reaction on the piles by a series of horizontal linear or non-linear springs; the piles are modelled as vertical beams. The deflected soil profile is then imposed on the bases of the soil springs to find the deflected shape of the piles. The resulting actions in the piles must be combined with dynamic loading from the superstructure. The two effects are unlikely to be perfectly correlated and therefore it would be rather conservative to design for a simple addition of the two effects. A dynamic analysis of the complete soil–pile–structure system would be needed for a more realistic combination.

Such analysis may indicate that plastic hinges are formed in the pile. Provided that appropriate detailing is present, plastic hinge formation may be acceptable. The detailing would take the form of closed spaced links or spirals in concrete piles and the use of compact sections able to develop full plasticity in steel piles. However, piles are difficult to inspect, and their capacity to resist further lateral loads may therefore need to be assumed to have been severely reduced by an extreme earthquake. Usually, locations of plastic hinges other than at the top of the pile are not considered acceptable. Further considerations for detailing of concrete piles are given in the next section.

7.6.2 Detailing concrete piles

Both EC8 and ASCE 7-10 (ASCE, 2010) require additional confinement steel in the form of hoops or spirals, both at the pile head and at junctions between soft and stiff soils, because these are potential plastic hinge points. Detailing of the vertical steel is also covered.

Particular regions where special detailing measures may be required are as follows.

- The junction between pile and pile cap is a highly stressed region where large curvatures may occur in the pile. Unless adequate confinement and good connection details are present, brittle failure may occur.
- Junctions between soft and hard soil strata may also impose large curvatures on piles; such junctions are likely to be potential points for the formation of plastic hinges.
- Piling through soil that may liquefy can pose special problems. In this case the pile may have a large unsupported length through the liquefied soil and should be reinforced as though it was an unsupported column. A reliable ductile behaviour will also be necessary in this situation.

A general review of the seismic assessment of piles for design is provided by Madabhushi and May (2009).

7.6.3 Raking piles

Raking piles pose a special problem, because they tend to attract not only the entire dynamic load from the superstructure, but also the horizontal load from the soil attempting to move past the piles (Figure 7.5), which is a particularly severe example of the kinematic interaction

Figure 7.5 The effect of raking piles on pile group deformation (reproduced from Pappin, 1991)

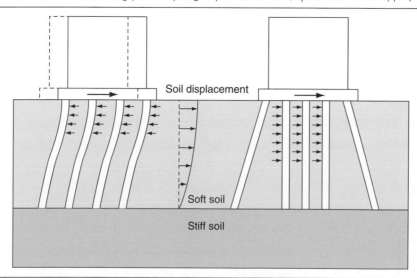

effect described in Section 7.6.1. Raking piles are found to be particularly susceptible to failure in earthquakes. They should therefore be used with care in seismic regions, with particular attention to the kinematic interaction effects shown in Figure 7.4, for which an explicit analysis is likely to be required.

7.7. Retaining structures

7.7.1 Introduction

During an earthquake, the soil behind a retaining structure may impose large inertia loads on it. When the soil is above the water table, complete collapse of the wall is unlikely to occur (Seed and Whitman, 1970) but large horizontal movements and/or rotations are often observed. In the case of bridge abutments, these movements have led to damage or loss of support to the bridge deck in many earthquakes.

Where liquefaction can occur, complete collapse is more common; it was associated with the quay wall collapse shown in Figure 7.6. The condition is severe because the liquefied soil imposes a lateral pressure much greater than the normal active soil pressure, while the liquefaction may also reduce the shear strength of the soil and weaken the restraint offered by the soil to the base of the wall.

Steedman (1998) provides a general review of methods for the seismic design of retaining walls and Anderson *et al.* (2008) also provide advice.

Figure 7.6 Failure of a quay wall, San Antonio, Chile, 1985 (© E Booth)

7.7.2 Analysis of earth pressures during an earthquake

Active and passive soil pressures from granular soils on retaining walls arising from earthquakes are still commonly assessed by the Mononobe–Okabe equations, originally developed in Japan in the 1920s; they are quoted in annexe E of EC8 part 5. The equations assume that the wall movement is large enough for an active state to develop. However, for rigid structures, such as basement walls or gravity walls founded on rock or piles, higher pressures will develop and EC8 part 5 equation E.19 provides an alternative basis. They also apply to non-cohesive soils; while most retaining walls will be backfilled with sandy or gravelly material, in many cases the failure planes will also pass through clayey or silty material further behind the wall. Anderson *et al.* (2008) provide expressions for seismic loads on walls retaining cohesive soils.

Many retaining walls, for example, in road cuttings or harbour walls, are designed to be able to move forward slightly during earthquake loading, either due to sliding or rotation. This results in a reduction in the required design strength of the retaining structure, in just the same way as ductility factors reduce design forces in ductile superstructures. EC8 part 5 table 7.1 provides for reduction factors of between 1 and 2, depending on the circumstances, and acceptability of permanent movement. In some circumstances, significant movement is either not possible (e.g. in the retaining walls of basements in buildings) or not acceptable, in which case the full force needs to be accommodated. Steedman (1998) provides the theoretical basis for allowing for permanent movements.

7.7.3 Fluid pressures

For retaining walls with one face in contact with water, the hydrodynamic interaction of the water and the wall must be accounted for. Westergaard (1933) demonstrated that, for a rigid wall retaining a water reservoir, the hydrodynamic interaction could be visualised as a portion of the water mass moving in phase with the wall. Based on his solution, EC8 part 5 (CEN, 2004) equation E.18 provides a simplified design pressure distribution, corrected for the wall's restraint conditions. More recent research compares the results of Westergaard and the more commonly used Chopra's method of estimating the hydrodynamic pressures with centrifuge-based experimental data – see Saleh and Madabhushi (2010a, 2010b, 2010c).

In addition, there is also likely to be water in the retained soil on the other face of the wall. This may move in phase with the soil, and in that case would merely add inertia to the soil. This can be accounted for by taking the total wet density of the soil in the Mononobe–Okabe equations. This is the most common situation; EC8 part 5 recommends it can be assumed for soils with a coefficient of permeability of less than 5×10^{-4} m/s and gives the relevant equations in section E.6.

However, in highly permeable soils the water within the soil has some freedom to move independently, and will give rise to additional hydrodynamic effects. In this case, the Eurocode recommends assuming that the water is totally free. The pressures on the soil face of the wall are then the soil pressure calculated from vertical effective stress, plus an additional hydrodynamic term based on Westergaard, but reduced by the porosity of the soil; the relevant equations are given in section E.7 of EC8 part 5. The soil strength properties should be taken as their undrained values. Steedman (1998) more conservatively recommends that the walls retaining permeable soils should be checked under both assumptions – that is, either that the water is both fully restrained by the soil or that it is totally free. For some case histories see Madabhushi and Zeng (1998, 2007), who compared finite element analysis results with centrifuge test data for gravity and flexible, cantilever retaining walls.

7.8.   Design in the presence of liquefiable soils

Two types of countermeasure are possible in the presence of liquefiable soils. Either the structures can be modified to minimise the effects of liquefaction, or the soils can be modified to reduce the risk of their liquefying.

If liquefaction is expected to be limited in extent, causing only minor local settlements, structural modification could take the form of local strengthening to cope with the settlement stresses. More radically, foundations can be moved to avoid the liquefiable soils. For example, the foundation depth can be increased to found below the levels at risk. Movement of the entire structure may also be worth considering; for example, a river bridge may be moved to a different crossing point or its span might be increased if the liquefiable material is confined to the river banks.

Foundations may also be designed that minimise the consequences of liquefaction. Possible options are as follows.

■   Provision of a deep basement, so that the bearing pressures due to vertical loads are greatly reduced. Essentially, the structure is designed to float in the liquefied soil. This may be less effective in countering soil pressures due to overturning forces, and so this option is likely to be confined to relatively squat structures.
■   Provision of a raft with deep upstands. The structure is designed to sink if liquefaction occurs until vertical equilibrium is regained. The solution may imply large settlements and again is most applicable to relatively squat structures.
■   Provision of end-bearing piles founded below the liquefiable layers. Although this will counter vertical settlements due to gravity loads and overturning moments, the piles may be subject to large horizontal displacements occurring between the top and bottom of the liquefying soil layer, and the piles must be designed and detailed to accommodate this.

The alternative strategy is to reduce the liquefaction potential of soils. A number of methods are possible and consist of four generic types, as follows. Further information is provided by Idriss and Boulanger (2008).

■   Densification, for example, by vibrocompaction, which produces a more stable configuration of the soil particles. This may not be an option for existing structures, because of the settlements induced by the process.
■   Soil stabilisation, for example, by chemical grouting, which makes the soil less likely to generate increases in pore water pressure.
■   Provision of additional drainage, for example, by provision of sand drains, which helps reduce the rise in pore water pressure.
■   Removal of soil and replacement with more suitable material.

These tend to be expensive solutions, although experience from the 1989 Loma Prieta earthquake in California (EERI, 1994), in various Japanese earthquakes (Ohbayashi *et al.*, 2006) and in Mexico at the port of Manzanillo suggests they are effective.

REFERENCES

Anderson DG, Martin GR, Lam I and Wang JN (2008) *Seismic analysis and design of retaining walls, buried structures, slopes and embankments.* NCHRP Report 611, Transportation Research Board, 148 pp.

ASCE (2010) ASCE/SEI 7-10. Minimum design loads for buildings and other structures. American Society of Civil Engineers, Reston, VA, USA.

CEN (2004) EN 1998: 2004: Design of structures for earthquake resistance. Part 1: General rules, seismic actions and rules for buildings. Part 5: Foundations, retaining structures and geotechnical aspects. European Committee for Standardisation, Brussels, Belgium.

EERI (Earthquake Engineering Research Institute) (1994) *Earthquake basics: liquefaction – what it is and what to do about it.* EERI, Oakland, CA, USA.

Idriss IM and Boulanger RW (2008) *Soil liquefaction during earthquakes.* Earthquake Engineering Research Institute, Oakland, CA, USA.

Madabhushi SPG and May R (2009) Pile foundations. In *Design of buildings to Eurocode 8* (Elghazouli A (ed.)). Taylor & Francis, London and New York.

Madabhushi SPG and Zeng X (1998) Behaviour of gravity quay walls subjected to earthquake loading. Part II: numerical modelling. *Journal of Geotechnical Engineering – ASCE* **124(5)**: 418–428.

Madabhushi SPG and Zeng X (2007) Simulating seismic response of cantilever retaining walls with saturated backfill. *ASCE Journal of Geotechnical and Geoenvironmental Engineering* **133(5)**: 539–549.

NZS (New Zealand Standards) (2006) NZS 3101: Part 1: The design of concrete structures. NZS, Wellington, New Zealand.

NZSEE (New Zealand Society for Earthquake Engineering) (2009) *Seismic design of storage tanks: 2009. Recommendations of a study group of the New Zealand Society for Earthquake Engineering.* NZSEE, Wellington, New Zealand.

Ohbayashi J, Harada K, Fukada H and Tsuboi H (2006) Trends and developments of countermeasure against liquefaction in Japan. *Proceedings of the 8th US National Conference on Earthquake Engineering, San Francisco, CA, USA.*

Pappin JW (1991) Design of foundation and soil structures for seismic loading. In *Cyclic Loading of Soils* (O'Reilly MP and Brown SF (eds)). Blackie, London, UK pp. 306–366.

Pappin JW, Ramsey J, Booth ED and Lubkowski ZA (1998) Seismic response of piles: some recent design studies. *Proc. ICE – Geotechnical Engineering* **131**: 23–33.

Pecker A (2004) Design and construction of the Rion Antirion Bridge. In *Geotechnical Engineering for Transportation Projects* (Yegian M and Kavazanjian E (eds)). American Society of Civil Engineers Geo Institute, Geotechnical Special Publication no. 126. ASCE, Reston, VA, USA.

Saleh S and Madabhushi SPG (2010a) Hydrodynamic pressures behind flexible and rigid dams. *ICE – Dams and Reservoirs Journal* **20(2)**: 73–82.

Saleh S and Madabhushi SPG (2010b) An investigation into the seismic behaviour of dams using dynamic centrifuge modelling, *Bulletin of Earthquake Engineering* **8**: 1479–1495.

Saleh S and Madabhushi SPG (2010c) Response of concrete dams on rigid and soil foundations under earthquake loading. World Scientific Press. *Journal of Earthquake and Tsunami* **4(3)**: 251–268.

Seed HB and Whitman RV (1970) *Design of earth retaining structures.* Cornell University, New York, UK.

Steedman RS (1998) Seismic design of retaining walls. *ICE – Geotechnical Engineering* **113**: 12–22.

Westergaard HM (1933) Water pressures on dams during earthquakes. *Transactions ASCE* **98**: 418–472.

Earthquake Design Practice for Buildings
ISBN 978-0-7277-5794-4

ICE Publishing: All rights reserved
http://dx.doi.org/10.1680/edpb.57944.163

Chapter 8
Reinforced concrete design

> The art of detailing reinforced concrete components for ductility comprises the skilful combination of the two materials, one inherently brittle, the other very ductile.
>
> Paulay (1994)

This chapter covers the following topics.

- The behaviour of reinforced concrete under cyclic loading.
- Ductility in reinforced concrete, and how to achieve it.
- Material specification of concrete and reinforcing steel.
- Special considerations for analysis.
- Design and detailing: frames, walls and diaphragms.
- Prestressed and precast concrete.

Reinforced concrete is often the structural material of choice for engineered structures, particularly in the developing world, yet even the static design of this complex composite is not straightforward and where earthquake loading may apply, the structural engineer needs to understand its response to cyclic excitation.

8.1.   Lessons from earthquake damage
Field evidence from past earthquakes shows that although reinforced concrete on average performs significantly better than unreinforced masonry, it can still be very vulnerable. Moreover, the high weight of concrete buildings means that collapse often proves lethal to its occupants. Buildings consisting of frames built from reinforced concrete beams and columns and that are not braced by walls have proved particularly vulnerable to earthquakes, unless special design and detailing measures are in place to resist earthquakes. The main points of vulnerability are the following.

- Beam–column joints (Figure 8.1).
- Bursting failures in columns (Figure 8.2).
- Shear failures in columns (Figure 8.3).
- Anchorage failure of main reinforcing bars in beams and columns.

Many of these defects are found in the ubiquitous 'soft storey' failures that are by far the most common cause of collapse of concrete frame structures. Most usually, it is the ground floor that fails, as occurred in thousands of buildings in the Kocaeli, Turkey earthquake of 1999

Figure 8.1 Failure of a beam–column joint in Erzincan, Turkey, 1992. Note failure of the concrete in the joint, and bursting out of column steel (© E Booth)

Figure 8.2 Bursting failure in a column, Kobe, Japan, 1995. Inadequate horizontal steel has caused the heavy main bars to buckle, and allowed the concrete in the column to shatter (courtesy Earthquake Engineering Field Investigation Team (EEFIT), UK)

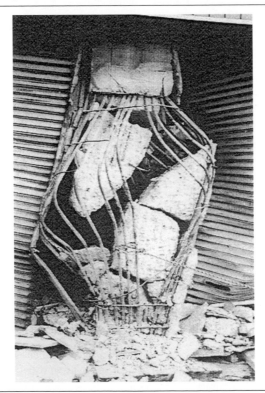

Figure 8.3 Shear failure of a lightly reinforced concrete column and the adjacent masonry, opposite a window opening in St Johns, Antigua, 1974. The masonry wall stopping short of the top of the column creates a 'short column' liable to fail in shear before bending (© David Key)

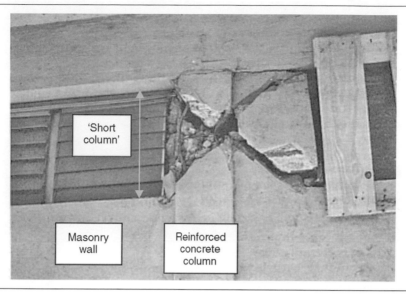

(Figure 8.4). However, intermediate or top storeys can also fail (Figure 8.5). When the failure spreads to more than one storey, perhaps due to the impact of one floor collapsing onto the next below, a general 'pancake' failure can occur, involving the loss of load-carrying capacity in all columns, not just those in a particular storey (Figure 8.6).

Frame buildings often fail due to interactions with inappropriately designed masonry infill walls (Figure 8.7, see also Figure 8.3) or due to impact between the columns of one building and the floor or roof diaphragm of an adjacent one (Figure 8.8). Torsional response due to differences between centres of stiffness and mass are also a common form of failure (Figure 8.9). Note that all the failures shown in Figures 8.2 to 8.9 are primarily those of columns, not beams or floor diaphragms.

Buildings with shear walls that provide a significant contribution to lateral resistance have generally proved less vulnerable; they protect the columns from the job of both resisting lateral forces as well as supporting the gravity loads (and hence preventing collapse), and the walls also limit storey drifts. Perhaps the most dramatic example was in the Spitak Armenia earthquake of 1988, where many precast concrete frame buildings suffered total collapse, while the precast concrete wall buildings, although equally poorly constructed, survived without endangering their occupants (Figure 8.10). Another notable example is Chile, where shear walls have for many years been used in the majority of medium to high-rise buildings and generally did an excellent job of protecting their occupants in the 1985 and 2010 earthquakes. Shear walls are not, however, immune from damage; for example,

Figure 8.4 Ground storey (soft storey) collapse of buildings in Kocaeli, Turkey, 1999 (courtesy Earthquake Engineering Field Investigation Team (EEFIT), UK)

collapses occurred both in Chile in 2010 and in Christchurch, New Zealand in 2011 (Figure 8.11). Potential failure triggers include compression failure of their outer edges (Figure 8.12(a)), diagonal shear and bending failure at their bases, and shear failure in wall piers (Figure 8.12(b)). Moreover, the intrinsic in-plane stiffness of shear walls does not necessarily protect other structural and non-structural elements from damage due to drifts

Figure 8.5 (a) Intermediate and (b) upper storey collapse of a multi-storey reinforced concrete structure in Mexico City, 1985 ((a) © E Booth; (b) photograph: M Winney)

(a) Intermediate floor   (b) Top floor

Figure 8.6 Total collapse of a multi-storey reinforced concrete building in Baguio, Philippines, 1990 (© E Booth)

Figure 8.7 Incipient soft storey collapse of a building in Erzincan, Turkey, 1992 (© E Booth)

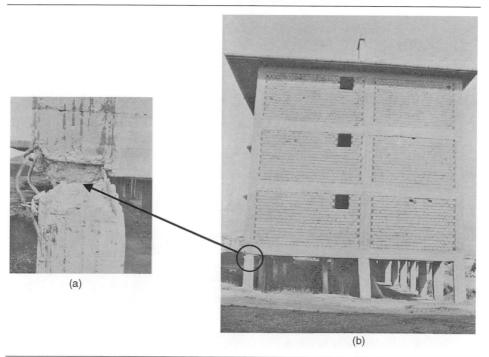

(a)

(b)

Figure 8.8 Failure in Mexico, 1985, caused by impact between adjacent buildings (© E Booth)

Figure 8.9 The nearer end of this building is restrained by stiff shear walls but the far end was supported by slender columns. The resulting torsional movement has led to collapse in Mexico City, 1985 (© E Booth)

caused by rocking of the walls on their foundations. Failure of the beam column frames in the Pyne Gould Corporation Building (Figure 8.11(b)) has been partly attributed to this mechanism (Canterbury Earthquakes Royal Commission, 2012).

8.2. Behaviour of reinforced concrete under cyclic loading

Reinforced concrete is composed of a number of dissimilar materials. Its complex response to dynamic cyclic loading is highly non-linear and depends on the interaction between its various parts, and in particular the concrete/steel interface. Some understanding of this behaviour is necessary for the design of concrete structures in earthquake country; for a fuller description than the outline that follows, see Fenwick (1994). Other standard texts on the seismic behaviour of reinforced concrete included Paulay and Priestley (1992) and (for a European approach) Fardis (2006).

8.2.1 Cyclic behaviour of reinforcement

When a reinforcing bar is yielded in tension or compression and the direction of the stress is reversed, the distinct yield point is lost and the stress–strain relationship takes the curvilinear form shown in Figure 8.13. This change in the stress–strain relationship is known as the Bauschinger effect. An important result is that the stiffness of the steel is lowered as it approaches yield, compared with the initial loading cycle, which means it is more prone to buckle in the compression cycle.

Figure 8.11 Collapse of shear wall buildings ((a) courtesy Earthquake Engineering Field Investigation Team (EEFIT), UK; (b) reproduced from Canterbury Earthquakes Royal Commission, 2012, © Crown Copyright, New Zealand)

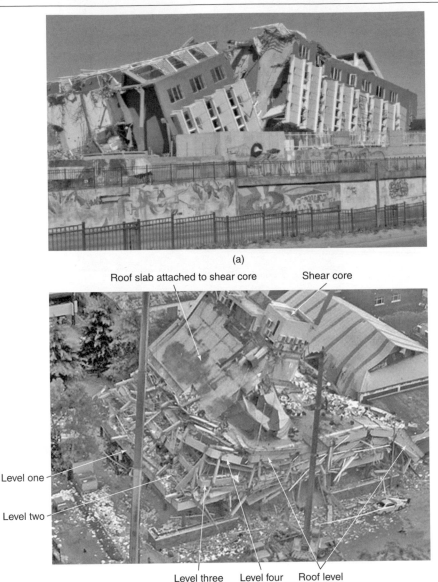

(a)

Roof slab attached to shear core   Shear core

Level one

Level two

Level three   Level four   Roof level

(b)

Another effect of loading and unloading steel to yield is known as 'strain ageing'. This takes days or weeks to develop, and results in an embrittlement of the reinforcement. In principle, this might reduce the capacity of concrete buildings stressed beyond yield in one earthquake to resist future ground motions of a similar or greater magnitude, although it appears there is no direct evidence of this having happened in past earthquakes.

Figure 8.12 Failure modes in shear walls ((a) and (b) reproduced from Westenenk *et al.*, 2013, with kind permission from Springer Science and Business Media)

(a) Compression failure of outside edge of shear wall, Chile 2010

(b) Failure of wall piers, Chile 2010

The high rates of loading that occur during earthquakes may lead to increases in initial yield stress of approximately 20% in mild steel, although the increase is lower in high yield steel, and the increase has also been found to be much lower in subsequent yielding cycles. These strain rate effects in reinforcement are therefore likely to be relatively minor for seismic loading. Rate effects in concrete may be much more significant (see Section 8.2.2).

Figure 8.13 Stress–strain relationship for mild steel reinforcement subjected to inelastic load cycles

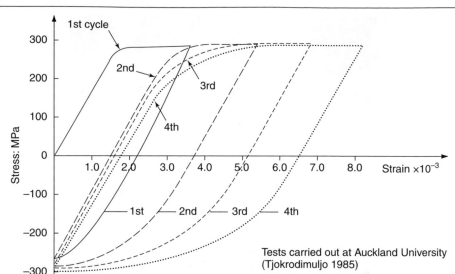

When the yield strength of reinforcement exceeds around 500 MPa, the margin between yield and fracture strain is reduced, and the use of high strength steel as passive reinforcement results in structures with limited overall ductility. High strength steel is used in prestressed concrete, discussed in Section 8.2.10.

8.2.2 Stress–strain properties of plain concrete

Concrete on its own is weak and brittle in tension. In uniaxial compression there is some ductility, but this reduces as the concrete strength increases (Figure 8.14). When there is a lateral confining pressure, the properties are very different, as discussed in the next section.

Strain rate effects in concrete can lead to strength increases in the order of 20% or more, and may be significant in columns with a high axial load, where flexural response is dominated by the concrete rather than the steel.

8.2.3 Stress–strain properties of confined concrete

It has long been recognised that a lateral confining pressure, when applied to concrete, can greatly increase both its compressive strength and compressive strain at fracture. Subsequently, it was recognised that the confinement need not come from a hydrostatic pressure, but could result from the confining effect of properly anchored rectangular or spiral reinforcement in the plane at right angles to the applied compressive stress. The mechanism is due to the tendency of concrete to expand in directions normal to an applied compressive stress. This expansion is due to Poisson's ratio effects, which are enhanced (once the compressive stress reaches 70% of the cylinder strength) by extensive microcracking. This expansion causes the confinement steel to stretch and hence develop tensile forces tending to resist the expansion.

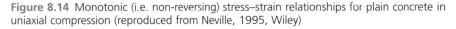

Figure 8.14 Monotonic (i.e. non-reversing) stress–strain relationships for plain concrete in uniaxial compression (reproduced from Neville, 1995, Wiley)

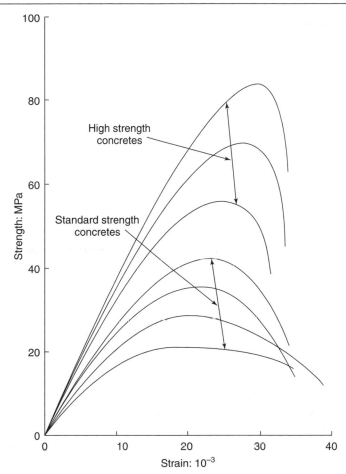

The effect of different quantities of confining steel on the stress–strain properties of concrete are illustrated in Figure 8.15. It can be seen that with even small amounts of confinement, there can be a dramatic improvement in the ductility of concrete in compression and also a significant strength increase. The confined concrete can sustain a substantial additional strain after reaching its maximum strength; failure occurs when the tensile strains in the confining steel reach a point at which the reinforcing bars break in tension.

Fenwick (1994) provides design equations for calculating the ultimate strength and fracture strain as a function of the degree and efficiency of steel confinement, based on the work of Mander *et al.* (1988). These are useful when a calculation is required of the rotational capacity of flexural hinges in reinforced concrete. In most cases, however, the design engineer will rely on rules for the quantity of confining steel given in codes of practice.

Figure 8.15 Idealised stress–strain graph for a rectangular section with varying confinement (cylinder strength $f'_c = 30$ MPa)

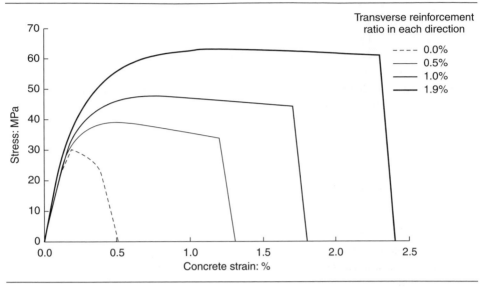

Figure 8.15 shows typical results for a moderate strength concrete. The mechanism by which confinement works implies that improvements are less dramatic in high strength concrete, which Figure 8.14 shows is already more brittle in its unconfined state. This is because the effective confining pressure applied by the steel is relatively lower compared to the strength of the concrete. High strength concrete, with its high strength to mass ratio, may have applications in seismic design, especially for tall buildings, but it must be used with care. An American Concrete Institute (ACI) report (ACI, 2007a) provides design advice.

Caution must also be exercised with concrete made from lightweight aggregates. In this case, the aggregates tend to crush when the confining reinforcement bears against them. The result is that the confining stress is reduced and the enhancement in strength and ultimate strain is considerably less than for normal-weight concrete of equivalent strength. An ACI special publication (Kowalsky and Dwairi, 2003) provides further information on lightweight concrete.

8.2.4 Bond, anchorage and splices

During earthquake loading, reinforcing bars are subject to reverse cycle loading, which in structures designed to be ductile will cause the bars to yield in both tension and compression. This places a much more severe demand on the bond between concrete and reinforcement than is the case for monotonic loading. If this bond is not maintained, bars will lose their anchorage and not be able to develop the forces needed to resist earthquake effects, and they will also lose continuity at lapped splices. Bond strength under cyclic loading is improved where the concrete is confined by closely spaced hoops or spirals or enclosed by ties or stirrups. Lapping by means of bar couplers avoids reliance on the concrete to transfer forces; however, the couplers must be capable of sustaining cyclic yielding of the bars they connect. ACI 318 (ACI, 2011) gives specific rules. Welded splices are not permitted where plastic hinges might form.

Some implications for design are as follows.

- Anchorage of bars in earthquake-resisting structures needs special attention. When possible, bars should terminate in a bend or hook to provide mechanical anchorage.
- Bars must be anchored into concrete that will not spall or crack significantly during a severe earthquake. The concrete in which bars are anchored or spliced should therefore be well confined with hoops or spirals.
- Anchorage and splicing of bars should be avoided in areas where plastic hinges are expected to form. One advantage of capacity design procedures (Section 3.5) is that they provide the designer with some confidence in identifying non-yielding areas of the structure where anchorage and splicing may take place.

8.2.5 Flexure and shear in beams: reversing hinges

8.2.5.1 Introduction

The potential plastic hinge regions in ductile concrete frames must be able to sustain large plastic rotations, without significant loss of flexural strength and without shear failure. These potentially yielding regions will primarily be in the beams, but yielding in columns may also occur, as discussed in Section 8.2.7 below.

8.2.5.2 Cyclic degradation of shear strength

In beams with relatively low levels of gravity loading, plastic hinges will form at the ends. These will yield first in one direction and then the other as the frames sway to and fro during a large earthquake. Under these conditions of reversing load, diagonal shear cracks form in the plastic hinge region, which widen progressively with the number of loading cycles as both flexural and shear steel accumulate plastic tensile strains. This tends to destroy the contribution of aggregate interlock and dowel action to shear resistance. Under static loading conditions, this contribution can be safely included, but under seismic loading, some or all of this contribution will be lost, and generally codes specify that it should be discounted in beams, unless the shear stresses or ductility demands are low or an axial compressive load is present. Eurocode 8 (EC8) (CEN, 2004) is an exception, in that there is no distinction between the shear strength calculation for static and seismic loading of ductility class medium (DCM) beams, although there is for ductility class high (DCH) beams; this may lead to an overestimate by EC8 of shear strength in DCM beams.

The loss of 'concrete contribution' means that shear has to be resisted entirely by a truss action formed by the steel flexural and shear steel as tension members and diagonal concrete compression struts. The widening diagonal cracks then lead to another consequence; in order for the compression strut to take its load after a stress reversal, the diagonal crack across it must first close (Figure 8.16), increasing the shear deformation. This leads to a situation in which there is very little resistance to shear and hence stiffness around the point of load reversal in the loading cycle. When the effect of this shear deformation from yielding of the shear reinforcement is added to the flexural deformation from yielding of the main bars, the characteristic pinched shape of hysteresis loops is obtained (Figure 8.17).

Figure 8.16 Shear deformation in reversing hinge zones

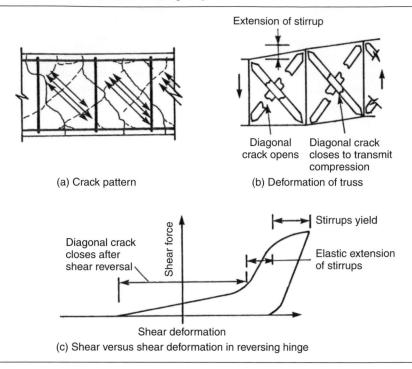

(a) Crack pattern

(b) Deformation of truss

(c) Shear versus shear deformation in reversing hinge

Strength degradation and eventual failure in a reversing hinge can occur in a number of ways. First, the longitudinal bars may fracture in tension or they may buckle in compression; the tendency to buckle increases for a number of reasons under cyclic loading, including softening due to the Bauschinger effect (Section 8.2.1) and the reduction in restraint from yielding of the shear steel.

Second, the opening, closing and widening of the diagonal cracks in the web causes loss of the concrete contribution to shear strength, as explained above. It can also lead to strength deterioration of the concrete in the diagonal compression strut, eventually leading to failure of the strut at stress levels considerably lower than those that can be sustained under monotonic conditions. Failure in diagonal compression then generally occurs close to one of the major cracks in the plastic hinge, and it is accompanied by high shear displacement (Figure 8.18) in what is known as 'sliding shear'. Sliding shear failure can be prevented by the addition of diagonal shear steel – for example, EC8 (CEN, 2004) requires such steel in DCH beams where reversing shear forces exceed a given threshold.

Finally, failure might occur due to fracture of the concrete in the flexural compression zone, particularly if there is inadequate confinement steel, but this is less likely in the absence of an overall compressive force in the beam.

Figure 8.17 Load deflection test results for a beam developing a reversing plastic hinge (reproduced from Fenwick et al., 1981, with permission)

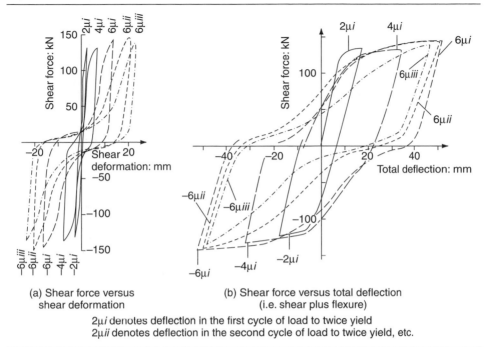

(a) Shear force versus shear deformation

(b) Shear force versus total deflection (i.e. shear plus flexure)

$2\mu i$ denotes deflection in the first cycle of load to twice yield
$2\mu ii$ denotes deflection in the second cycle of load to twice yield, etc.

Figure 8.18 Sliding shear failure in a reversing hinge (photograph courtesy of Richard Fenwick)

8.2.5.3 Beam and wall elongation

Crack widths under reversing cyclic loads become progressively larger. This is because part of the yielding tension force in the main steel is resisted by the concrete, so that for equal areas of top and bottom steel, there is never enough tension to force the compression steel to yield in compression and recover some of the plastic tension yielding from previous cycles. Tensile plastic strains in the steel therefore accumulate and the overall length of the beam increases. Even for unequal areas of top and bottom steel, the side with the most steel will tend to accumulate tensile strain. As a result, the beams elongate in a severe earthquake, and this imposes additional rotations on the lowest columns, particularly at the two ends of a frame (Figure 8.19). Thus, plastic hinges may form at both the top and bottom of some of the lowest columns, even when a capacity design has been carried out to ensure a 'strong column/weak beam' system, although this can never on its own lead to a soft storey failure. Moreover, the elongation imposes severe conditions on attached elements such as cladding panels and floor diaphragms, which have implications for the design of their connection and bearing arrangements. Elongation also occurs in the plastic hinges forming at the base of shear walls.

These deleterious effects have been observed in real earthquakes. The Royal Commission reporting on the Canterbury New Zealand earthquakes of 2010 and 2011 (Canterbury Earthquakes Royal Commission, 2012) identified buildings in which elongation of plastic hinges in beams and shear walls had led to significant damage. In particular, large tensile cracks forming in the in situ toppings to precast floors as a result of beam elongation appeared to have reduced the ability of the floor to distribute seismic shears to lateral resisting elements such as shear walls; floor beams also lost their support.

Current US and European codes do not refer explicitly to beam elongation, although the New Zealand standard NZS 3101: 2006 does (NZS, 2006). Note that even quite sophisticated

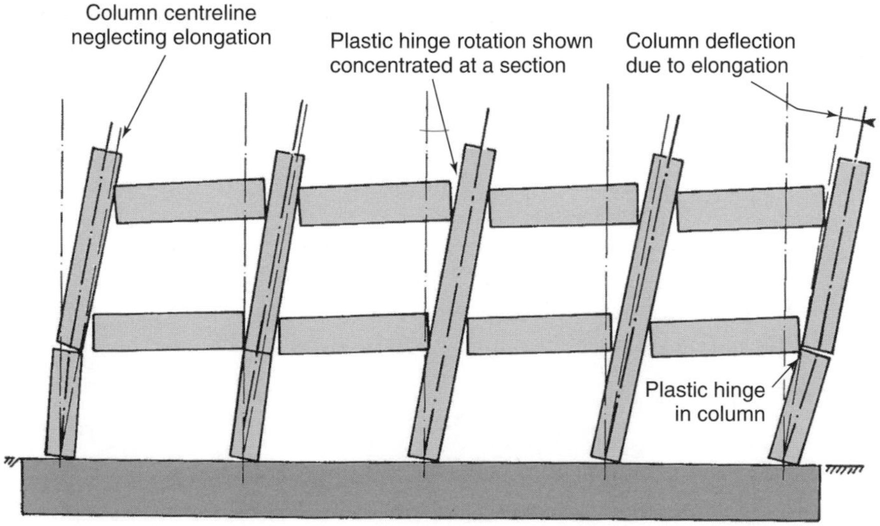

Figure 8.19 Effect of beam elongation on deflections and rotations in a ductile frame

non-linear dynamic seismic analysis will not generally model elongation effects and their consequences. Peng *et al.* (2007) cite the literature describing the extensive research on the subject in New Zealand over the last 30 years, and describe an analytical model for elongation in reversing hinges.

8.2.5.4 Implications for design

Some implications for the design of the response of concrete to cyclic shear are as follows.

■ Closely spaced transverse steel is required at potential plastic hinge points for four reasons
  – to provide adequate shear strength
  – to limit shear deformations
  – to provide buckling restraint to main steel
  – to confine the concrete in the flexural compression zone, in order to ensure its integrity.
■ The concrete contribution to shear strength in beams tends to be lost under conditions of cyclic loading.
■ Diagonal shear steel is needed at potential plastic hinge regions under conditions of high levels of reversing shear.
■ Elongation of beams occurs under cyclic loading and may require special detailing of lower columns and of restraint and bearing details to precast floors. Beam elongation may also lead to additional reinforcement requirements for the in situ toppings to precast floors.

8.2.6　Flexure and shear in beams: unidirectional hinges

The previous section considered beams in which two plastic hinges form under extreme earthquake loading, one at each end of the beam. With successive cycles of earthquake loading, the plastic hinges rotate first in one direction and then the other (Figure 8.20(a)). This is the situation likely to apply when the seismic resistance is provided by a perimeter frame that takes relatively low gravity loads, and the main vertical load bearing system is formed from gravity-only internal frames.

A different situation occurs when the beams in a seismic frame carry significant gravity loads. In this case, the positive (sagging) moments due to the vertical loads may be sufficient to cause plastic hinges to form in the beam span (Figure 8.20(b)). Four plastic hinges will then form during earthquake loading; a left-hand support hinge and a right-hand span hinge during one direction of loading, and a right-hand support hinge and left-hand span hinge during the other. For beams that have unvarying bending strength along the beam, it can easily be shown that unidirectional hinges will form if the shear forces at the two ends of the beam differ in sign when the first hinge forms at a support. This condition corresponds to

$$w > 2(M_A + M_B)/(L')^2 \tag{8.1}$$

where w is the vertical loading per unit length on the beam (gravity plus vertical seismic accelerations, assumed uniform along the beam), M_A and M_B are the positive and negative flexural strengths of the beam and L is the clear span. Where w exceeds this limit, the distance

Figure 8.20 Reversing and unidirectional hinge beams

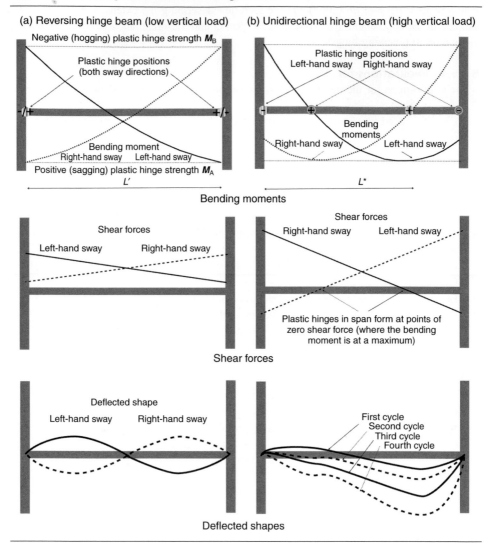

between the span and support hinges forming in any cycle is L^*, where

$$L^* = \sqrt{2(M_A + M_B)/w} \qquad (8.2)$$

The important consequence is that each hinge rotates in one direction only. While this avoids some of the degradation effects due to hinge reversals noted previously, the plastic rotations increase cumulatively instead of alternating between two extremes, and the ultimate rotation of unidirectional hinges may be quickly reached. Hence the overall ductility supply will be less than for frames with reversing hinges, and becomes linked more critically to the duration of earthquake (i.e. the number of loading cycles) as well as its maximum intensity. Elongation of the beams is also more severe (Fenwick and Megget, 1993).

Another consequence is that the maximum shear in the beam increases because the two plastic hinges are separated by a distance less than the clear shear span L'. For beams with unidirectional hinges, the equation in Figure 3.20 therefore needs to be amended by replacing L' with L^*, the distance between span and support hinge points.

Some implications for design are as follows.

- Where the beams of frames take significant vertical as well as lateral loads, unidirectional hinges may form, which will reduce significantly the effective frame ductility. The reduction in ductility is greater for large magnitude, long duration earthquakes.
- Peak shear forces may be somewhat greater in beams in which unidirectional rather than reversing hinges form.
- The formation of plastic hinges within the beam span – and hence the undesirable unidirectional effects described – could be prevented by increasing the relative bending strength of the beam in the span to the bending strength at supports. For example, additional bottom steel could be placed that stops at least a beam depth short of the plastic hinge region at the beam supports.

8.2.7 Flexure and shear in columns

Column failure is likely to have more disastrous consequences than beam failure, because the loss of support will extend to all floors above the failed column. Columns therefore need additional protection to guard against flexural or shear failure.

The differences between beam and column behaviour under cyclic loading arise from the compressive load that a column carries. This has two major consequences.

1 Shear and flexural cracks opening under one cycle of loading are likely to close under the influence of the compressive load in the reverse cycle. The pinching of the hysteretic loops found in beams (Figure 8.17) is therefore less pronounced, and the loss of 'concrete contribution' to shear strength (Section 8.2.5) is less severe. ACI 318 therefore allows the full shear strength under static loading conditions to be assumed when a significant compressive stress is present, as does EC8 (probably unconservatively) in all circumstances.
2 The additional compressive stress increases the cyclic compressive strain that the concrete must sustain, and as a consequence the concrete strength will quickly degrade at plastic hinge locations unless adequate confinement steel is present.

Flexural hinges may form columns, even when capacity design procedures have been applied, for reasons discussed in Section 3.5. However, as columns under seismic loading are rarely subject to significant lateral load, the hinges will almost always form as reversing hinges at the ends of the column, as discussed for beams in Section 8.2.5, rather than uniaxial hinges (Section 8.2.6).

8.2.8 Flexure and shear in slender shear walls

A slender shear wall is defined as one in which the height exceeds twice the width. Under these conditions, a suitably designed wall can form a ductile flexural hinge at the base,

Figure 8.21 Bending moment versus lateral displacement in a shear wall with a flexurally dominated response (reproduced with permission from Paulay and Priestley, 1992, with acknowledgement to Dr W. J. Goodsir)

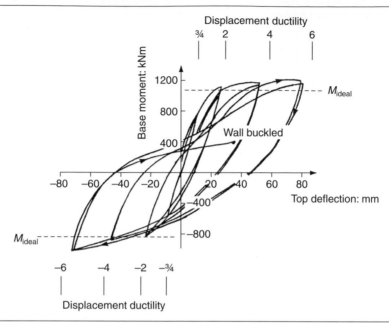

which achieves a level of ductility only slightly less than that of a well detailed frame (Figure 8.21). In New Zealand practice, the term 'shear wall', with its connotations of brittle shear failure, is felt to be a misnomer, and the term 'structural wall' is preferred.

Note the absence of stiffness and strength degradation in Figure 8.21 and the absence of significant pinching in the hysteresis loops. In order to achieve this ductile flexural behaviour, a number of conditions must be met.

- The vertical reinforcement in the plastic hinge region of the hinge must be restrained against buckling by closely spaced links. The Canterbury Earthquakes Royal Commission (2012) found evidence that this restraint needs to extend over the full wall length, not just at the ends, as currently required by codes.
- The compression edge of the wall must not be so slender as to suffer a buckling failure. A thickening of the edge, or the bracing provided by a transverse wall, will help prevent this.
- The concrete must be well confined at the ends of the wall where it is required to sustain high compressive cyclic strains.
- Plastic hinge formation must occur in a location (usually the base) where there is adequate detailing to sustain large plastic deformations. In order to achieve this, both EC8 (CEN, 2004) and the New Zealand concrete code (NZS, 2006) stipulate a capacity

design procedure to ensure that flexural yielding occurs only at the base of the wall. US practice (ACI, 2011) does not include this requirement.

■ The shear strength of the wall throughout its height must be sufficient to sustain the chosen plastic flexural hinge mechanism. Once again, EC8 and the New Zealand code have special requirements to achieve this which do not appear in US practice.

Shear failure in shear walls can occur in diagonal tension or compression in a similar way to beams. Another form of shear failure is sliding shear at horizontal planes; this can be resisted by shear friction across any horizontal crack and by dowel action. Distributed vertical reinforcement plays several roles in this: it helps to distribute cracking, provides dowel resistance and also helps clamp concrete surfaces together. Construction joints are clearly potential sliding shear planes; they should be well roughened, cleaned of loose debris and checked for strength using a shear friction calculation.

Anchorage failure of the main reinforcement steel leads to loss of strength and must be detailed against by providing generous anchorage. Under cyclic loading, yielding of the steel will occur progressively further down the reinforcing bars, and may penetrate into the wall foundation by around 20 bar diameters for a displacement ductility of 6. Full tension anchorage of the bars is therefore required beyond this point.

8.2.9   Squat shear walls

When the height to width ratio of a wall is less than about 2, the shear force necessary to develop a flexural hinge becomes relatively large, and ductile flexural behaviour may be hard to achieve. Often, this is not of concern because the inherent strength of a squat shear wall enables seismic action to be resisted without the need to develop much ductility. However, provided sliding shear failure is prevented, some ductility can be achieved through yielding of vertical reinforcement. The provision of diagonal reinforcement anchoring the base of the wall to its foundation greatly improves resistance to sliding shear, and is required by EC8 for DCH squat shear walls.

8.2.10   Prestressed concrete

The behaviour of prestressed concrete beams has similarities with that of passively reinforced columns; thus, the prestressing force improves shear resistance and reduces the tendency for stiffness degradation but increases concrete compressive strain demands, and the steel itself has a lower fracture strain, which in itself tends to reduce ductility. Figure 8.22 compares the cyclic response of prestressed and passively reinforced members; the lower ductility and hysteretic energy dissipation in the former is evident.

There is limited codified information on the seismic design of prestressed concrete buildings, and it is not covered by EC8 part 1 (CEN, 2004), although part 2 gives some advice for bridges. ACI 318 (ACI, 2011) permits partial prestressing, which can be used to provide not more than 25% of the flexural strength at critical sections and must be unbonded through these sections. ACI 318 also gives rules for the maximum strain in the prestressing tendons, the maximum level of prestress and the cyclic performance required of prestress anchorages. More extensive advice is provided in New Zealand code NZS 3101 (NZS, 2006).

Figure 8.22 Cyclic response of prestressed and reinforced concrete

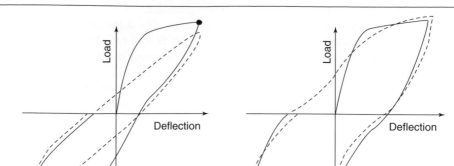

8.2.11 Non-ferrous reinforcement in seismic resisting structures

In non-seismic applications, increasing use is being made of non-ferrous reinforcement to provide tensile strength both for concrete and for non-cementitious resins. The reinforcement takes the form of fibres made of carbon, various types of plastic (aramid, polyethylene) or glass. They are characterised by high tensile strength and good corrosion resistance, compared to steel, but possess little or no ductility. Their main use in seismic applications has been as external jacketing applied as confinement to existing concrete columns.

8.3. Material specification

Table 8.1 shows the material specifications of EC8 (CEN, 2004) and ACI 318 (ACI, 2011). The Eurocode requirements are shown for the regions of DCM structures expected to form plastic hinges; more stringent rules apply to DCH structures.

The rationale behind the main requirements of Table 8.1 is as follows. Minimum concrete strength is specified to ensure a reasonable level of strength and ultimate strain, while the reasons for restrictions on maximum strength were discussed in Section 8.2.2.

Reinforcement with reasonable ultimate tensile strain is an obvious requirement to ensure ductility. The restrictions on the difference between actual and specified yield strength arise from capacity design considerations. Thus, for example, the required shear strength of a beam or column should be based on the actual flexural strength that the beam achieves, and if on site this exceeds the designer's assumptions, the shear strength provided may be insufficient to develop the actual flexural strength. The minimum ratio between ultimate tensile and yield strength is to ensure that yielding in regions of rapidly changing moment spreads over a reasonable length of the beam, thus producing good ductility. This is because if a plastic hinge increases in strength as it rotates, it will force adjacent sections further down the beam to yield. Without such an increase in strength with rotation, all the plastic yielding, and hence plastic rotation, becomes concentrated at the point of maximum moment, leading to high plastic strains.

Table 8.1 Concrete and steel specifications for high ductility seismic resisting structures

	EC8 (CEN, 2004) for DCM	ACI 318 (ACI, 2011) for special moment frames and walls
	Concrete cylinder strength	
Minimum	16 MPa	21 MPa
Maximum	a	b
	Reinforcement resisting flexural and axial loads	
General	Plain round bars are only acceptable as hoops or ties; otherwise deformed or ribbed steel must be used	Plain round bars are only acceptable as spirals or for prestress; otherwise deformed reinforcement must be used
Yield strength	400–600 MPa	A706M: 420 MPa A615M: 280 MPa or 420 Mpa
Min. tensile strain (on 200 mm)	Min. strain at ultimate tensile strength: 5%	Min. strain at fracture: A706: 10–14% A615: 7–12% (depending on bar diameter)
Ultimate tensile strength Yield or 0.2% proof strength	≥1.08	≥1.25 (A615M)
Actual yield strength less specified yield strength	Not specified	≤125 MPa (A615M)

[a] Concrete with cylinder strength exceeding 50 MPa is not covered by EC8
[b] The cylinder strength of lightweight concrete may not normally exceed 35 MPa, but this may be increased if justified by tests

8.4.   Analysis of reinforced concrete structures

Concrete structures may be analysed by any of the methods discussed in Chapter 3. This section discusses some particular aspects with respect to modelling.

8.4.1   Damping in concrete structures

In a structure responding plastically to an earthquake, most of the damping is hysteretic and in a ductility modified response spectrum analysis, this is represented by the ductility reduction factor (e.g. the q factor in EC8 or R in US codes). Therefore, no separate allowance needs to be made for the level of viscous damping. However, in an elastically responding structure, the usual assumption of 5% modal damping may need to be adjusted. ASCE 4-98 (ASCE, 1998) recommends modal damping values in reinforced concrete structures of 4–7%, the former being applicable when stresses are generally below half yield, and the latter when stresses are approaching yield. For prestressed members, these values reduce to 2% and 5%, respectively. These values would be appropriate in an elastic response spectrum or time history analysis.

In a non-linear time history analysis, the hysteretic damping is accounted for explicitly, because yielding is taken directly into account. Viscous damping to account for energy dissipation in the elastic range needs to be used rather carefully, and the ASCE 4-98 values quoted above may be unconservative. This is because although they are appropriate while the structure is in its elastic range, they may substantially overestimate the dissipated energy when the structure yields. A full discussion is given by Priestley *et al.* (2007: section 4.9.2(g)). Key points are first that the damping force taken in many computer programs is related to the initial stiffness, but this is likely to be unconservative, and relating to the secant stiffness gives more representative results. Second, programs often require viscous damping to be specified as 'Rayleigh damping'. This is a computationally simple method that can achieve approximately constant damping in all modes of interest, and is usually satisfactory for elastic responding structures, but needs care to avoid overestimating damping in the post-yield phase.

8.4.2   Assessing the rotational capacity of concrete elements

In a well-designed ductile concrete frame or shear wall structure, yielding and energy dissipation under extreme seismic loading normally takes place through rotation of plastic hinges forming within the structure. The ductility available then depends on the ultimate rotational capacity of those plastic hinge regions.

In conventional code design using force-based methods and elastic analysis, the adequacy of the plastic hinge regions is obtained by designing for a strength depending on a structural factor (e.g. the q factor in EC8 or R factor in ASCE 7) and then applying detailing rules given in the code that correspond to the structural factor adopted.

Non-linear static (Section 3.7.9) or time history analysis (Section 3.7.8) enables the local ductility demand at plastically yielding regions to be determined directly, which in principle is a much more satisfactory procedure. Where these regions are modelled as discrete plastic hinges, the analysis will result in values of plastic rotation, which must then be compared with the available capacity.

The most direct advice is given by ASCE 41-06 (ASCE, 2006) and EC8 (CEN, 2004) part 3, which provides limiting rotations corresponding to different performance goals for a variety of elements. This information is not currently provided by US or European codes for the seismic design of new buildings.

A calculation is also possible from first principles. This calculation starts by establishing the relationship between curvature of a section and moment, taking into account any axial load that may be present. With the knowledge of the stress–strain characteristics of both concrete and steel, and on the assumption that plane sections remain plane, this is in principle straightforward to do (Figure 8.23), and a number of programs exist to perform the calculation. Note that the stress–strain characteristics of concrete depend on the amount of confining steel (Figure 8.15). Note also the limitations of the assumption of plane sections remaining

Figure 8.23 Calculation of moment curvature relationship under uniaxial bending, assuming plane sections remain plane

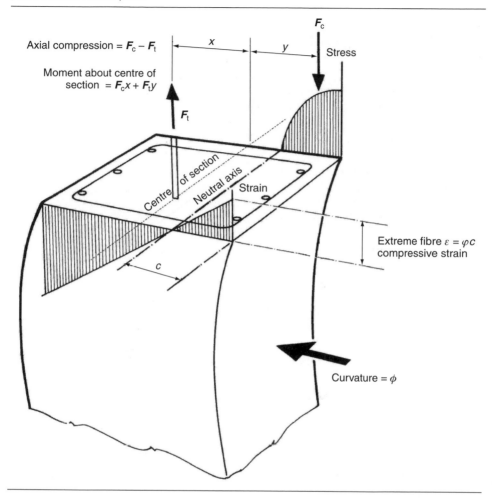

planc. It is strictly only true for monolithic sections with zero shear, although the shear stress needs to be high for significant error to occur. More importantly for reinforced concrete sections, slip between steel and concrete is not included. Therefore, under the conditions of high shear and bond slip likely in plastic hinge regions with high ductility demand, the results are approximations.

The curvature–moment relationship can be transformed to a rotation–moment relationship by multiplying the curvature by an effective plastic hinge length L_{pl} (Figure 8.24). Priestley *et al.* (2007) provide the following expressions for L_{pl}.

$$L_{pl} = kL_v + L_{SP} \geq 2L_{SP} \qquad (8.3)$$

$$L_v = M/V, \text{ the moment to shear ratio at the plastic hinge section} \qquad (8.4)$$

187

Figure 8.24 Calculation of effective plastic hinge rotation

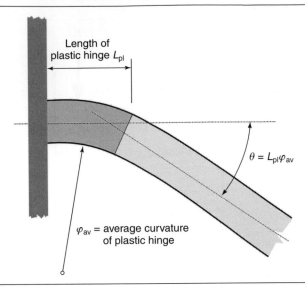

φ_{av} = average curvature
of plastic hinge

$$k = 0.2\left(\frac{f_{ult}}{f_y} - 1\right) \leq 0.08 \text{ where } \frac{f_{ult}}{f_y} \text{ is the ratio of ultimate to yield strength}$$

of the reinforcement $\hspace{10cm}$ (8.5)

$L_{SP} = 0.022 f_y d_{bl}$ where d_{bl} is the diameter of the longitudinal bars $\hspace{1cm}$ (8.6)

The term L_{SP} in Equation 8.3 allows for the slippage of reinforcement at the end of a plastic hinge, for example, of beam reinforcement into a beam–column joint, or column reinforcement into a column base. Equation 8.5 emphasises the importance of using reinforcement where the ultimate strength exceeds the yield strength by a sufficient margin, as required by codes (see Table 8.1). Note that different expressions for L_{pl} are given in EC8 part 3 annexe A, and in EC8 part 2 annexe E.

The plastic rotation θ_p of the hinge corresponding to a curvature φ_p can then be estimated as

$$\theta_p = (\varphi_p - \varphi_y)L_{pl} \hspace{9cm} (8.7)$$

where φ_y is the curvature at first yield.

The simplest approach for calculating the ultimate rotation would be to substitute $\varphi_u = \varphi_p$ in Equation 8.7, where φ_u is the curvature at concrete or steel fracture. However, that overestimates capacity, because it assumes that the curvature over the plastic hinge is constant at its ultimate value and also neglects the interaction between shear and rotation capacity. EC8 part 2 annexe E provides the following expressions, which account for these factors.

$$\theta_{pu} = (\theta_u\theta_y)L_{pl}(1 - L_{pl}/L_v)\alpha \hspace{6cm} (8.8)$$

$$\alpha = (L_v/d)^{1/3} \leq 1 \hspace{8cm} (8.9)$$

where θ_{pu} is the ultimate hinge rotation, d is the depth of the member and other symbols have the same meanings as before. Yield refers to conditions when the reinforcement at the most critical section first reaches its yield stress, while the ultimate rotation corresponds to a near collapse limit state. The plastic rotation corresponding to significant damage is taken by EC8 part 3 as three-quarters of this value.

Equation 8.8 takes no direct account of the degradation of plastic hinge capacity on the number of cycles to which they are subjected – the 'low cycle fatigue' effects. Sivaselvan and Reinhorn (1999) discuss models that attempt to include such effects.

Another important proviso is that there should be sufficient flexural reinforcement to ensure well distributed cracking in the plastic hinge region. With low steel ratios, the plastic hinge will form as one large crack with high associated plastic strains in the reinforcement and Equation 8.8 will greatly overestimate the available ductility.

8.4.3 Modelling the stiffness of reinforced concrete members
8.4.3.1 The ratio of cracked to uncracked flexural stiffness
The stiffness of concrete structures reduces under severe ground shaking. One reason for this is the onset of plasticity, and this is accounted for in non-linear analyses in which the plastic hinges are explicitly modelled and also (in an approximate way) in ductility modified response spectrum analyses. However, cracking of concrete also takes place away from plastic hinge positions, and therefore basing member properties on the uncracked concrete section tends to overestimate the stiffness and underestimate the natural periods of the structure prior to yield. For response to serviceability level events, this will lead to an underestimate of deflections and also probably an overestimate of member forces. Moreover, there are important consequences for response at ultimate limit states; because deflections and curvatures at yield are underestimated, effective ductility demands are also miscalculated.

ACI 318 (ACI, 2011) allows the flexural stiffness of concrete members to be based on half of the gross section properties, but also provides alternative values that depend on whether members are cracked or uncracked. The amount of reinforcement may also be taken into account. EC8 (CEN, 2004) also allows the stiffness of concrete (and masonry) members to be based on half the gross stiffness, in the absence of a more detailed analysis. The EC8 manual (ISE/AFPS, 2010) recommends ratios of between half and the full gross stiffness, depending mainly on the axial compressive load in the member, as shown in Table 8.2.

The approach of using half the gross section flexural stiffness in a seismic analysis is conveniently simple, but unfortunately it describes the real behaviour of concrete structures rather poorly. In fact, the ratio of effective to gross section stiffness may vary between about 0.1 and 0.9 (Figures 8.25 and 8.26) and is highly dependent on both the amount of longitudinal reinforcement and axial load; that is to be expected because both affect the amount of cracking. Rather than basing the stiffness ratio on simple tables such as Table 8.2, an alternative approach is to relate flexural stiffness to yield curvature, in a way that allows for the influence of both the quantity of reinforcement and axial load. This approach, which was developed by Nigel Priestley and his co-workers, is outlined in the sections that follow; for a more complete explanation, see Priestley et al. (2007).

189

Table 8.2 Effective stiffness of reinforced concrete (reproduced from ISE/AFPS, 2010 © The Institution of Structural Engineers and Association Française du Génie Parasismique)

Member	Flexural stiffness	Shear stiffness	Axial stiffness
Beams: non-prestressed	$0.5E_{cd}I_g$	$G_{cd}A_w$	
Beams: prestressed	$E_{cd}I_g$	$G_{cd}A_w$	
Columns in compression	$0.7E_{cd}I_g$	$G_{cd}A_w$	$E_{cd}A_c$
Columns in tension	$0.5E_{cd}I_g$	$G_{cd}A_w$	$E_{sd}A_s$
Walls and diaphragms: uncracked	$E_{cd}I_g$ ($\sigma_t < f_{ctm}$)	$G_{cd}A_w$ ($V_{Ed} < V_{Rd,c}$)	$E_{cd}A_c$
Walls and diaphragms: cracked	$0.5E_{cd}I_g$ ($\sigma_t > f_{ctm}$)	$0.5G_{cd}A_w$ ($V_{Ed} > V_{Rd,c}$)	$E_{cd}A_c$

A_c, cross-sectional area of the concrete section; A_s, cross-sectional area of the reinforcing steel; A_w, web cross-sectional area; E_{cd}, design value of concrete compressive modulus; E_{sd}, design value of reinforcing steel modulus; f_{ctm}, mean value of axial tensile strength of concrete; G_{cd}, design value of concrete shear modulus = $0.4E_{cd}$; I_g, gross moment of inertia of concrete section, based on uncracked properties; σ_t, maximum tensile stress in concrete due to bending, assuming an uncracked section; V_{Ed}, design value of shear force in wall in the seismic design situation; $V_{Rd,c}$, design shear resistance of the wall without shear reinforcement
This table gives tentative values for concrete elements interpolated from ASCE 43-05

8.4.3.2 The relation between flexural stiffness and yield curvature

Figure 8.27 shows the relationship between curvature and moment of a typical concrete section. In addition to the experimental line, a bi-linear approximation is shown, as is commonly assumed in analysis. The initial slope of this line is the secant stiffness up to the point of yield, and is appropriate for describing the stiffness of members away from plastic hinge locations. It can therefore be used to set member properties for linear frame elements in an analysis in which the non-linearity is modelled by the use of discrete plastic hinges. It is also appropriate for a ductility modified analysis of a frame. The second, plastic portion of the bi-linear approximation then sets the yield strength of the plastic hinge and its strain hardening characteristics.

A standard result is that the moment M developing in an elastic beam equals the flexural stiffness, EI times the curvature φ

$$M = \varphi EI \tag{8.10}$$

Therefore, if the curvature at flexural yield of a concrete section is known, its stiffness at the point of effective yield EI_{eff} can easily be found

$$E_{cm}I_{eff} = M_{Rk}/\varphi_y \tag{8.11}$$

Changing the amount of reinforcement and the axial load of course has a large effect on M_{Rk} the characteristic flexural resistance, but it turns out that it changes the curvature at yield φ_y much less. Extensive analysis, coupled with some experimental tests, has shown that for a given section, a reasonable approximation is to take the yield curvature of a concrete section as a constant ratio of ε_{yk}/D where ε_{yk} is the characteristic yield strain of the reinforcement, and D is the section depth. Table 8.3 shows this relationship for various types of section.

Figure 8.25 Flexural stiffness ratios for large concrete columns (reproduced from Priestley *et al.*, 2007, courtesy of IUSS Press)

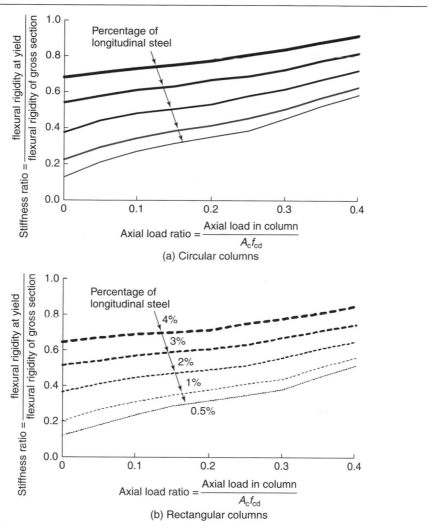

(a) Circular columns

(b) Rectangular columns

8.4.3.3 Approximate derivation of yield curvature expressions

The expressions for φ_y in Table 8.3 are not exact, but it is no coincidence that they are all equal to about $2\varepsilon_y/D$. For a beam or column made of an elastoplastic material like steel, this can be shown rather easily, as follows. Equation 8.11 can be rewritten

$$\varphi_y = M_{Rk}/(IE_{sm}) = f_{yk}S/(IE_{sm}) = \varepsilon_{yk}E_{sm}S/(IE_{sm}) = \varepsilon_{yk}S/I \qquad (8.12)$$

where M_{Rk} is the characteristic flexural resistance, ε_{yk} is the characteristic yield strain of the steel, I is the moment of inertia of the section and S is its plastic section modulus. For

Figure 8.26 Flexural stiffness ratios for rectangular beams (see note to Figure 8.25)

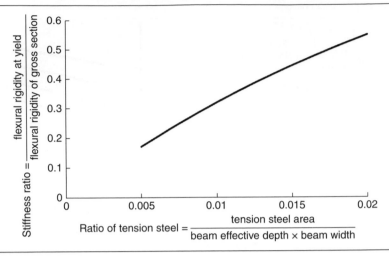

symmetrical sections, S/I equals (S/Z) $(2/D)$, where (S/Z) is the ratio of plastic to elastic modulus, or plastic shape factor. For I sections bending about their major axis, a well-known result is that the plastic shape factor is approximately 1.15, making S/I equal to $2.3/D$. In fact, for steel universal beams and column, joists and channels available in the UK, S/I ranges from $2.2/D$ to $2.46/D$, with an average of $2.3/D$ giving a value for $\varphi_y = \varepsilon_{yk}S/I$ comparable to that for concrete columns in Table 8.3. This is for major axis bending; the values for bending about the minor axis are approximately 25% greater.

Figure 8.27 Yield curvature relationship

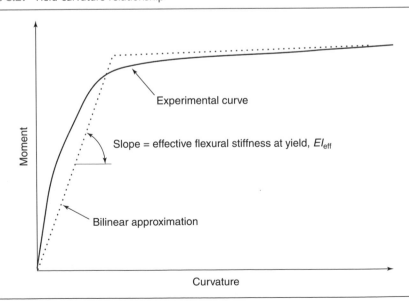

Table 8.3 Yield curvatures of concrete sections ($f_{cu} \leq 50\,\text{MPa}$, $f_{yk} \leq 500\,\text{MPa}$) (Priestley *et al.*, 2007)

Section type	φ_y Curvature at yield	
Circular column	$\dfrac{2.25\varepsilon_{yk}}{D}$	
Rectangular column	$\dfrac{2.10\varepsilon_{yk}}{h_c}$	
Rectangular wall	$\dfrac{2.10\varepsilon_{yk}}{l_w}$	
Flanged concrete walls	$\dfrac{1.50\varepsilon_{yk}}{l_w}$	
Rectangular beam	$\dfrac{1.85\varepsilon_{yk}}{h_b}$	
T-section beam	$\dfrac{1.70\varepsilon_{yk}}{h_b}$	

Figure 8.28 Idealisation of a beam at yield

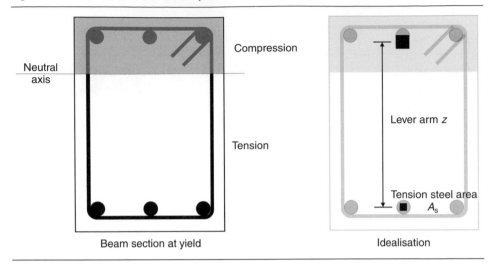

| Beam section at yield | Idealisation |

Reinforced concrete presents a much more complex case, because it consists of two dissimilar materials, one of which cracks in tension. However, an approximate equivalent to Equation 8.12 can be derived for rectangular sections as follows. The concrete in compression, plus any compression steel, has been idealised as a point area, and so has the reinforcement in tension, with the two areas separated by the lever arm z (Figure 8.28). Taking I_{cr} as the moment of inertia of the cracked section, allowing both for the reinforcement and also for the concrete in compression (but not tension), gives a transformed version of Equation 8.12 as

$$\varphi_y = M_{Rk}/(I_{cr}E_{cm}) = zf_{yk}A_s/(I_{cr}E_{cm}) = z\varepsilon_{yk}E_{sm}A_s/(I_{cr}E_{cm}) \tag{8.13}$$

With the assumptions of Figure 8.28, it can easily be shown that

$$I_{cr} = A_s z^2 \frac{E_{sm}}{E_{cm}} \bigg/ \left(1 + \frac{f_{ck}}{f_{yk}}\frac{E_{sm}}{E_{cm}}\right) \tag{8.14}$$

Combining Equations 8.13 and 8.14 gives

$$\varphi_y = \frac{\varepsilon_{yk}}{D}\left\{\frac{D}{z}\left(1 + \frac{f_{ck}}{f_{yk}}\frac{E_{sm}}{E_{cm}}\right)\right\} \tag{8.15}$$

For values of f_{ck} from 30 MPa to 50 MPa and f_{yk} from 400 MPa to 500 Mpa, $(1 + (f_{ck}/f_{yk}) \times (E_{sm}/E_{cm}))$ is approximately 1.5, with a variation of $\pm12\%$. This suggests that although concrete and steel strengths will affect the Table 8.3 values, the influence is relatively small, and can be neglected for normal concrete and steel strengths, because other uncertainties are substantially greater. The lever arm ratio to total depth z/D for a rectangular reinforced concrete beam is typically approximately 0.75, making the expression $\{(D/z)(1 + (f_{ck}/f_{yk}) \times (E_{sm}/E_{cm}))\}$ equal to about 2.0, comparable to the value for rectangular beams in Table 8.3.

Note that z/D tends to increase in flanged sections, compared with rectangular ones, and hence its inverse is less, explaining why the Table 8.3 values are smaller for flanged sections.

8.4.3.4 Implications for designers and analysts for the interdependence of stiffness and reinforcement ratio

Figures 8.25 and 8.26 show that a realistic analytical representation of the stiffness of a concrete structure is only possible when both the axial load and (more significantly) the amount of reinforcement are known. This in principle is the case for an existing structure, but of course for new structures it is not. If the reinforcement of the new structure is designed using a traditional force-based method based on strength requirements, an iterative approach is therefore needed until the reinforcement ratios and consequential stiffness match the force demands. A major advantage of the displacement-based method of seismic design proposed by Priestley *et al.* (2007) and outlined in Section 3.6 is that the amount of reinforcement is determined not from strength requirements but from the need to provide sufficient stiffness to achieve a target displacement. While the procedure involves a number of approximations, it enables efficient section sizes and reinforcement ratios to be determined much more rapidly than does the use of a force-based method procedure; these can then be confirmed by a detailed analysis using stiffnesses based on Table 8.3.

There is another important implication of the strong link between flexural stiffness and reinforcement ratio. Consider a structure that is designed on a force-based method and the analysis has assumed that flexural stiffnesses are a fixed ratio of gross section stiffness, using, for example, Table 8.2. The forced-based procedure leads to the most highly stressed members being designed with the highest reinforcement ratios, and the most lightly stressed with the lowest ratios. In reality, this will lead to very different stiffnesses than those assumed in the analysis. The high level of reinforcement in the highly stressed members will stiffen them by reducing cracking, thus attracting more load than the analysis suggests, while the lightly stressed members will crack sooner and hence be more flexible and lightly loaded. An iterative procedure is needed to get a more rational distribution of steel. Alternatively, as previously discussed, a displacement-based approach leads directly to a more uniform distribution of ductility demands and hence a more efficient use of reinforcement.

8.4.3.5 Limitations of the recommended flexural stiffness method

Equation 8.11 predicts the flexural stiffness of a concrete section at the point of yield, on two assumptions

1 the concrete is completely cracked in tension, and
2 plane sections remain plane.

It therefore corresponds to the cracked section stiffness I_{cr} at yield. Assumption 1 leads to an underestimate of stiffness, because of the stiffening effect of the uncracked concrete between cracks. This will become more significant the closer the cracking moment of the section, M_{cr} is to the plastic yield moment M_{Rk}. As noted in Section 8.4.2, ratios of M_{cr}/M_{Rk} approaching unity will also lead to a severely reduced ductility supply in the plastic hinge.

Assumption 2 – plane sections remain plane – is likely to result in an overestimate of stiffness near the plastic hinge, because of shear deformations and bond slip between the reinforcement and concrete; these effects will get progressively greater with the number of reverse yielding cycles that the structure experiences.

Overall, it is likely that using the Table 8.3 values will result in an underestimate of stiffness at serviceability level, and a sensitivity check using gross concrete section properties is recommended if the cracking moment M_{cr} exceeds 0.35 M_{Rk}. For post-yield response under ultimate limit state conditions, the limited experimental evidence suggests that the Table 8.3 values gives reasonable results (Priestley, 1998). The values are generally recommended as the best option for a simple frame analysis, given the limitations discussed in Section 8.4.3.6 below. However, when a low reinforcement ratio leads to $M_{cr} > 0.5 M_{Rk}$, a sensitivity check using gross concrete section properties should also be performed. These recommendations, which do not appear in codes, are tentative and research is needed. M_{cr} may be estimated from Equation 8.16.

$$M_{cr} = f_{ctm} I_g / (h - x_u) \tag{8.16}$$

where $(h - x_u)$ is the distance from the compression face of the section to the neutral axis and other symbols are as defined previously.

8.4.3.6 Summary of the recommended method for determining flexural stiffness

When the amount of reinforcement and level of axial load due to gravity are known, the following steps are involved.

- Determine the characteristic flexural resistance M_{Rk} of each member, based on the characteristic concrete strength and reinforcement without dividing by the material partial factor γ_m (or equivalently in US practice, determine the nominal strength without capacity reduction factor ϕ). The axial load due to gravity in the columns should be included when calculating the flexural strength, as should any prestress.
- From Table 8.3, determine the curvature at yield φ_y from the depth of the section and the characteristic strain in the reinforcement at yield, ε_{yk}.
- From Equation 8.11, determine the effective flexural stiffness $E_{ck} I_{eff}$.
- For serviceability limit state, when the cracking moment M_{cr} exceeds 0.35 M_{Rk}, a sensitivity check using the gross (uncracked) section properties should be carried out. For ultimate limit state, this check is recommended if $M_{cr} > 0.5 M_{Rk}$. These are tentative recommendations, subject to further research.

For new structures designed on a force-based method, an iterative procedure is needed, as the reinforcement is not initially known, and stiffnesses must be estimated for the first iteration – for example, using Table 8.2. Alternatively, initial estimates of reinforcement requirements can be made using a displacement-based method, and the flexural stiffnesses can be based on these estimates before detailed analysis is carried out.

8.4.3.7 Shear stiffness

Shear deformations in slender members usually represent only a small proportion of total deformations, and basing the shear stiffness on simple expressions such as those in

Table 8.2 in most cases will not cause unacceptable errors. However, shear deformations may become significant in members that are short compared to their depth, and the reduction in shear stiffness due to cracking may then need to be accounted for. Priestley *et al.* (2007) advise that, up to the point at which shear cracks develop, the reduction in shear stiffness is approximately the same as that of the flexural stiffness; from Equation 8.11, the reduction factor R_{sh} equals

$$R_{sh} = \frac{G_{cm}A_{sh,eff}}{G_{cm}A_{sh,g}} \approx \frac{E_{cm}I_{eff}}{E_{cm}I_g} = \frac{M_{Rk}}{\varphi_y E_{cm}I_g} \tag{8.17}$$

where φ_y is found from Table 8.3, G_{cm} is the mean concrete shear modulus, $A_{sh,eff}$ and $A_{sh,g}$ are the effective and gross shear areas of the section, and the other symbols are as previously defined. After shear cracking has developed, the shear stiffness reduces further; see Priestley *et al.* (2007) for details.

8.4.3.8 Axial stiffness

The axial stiffness of a concrete column is of course very different in compression than it is in tension after cracking. In an equivalent static or static pushover analysis, different axial stiffnesses can be applied to tension and compression members, using, for example, the values recommended in Table 8.2. For a time history analysis, members may alternate between tension and compression, so modelling the stiffness change is harder, and it is impossible in a response spectrum analysis, which does not allow for such non-linearity. Sensitivity analysis to investigate the effect of axial stiffness variations is one option for dealing with this.

8.4.4 Modelling of a concrete frame structure

The flexural and shear stiffnesses calculated from the procedures outlined in the previous section are appropriate for use in a ductility modified response spectrum analysis in which the superstructure consists of concrete beams, columns and (if present) shear walls, modelled by line elements. Figure 8.30 illustrates such a model for the perimeter frame with shear wall building shown in Figure 8.29. Note that response spectrum analysis assumes the stiffness in positive and negative directions are equal. Where this is not the case, the tendency of a structure to 'ratchet' in the more flexible direction will be missed, and non-linear analysis must be used. For example, a T-section shear wall is more flexible when the T-flange is in tension than when it is in compression and this might cause differential stiffness.

In the most straightforward, and common, type of non-linear analysis, the regions where plasticity is expected to develop are modelled by discrete plastic hinges, with the surrounding superstructure modelled by line elements (Figure 8.31). Stiffness properties of the line elements were discussed in the previous section; ASCE 41 (ASCE, 2006) provides data for the modelling of the plastic hinges. Sophisticated non-linear models for concrete are provided in a number of academic-based programs, including the freely distributed open source program OpenSees (see http://opensees.berkeley.edu) and the freely distributed program IDARC (http://civil.eng.buffalo.edu/idarc2d50/).

A number of further issues when creating this type of non-linear model are now discussed.

Figure 8.29 Example building with perimeter frames and shear walls. NB: Floor and roof slabs not shown for clarity

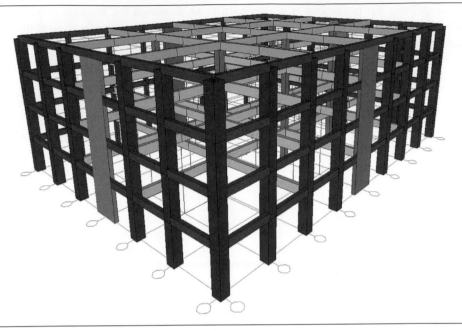

8.4.4.1 Modelling of beam–column joints

The shear forces in a joint under seismic loading are much higher than in the columns and beams that it connects (see Figure 8.37), and shear deformation of the joint may contribute significantly to overall deformation, despite the short length of the joint. Although it might be expected that flexural deformations of the joint could be neglected, they may also be significant; the strain in the reinforcement does not suddenly reduce to zero at the face of the joint, but on the contrary will be at yield when a plastic hinge forms. Therefore, the common assumption that line elements representing the joint can be modelled as effectively rigid in shear and flexure is not correct; it will underestimate flexibility and hence deflections.

When analysing a structure in which the beam–column joint is well confined with transverse reinforcement, the line elements may be modelled with the same shear and flexural stiffnesses as the beams and columns they join. For the analysis of an existing building where the joint is poorly confined, extensive shear cracking of the joint may occur, dramatically reducing its shear stiffness and the structural effectiveness of the frame. Linear elements cannot model this possibility realistically, and non-linear modelling of the joint is needed, for example as a series of shear and axial springs. Lowes *et al.* (2004) propose the arrangement illustrated in Figure 8.32; the assembly representing the beam–column joint has been implemented as a superelement in the OpenSees freeware suite of analysis programs maintained by the University of California at Berkeley (http://opensees.berkeley.edu/).

It is important to note that the modelling of beam–column joints shown in details 1 and 3 (Figures 8.30 and 8.31) will significantly underestimate the shear forces in the beam within

Figure 8.30 Elastic model of perimeter frame in example building

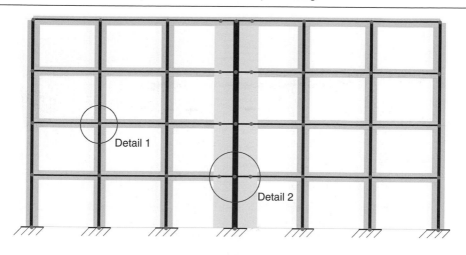

• Rigid node point   Elastic column, beam and
                             wall frame members

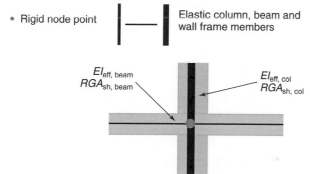

$EI_{\text{eff, beam}}$ $EI_{\text{eff, col}}$ Flexural stiffnesses from Equation 8.11
$RGA_{\text{sh, beam}}$ $RGA_{\text{sh, col}}$ Shear stiffnesses from Equation 8.17, based on web area
Axial stiffness of column – see Table 8.2

Detail 1 at beam column joint

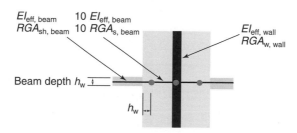

$EI_{\text{eff, beam}}$ $EI_{\text{eff, wall}}$ Flexural stiffnesses from Equation 8.11
$RGA_{\text{sh, beam}}$ $RGA_{\text{sh, wall}}$ Shear stiffnesses, from Equation 8.17, based on web area
Axial stiffness of wall – see Table 8.2

Detail 2 at beam wall joint

Figure 8.31 Non-linear model of perimeter frame in example building

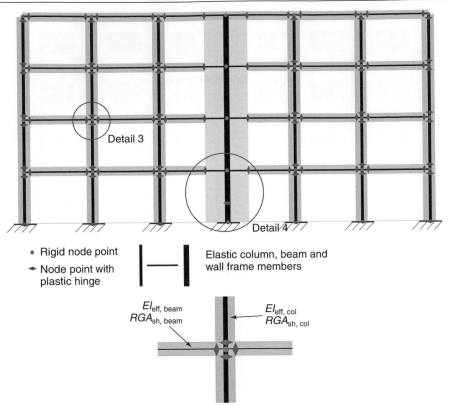

- • Rigid node point
- ➤ Node point with plastic hinge

Elastic column, beam and wall frame members

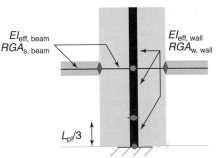

$EI_{\text{eff, beam}}$
$RGA_{\text{sh, beam}}$

$EI_{\text{eff, col}}$
$RGA_{\text{sh, col}}$

$EI_{\text{eff, beam}}$ $EI_{\text{eff, col}}$ Flexural stiffnesses from Equation 8.11
$RGA_{\text{sh, beam}}$ $RGA_{\text{sh, col}}$ Shear stiffnesses away from joint, from Equation 8.17, based on web area
In beam-column joint, shear forces are under estimated (see Figure 8.37) and shear stiffness may be reduced by shear cracking
Axial stiffness of column – see Table 8.2

Detail 3 at beam column joint

$EI_{\text{eff, beam}}$
$RGA_{\text{s, beam}}$

$EI_{\text{eff, wall}}$
$RGA_{\text{w, wall}}$

$L_{\text{pl}}/3$

$EI_{\text{eff, beam}}$ $EI_{\text{eff, wall}}$ Flexural stiffnesses from Equation 8.11
$RGA_{\text{sh, beam}}$ $RGA_{\text{sh, wall}}$ Shear stiffnesses from Equation 8.17, based on web area
Axial stiffness of wall – see Table 8.2 L_{pl} plastic hinge length from Equation 8.4

Detail 4 at beam wall joint

Figure 8.32 Non-linear model of beam–column joint (Lowes *et al.*, 2004)

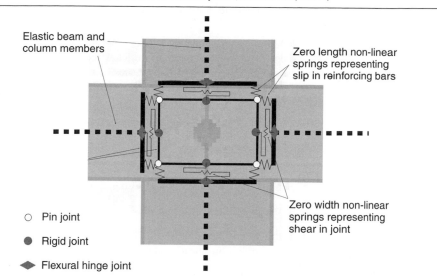

the joint, because the flexural restraint from the column is modelled as being concentrated at the centre of the joint, instead of being spread over the joint width. The column shears within the joint are similarly underestimated. By contrast with detail 3 in Figure 8.31, the shear springs in Figure 8.32 captures joint shear correctly.

8.4.4.2 Position of plastic hinges in beams

Generally, plastic hinges are expected to form at the faces of beam–column joints. However, this may not be the case for beams subject to substantial gravity loads, where plastic hinges may form at mid-span (Section 8.2.6). Where Equation 8.1 or other analysis shows that mid-span hinges could occur, the model needs to include additional plastic hinges at the appropriate positions.

8.4.4.3 Position of plastic hinges in columns

Plastic hinges are expected to form at the bases of columns as part of a ductile sway mechanism (Figure 3.19 – frame A), but capacity design measures are intended to prevent their formation elsewhere in columns. For the analysis of a new structure where capacity design is required, or for an existing one where it was applied, it would therefore seem unnecessary to model potential plastic hinge points in columns, other than at their base. However, there are three reasons for expecting hinges to form elsewhere.

1   As discussed in Section 3.5, capacity design only reliably prevents yielding simultaneously at the top and bottom of a column, and a single hinge may still form at one or other end.
2   Column hinging is difficult to prevent from occurring at the top of columns both in single-storey frames and in the top storey of multi-storey frames, and capacity design rules in codes do not apply at these locations.

3　Beam elongation leads to the possibility of hinges forming at the top of ground floor columns (Figure 8.19), although simple non-linear models will not capture this effect.

It is therefore recommended that, in all cases, potential hinge points are modelled at both ends of all columns. However, except in unusual cases of significant lateral loading, mid-span hinging of columns is never expected, unlike the case for beams.

8.4.4.4　Modelling of hinges at the base of shear walls
A plastic hinge forming at the base of a slender shear wall is likely to extend over most or all of the ground floor storey height. Modelling it as a discrete hinge at the base of the wall is therefore a rather crude approximation. In order to model more accurately the effective centre of rotation, it is tentatively suggested that the hinge position should be raised by one third of the assumed hinge length, L_{pl}, which can be estimated from Equation 8.3 (see Figure 8.31, detail 4).

8.4.4.5　Limitations of simple non-linear frame models
As discussed in Section 3.8.6, a simple non-linear model consisting of linear line elements and discrete plastic hinges provides far more information on the true post-yield behaviour of a concrete superstructure and the distribution of plastic demand, than can a ductility modified response spectrum analysis. However, in many ways it still provides a crude representation of the actual complex non-linear behaviour of a reinforced concrete structure, and this should be borne in mind in interpreting and using the results. Specific issues are as follows.

- The members modelled as linear will in fact vary in stiffness with curvature, and with fluctuating axial load due to seismic loading.
- Simple models of plastic hinges assume a load rotation characteristic that is independent of the level of shear and axial load and of loading effects in the axis orthogonal to the one being considered. However, all these effects can be significant.
- Simple plastic hinge models also take no account of the variation in hinge properties with the number and amplitude of loading cycles, which again may be very significant.
- The beam elongation effects described in Section 8.2.5.3 are not included, but may become important.
- Special measures, such as are shown in Figure 8.32, are needed to model correctly the forces in beam–column joints.

More sophisticated plastic hinge models are available or alternatively non-linear 'fibre elements' or non-linear plate or brick elements can be used; further discussion of these is provided in section 4.9 of Priestley et al. (2007) but they are currently primarily used in research rather than for design.

8.5.　Design of concrete building structures
The remaining sections of this chapter discuss design considerations for concrete structures, with particular reference to the provisions of EC8 and ACI 318. Further information on EC8 is given by Fardis et al. (2005), Fardis (2006) and the EC8 manual (ISE/AFPS, 2010) and on ACI 318 by Derecho and Kianoush (2001) and Mo (2003).

8.5.1   Design levels of ductility

EC8 recognises two classes of ductility in concrete structures designed for areas of high or moderate seismicity. DCH structures may be designed for lower lateral strength, but have stringent rules for detailing and strength assessment. These rules are relaxed (sometimes substantially) in DCM structures, but at the expense of an increase in lateral strength requirement of typically 50%. Because checks are less stringent, the design effort required for DCM structures is significantly less than that for DCH, as noted in the following sections. It appears that DCH has been little used, and some of its detailing provisions are found hard to achieve in practice. A third, 'low' ductility class (DCL) is also defined in EC8, requiring the highest lateral strength but with no special seismic rules so that design to the non-seismic Eurocode 2 for concrete suffices. However, DCL structures may only resist seismic forces in areas of low seismicity. Non-seismically detailed ('secondary') elements may be used in areas of high and moderate seismicity, but they can only be used for supporting gravity loads and their contribution to lateral resistance must be neglected, as discussed in Section 8.5.2.6.

Similar classifications apply in US practice in ACI 318, although the ductility levels are classified as 'special', 'intermediate' and 'ordinary', with the stringency of requirement increasing with both the level of seismic hazard and the importance of the building's function. However, there is not a one-to-one correspondence with the Eurocode ductility classes; DCH is generally more stringent than 'special', while DCM broadly lies somewhere between 'special' and 'intermediate'. Requirements for non-seismically designed secondary elements also differ between US and Eurocode practice.

8.5.2   Design of reinforced concrete frames

8.5.2.1   Introduction

Moment-resisting concrete frames rely on the rigidity of the beam–column joints to resist lateral loads, rather than on shear walls or cross-bracing. They are sometimes called unbraced frames.

8.5.2.2   Preliminary sizing

Codes place restrictions on the range of geometries permitted in ductile frames, as shown in Table 8.4. The rationale behind the main requirements shown is as follows.

The restrictions on beam to column width ratios are to ensure a flow of moment between beams and columns without undue stress concentrations, and to harness benefit from the improvement that column compression has on the bond of beam reinforcement passing through the joint region; the restrictions effectively prohibit the use of flat slab systems as ductile frames, because they perform poorly under earthquake loading. Depth to width ratios within individual elements are restricted to prevent buckling instability. Low beam span to depth ratios are likely to result in members governed by shear rather than flexure. This will restrict their ductility unless special measures are taken, such as the provision of diagonal steel.

Section 5.4.3 in Chapter 5 set out some of the factors influencing overall frame geometry. Preliminary design using conventional force-based methods then follows on an iterative basis using an equivalent static analysis, to establish that the chosen sections can be reinforced

Table 8.4 Limiting dimensions for special moment frames to ACI 318 or DCM frames to EC8 (data taken from ACI, 2011, and CEN, 2004)

	Columns
ACI 318	Shortest c/s dimension $\geq$300 mm
	$\geq$0.4 perpendicular direction
EC8	Shortest c/s dimension $\geq$250 mm
	C/s dimension $\geq$one tenth of larger distance between point of contraflexure in column and end of column, for bending in the plane of dimension considered (unless axial forces are low)

	Beams
ACI 318	Beam clear span $\geq$four times effective depth of beam
	Beam width to depth ratio $\geq$0.3
	Beam width $\geq$250 mm $\geq$width of supporting member (on plane perpendicular to beam axis) plus distances on each side of not greater than either 3/4 of width of supporting member, or depth of supporting member, if less
EC8	Beam width $\leq(b_c + h_w)$
	$\geq 2b_c$
	Centroidal axes of beam and column must not be more than $(b_c/4)$ apart
	b_c = largest c/s dimension of column perpendicular to beam
	h_w = depth of beam

for the strength required, and that the stiffness is adequate. Often, stiffness rather than strength may govern the design of tall buildings. The process needs to be iterative because changing the stiffness of the structure changes its period of vibration, and hence the seismic loads it attracts. Displacement-based design (Section 3.6) is an alternative procedure, which establishes member sizes and reinforcement ratios on the basis of limiting storey drifts and member ductility demands to acceptable levels under the design earthquake loading.

Capacity design considerations are also important, even at the preliminary planning stage, to ensure that favourable yielding mechanisms apply. In particular, a 'strong column/weak beam' frame should be assured to prevent soft or weak storey mechanisms forming during an earthquake, and the shear strength of an element should in most cases exceed that required to develop its flexural strength.

To satisfy the 'strong column/weak beam' condition, EC8 requires the sum of design column flexural strengths to exceed 130% of the sum of beam flexural strength framing into a joint. It is difficult to prevent hinges forming in the top of columns in the top storey of multi-storey frames, because each column must resist the plastic capacity of one or two beams, as there is no column above. Codes therefore exclude top storeys from column capacity design requirements for flexure, and (for the same reason) columns in single-storey frames are also excluded. In calculating the column flexural strength, due allowance must be made for the most unfavourable axial load that may be present. EC8 alternatively allows a non-linear static (pushover) analysis to check that the hierarchy of beam and column strengths is satisfactory, and that weak storey mechanisms or other brittle failure modes are avoided. More rigorous and complex procedures are required by the New Zealand concrete code, NZS 3101 (NZS, 2006).

To some extent, these capacity design considerations for relative flexural strength and shear strength can be satisfied by adjusting the amount of reinforcement. However, preliminary design needs to ensure that the section sizes chosen are sufficient to accommodate the required reinforcement without undue congestion.

8.5.2.3 Detailing of beams and columns

In order to ensure satisfactory seismic performance, careful detailing of reinforcing bars is essential, and codes of practice provide extensive guidance. Figures 8.33 and 8.34 show typical details for beams and columns, respectively, while Table 8.5 provides a summary of ACI and Eurocode requirements for ductile members.

Figure 8.33 Detailing notes for a ductile beam: (a) closed hoops; (b) stirrups with ties; (c) leg hoops for wide beam; (d) multiple layers of flexural steel

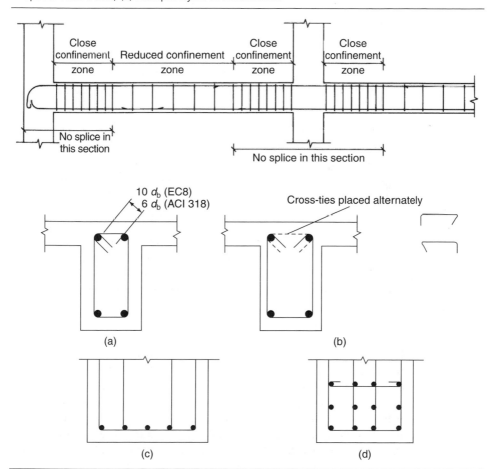

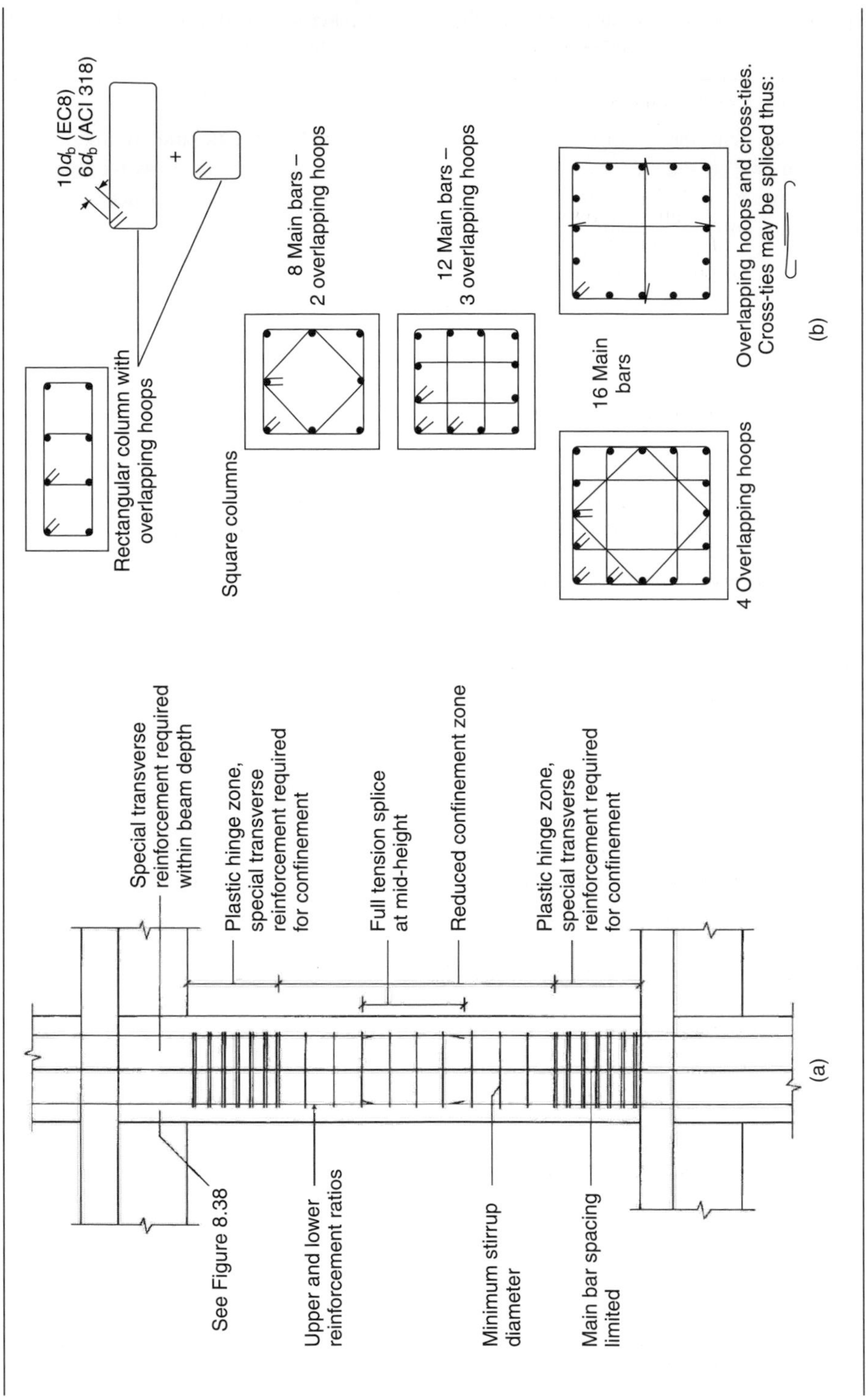

Figure 8.34 Detailing notes for a ductile column: (a) elevation; (b) sections through column

Special transverse reinforcement required within beam depth

Plastic hinge zone, special transverse reinforcement required for confinement

Full tension splice at mid-height

Reduced confinement zone

Plastic hinge zone, special transverse reinforcement required for confinement

See Figure 8.38

Upper and lower reinforcement ratios

Minimum stirrup diameter

Main bar spacing limited

(a)

$10d_b$ (EC8)
$6d_b$ (ACI 318)

Rectangular column with overlapping hoops

Square columns

8 Main bars – 2 overlapping hoops

12 Main bars – 3 overlapping hoops

16 Main bars

4 Overlapping hoops

Overlapping hoops and cross-ties. Cross-ties may be spliced thus:

(b)

Table 8.5 Detailing requirements for special moment frames to ACI 318 or DCM frames to EC8 (data taken from ACI, 2011, and CEN, 2004)

Beams: main reinforcement

ACI 318   Min area of main reinforcement top and bottom (see Figure 8.35)

$\geq (1.4 \, b_w \, d)/f_{yk}$

b_w = width of beam web (m)

d = effective depth of section (m)

f_{yk} = yield strength of main steel (MPa)

Max percentage of tension reinforcement $\leq 2.5\%$

At least 2 bars must be provided at top and bottom of the beam throughout its length

At joint faces, positive bending strength beam $\geq 50\%$ of negative bending strength

Positive and negative strength everywhere in beam $\geq 25\%$ max. bending strength at either joint

No lap joints are allowed:

(a) Within beam–column joints

(b) Within 2 times member depth from joint face

(c) Within anticipated plastic hinge zones

Laps must be confined by hoops or stirrups spaced at not more than $d/4$, or 100 mm, if less

On tension and compression faces of beam, at least alternate bars shall be laterally restrained by hoops or ties

Spacing of laterally supported bars should not exceed 350 mm

EC8   Max and min ratios of tension steel – see Figures 8.35 and 8.36

No lap joints are allowed:

(a) Within beam–column joints

(b) Within anticipated plastic hinge zones

Laps must be confined by hoops or stirrups spaced at not more than $h/4$, or 100 mm, if less, where h is the minimum cross-sectional dimension

Beams: transverse reinforcement

ACI 318   In the special confinement zone, which extends twice the member depth from the face of the supporting member, spacing of hoops must not exceed:

$d/4$

6 times diameter of smallest longitudinal bar

150 mm

Outside this zone, spacing may be relaxed to $d/2$

The first hoop not be more than 50 mm from the face of a supporting member

Hoops may consist of closed hoops, or stirrups and cross-ties (see Figure 8.33)

Hoops and stirrups must be terminated with a 135° hook, which extends 6 hoop bar diameters (or 75 mm if greater) into the confined core of the beam

EC8   In the special confinement zone, which extends the member depth from the face of the supporting member, spacing of hoops must not exceed:

Total beam depth/4

8 times diameter of smallest longitudinal bar

24 times diameter of hoop bars

225 mm

Hoops may consist of closed hoops, or stirrups and cross-ties (see Figure 8.33)

Hoops and stirrups must be terminated with a 135° hook, which extends 10 hoop bar diameters into the confined core of the beam

Table 8.5. Continued

Columns: main reinforcement

ACI 318   Ratio of main reinforcement $\geq 1\%$
$$\leq 6\%$$
Lap splices shall occur only in the centre half of the column
At least six bars in circular columns

EC8   Ratio of main reinforcement $\geq 1\%$
$$\leq 4\%$$
At least one intermediate bar shall be provided between column corners
Symmetrical sections shall be reinforced symmetrically

Columns: transverse reinforcement

ACI 318   Spacing of column hoops or spirals in special confinement zone not to exceed the least of the following, but need not be less than 100 mm

     1/4 quarter minimum dimension of column
     6 times diameter of smallest longitudinal bar
     $100 + (350 - h_x)/3$ or 150 mm, if less

where h_x is the largest horizontal spacing between hoop or cross-tie legs
$h_x \leq 350$ mm
Outside special confinement zone, spacing can be relaxed to 6 times diameter of smallest longitudinal bar or 150 mm if less
Height of special confinement zone
     $\leq$depth of column at joint face
     $\leq$1/6 of clear span of column
     $\leq$450 mm

EC8   Min diameter of hoops, ties or spirals is $0.4d_{bL}\sqrt{f_{ydL}/f_{ydw}}$ where d_{bL} is the diameter of the main column bars and f_{ydL}/f_{ydw} is the ratio of yield strength in the main bars to that in the hoops
Spacing of column hoops or spirals in special confinement zone
     $\leq$1/3 minimum dimension of confined core of column
     (to centre line of hoops or spirals)
     $\leq$6 times diameter of smallest longitudinal bar
     $\leq$125 mm
Distance between main bars engaged by hoops or cross-ties
     $\leq$150 mm
Outside special confinement zone, Eurocode 2 (ie non-seismic) rules apply
Height of special confinement zone
     $\leq$the largest c/s dimension of column
     $\leq$1/6 of clear span of column
     $\leq$450 mm
The entire column shall be treated as a special confinement zone where the ratio (clear column height)/(max column c/s dimension) is less than 3

8.5.2.4 Beam–column joints

The joint region between beams and columns in a moment-resisting frame is a highly stressed region, in which the shear stresses are many times greater than those in a frame subjected solely to gravity loads (Figure 8.37). These high shear forces lead to high concrete diagonal compressive forces, which require good confinement of the joint region to be sustainable,

Figure 8.35 EC8 and ACI318 requirements for minimum percentage of tension steel as a function of characteristic yield stress f_{yk} and concrete grade

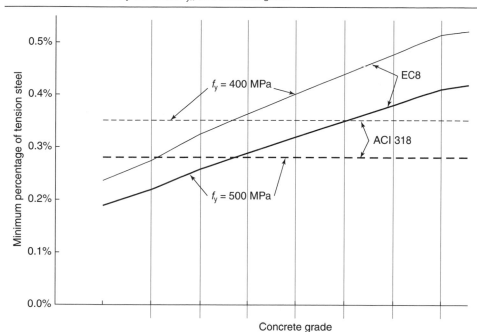

and the need for horizontal and vertical shear steel to transmit the diagonal tension. They also imply a high rate of change in bending moment and hence lead to rapid changes in the tension forces in the flexural steel. The resulting bond stresses between flexural steel and concrete in the joint zone are therefore also exceptionally high; bars passing through the joint are expected to be in full compressive yield on one side of the joint and in full tension yield on the other. This leads to both the need to restrict the diameters of such bars (because small diameter bars are more efficient in transferring their forces into the surrounding concrete) and also the need to provide good confinement to the bars, to sustain the high bond stresses that develop. Joints at the ends of beams also need special care, because the anchorage length for the beam steel on one side of the joint is restricted (Figure 8.38). A full and clear discussion of the complex transmission of forces within beam–column joints is provided by Paulay (1994).

In the rules for the design of beam–column joints, there is a clear distinction between the more rigorous approach of the New Zealand standard NZS 3101 (NZS, 2006) on the one hand and US practice, represented by ACI 318 (ACI, 2011), on the other. EC8 (CEN, 2004) stands somewhere in between, with rigorous rules for DCH structures, and much simpler ones for DCM structures. Debate on this issue is not entirely resolved. It is generally accepted that either approach should provide sufficient strength in the joint to allow plastic hinges to form in the beams, allowing the frame to develop its potential ductility. It is, however, argued by proponents of the more rigorous methods that the US approach may in

Figure 8.36 Approximate EC8 requirement for maximum value of percentage of tension steel less percentage of compression steel in a beam, as a function of q (behaviour) factor and concrete grade

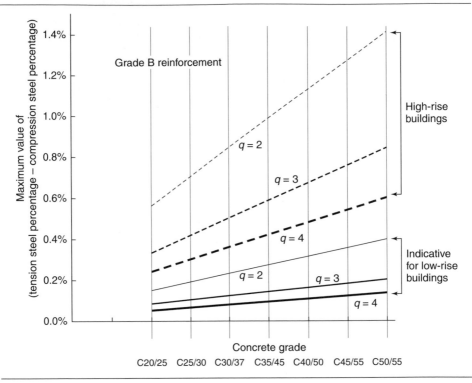

some cases lead to significant stiffness degradation in the joint under severe cyclic loading, leading to increased storey drift and associated damage.

8.5.2.5 Frames of 'low' or 'ordinary' ductility

The normal strength and detailing requirements of codes for structures designed to withstand wind and gravity loads in themselves provide some basic level of robustness and hence seismic resistance and ductility, which may be sufficient for areas of low seismicity. Such frames without seismic detailing or capacity design are recognised by EC8 as having 'low' (DCL) ductility and by ACI 318 as having 'ordinary' ductility. In EC8, they are designed for seismic forces calculated using the low behaviour factor q of 1.5 (compared with up to 5.9 for DCH frames), but are only recommended for areas of low seismicity. Similarly, ACI 318 specifies increased seismic forces for 'ordinary' frames, but only allows them in low seismicity areas.

8.5.2.6 'Secondary' frames not proportioned to resist lateral loads

Some frames may be designed to resist only the gravity loads in a building, with their contribution to lateral resistance neglected. For example, this would be the case for the internal structure in which lateral resistance is provided by a separate perimeter frame

Figure 8.37 Shear in beam–column joints: gravity and sway frames compared

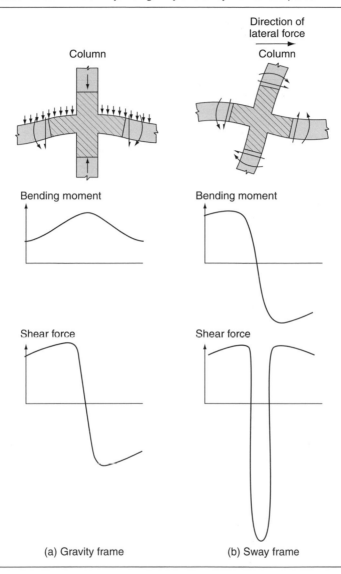

(a) Gravity frame   (b) Sway frame

(Figure 5.2(b)) or where shear walls take all the seismic loads. The design and detailing of gravity-only frames can clearly be less demanding than for the moment-resisting frames discussed so far; their only requirement is to maintain their load-carrying capacity under the maximum deflections imposed on them during an earthquake. The stiffer the seismic load-resisting system, the lower the deflection demand on these 'secondary' elements, which is one reason why shear walls generally offer good seismic protection.

Where the gravity-only members (and in particular the columns) are not expected to exceed their design flexural and shear strength under the imposed seismic deflections, less stringent

Figure 8.38 Anchorage of flexural steel in beam–column joints

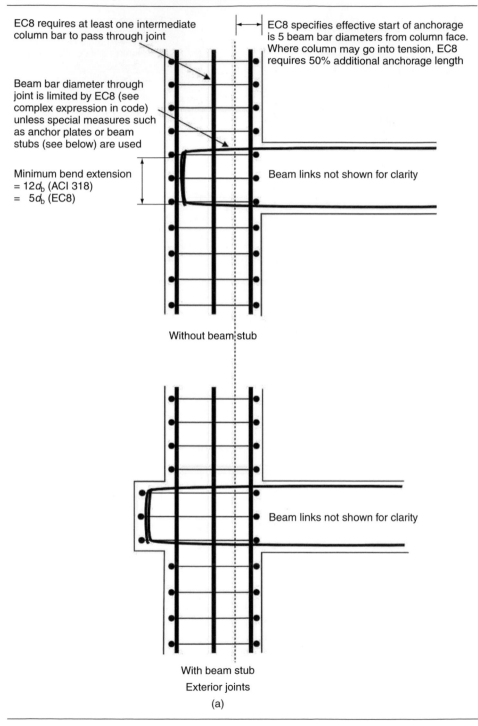

EC8 requires at least one intermediate column bar to pass through joint

EC8 specifies effective start of anchorage is 5 beam bar diameters from column face. Where column may go into tension, EC8 requires 50% additional anchorage length

Beam bar diameter through joint is limited by EC8 (see complex expression in code) unless special measures such as anchor plates or beam stubs (see below) are used

Minimum bend extension
$= 12d_b$ (ACI 318)
$= 5d_b$ (EC8)

Beam links not shown for clarity

Without beam stub

Beam links not shown for clarity

With beam stub

Exterior joints

(a)

Figure 8.38 Continued

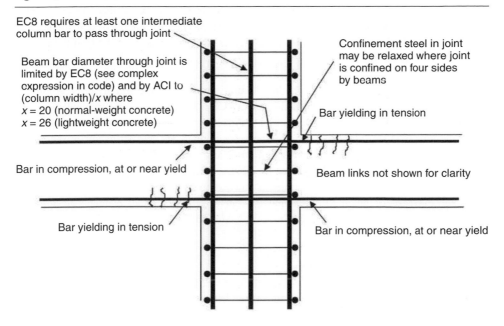

EC8 requires at least one intermediate column bar to pass through joint

Beam bar diameter through joint is limited by EC8 (see complex expression in code) and by ACI to (column width)/x where
$x = 20$ (normal-weight concrete)
$x = 26$ (lightweight concrete)

Confinement steel in joint may be relaxed where joint is confined on four sides by beams

Bar yielding in tension

Bar in compression, at or near yield

Beam links not shown for clarity

Bar yielding in tension

Bar in compression, at or near yield

Beam bars continuous through joint

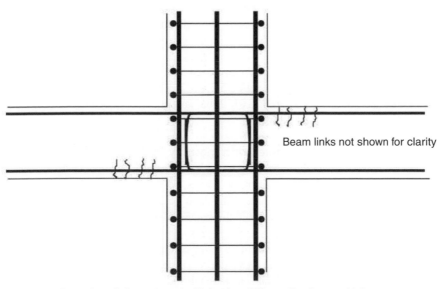

Beam links not shown for clarity

Beam bars fully anchored with hooks within confined area of joint

Interior joints

(b)

213

measures are needed. For this case, ACI 318 places some restrictions on spacing of confinement steel in columns, which increase with the level of axial load. In cases when the compressive stress due to the axial load exceeds $0.1f_c'$, a capacity design for shear is needed, whereby the shear strength must be greater than that needed to develop the flexural strength at the ends of the column.

For the more critical case in which the flexural strength of the gravity-only column is exceeded under the design seismic deflections, ACI 318 requires more stringent measures, which amount to full confinement steel as for ductile moment-resisting frames where the axial compressive stress exceeds $0.1f_c'$. The rules in EC8 are less stringent; no additional detailing beyond the non-seismic rules of Eurocode 2 are needed if the design deflections do not cause the gravity-only columns to yield, but full ductile detailing is (apparently) required if yielding does occur. Caution is needed in applying these rather relaxed EC8 rules for the case of no yield. Note that the lateral stiffness of all elements designed as 'secondary' must not exceed 15% of the total lateral stiffness of the structure, according to EC8.

8.5.2.7   Precast concrete frames

The major potential source of weakness in precast frames lies in their connections. If capacity design procedures are carried out to ensure that yielding does not occur here, the rest of the structure is to be designed to rules for cast in situ structures. Alternatively, the connections can be specially designed to yield and provide energy dissipation under extreme seismic loading (although this is more likely to be practical in precast wall systems). Both approaches are recognised by EC8, which provides detailed design rules. Design guidance is also provided by a New Zealand set of guidelines (New Zealand Concrete Society and Society for Earthquake Engineering, 1991). Precast concrete frames are permitted in ACI 318 but detailed design rules are not given.

Precast post-tensioned frames have been developed as a system designed to be self-centering after an earthquake, as described in Section 5.5.8.5. Design rules are given in an appendix to the New Zealand concrete code NZS 3101: 2006, and in an ACI document (ACI, 2003), but they are not currently referred to in EC8.

8.5.2.8   Moment-resisting frames with masonry infill panels

EC8 recognises three situations in which external frames have been infilled with masonry walls. In the first, the walls are separated from the frames, so that there is no structural interaction between them. In practice, this is quite difficult to achieve, because the walls need lateral restraint to prevent them from falling out of the frame under strong wind or earthquake loading.

In the second situation, the walls are built up to the column members but are not connected to them. It must then be ensured that no weak storeys (Figure 8.7) or short columns (Figure 8.3) are formed, and that the concrete frame can take the additional forces that the infill panel may subject it to. EC8 provides rules for this check, but the situation is not covered by ACI 318.

The third situation is when the walls are built first, and the beam–column frame is cast directly against the masonry. The system is then treated as a 'confined masonry' structure, in which all the seismic forces are considered to be resisted by the masonry, but enhanced resistance may be assumed, provided the reinforced concrete elements conform to certain minimum requirements (see Chapter 10).

In US practice, new buildings must separate the masonry infill from the concrete frame so only the first, and not the second, situation described above is allowed. Moreover, the infill masonry must be reinforced. However, existing buildings in the USA do have infill masonry that acts with the frame, and recent research has investigated its performance. Koutromanos *et al.* (2011) discuss analytical models for masonry infill, while a proposed amendment to ASCE 41 (ASCE, 2006), due for final publication in 2014, gives design rules.

8.5.3   Shear walls

8.5.3.1   Preliminary sizing

Shear walls must be thick enough to prevent buckling instability occurring under extreme seismic loading, and must also usually be able to accommodate two horizontal and two vertical layers of reinforcement. EC8 requires a minimum web thickness of 150 mm, or $(h_s/20)$ where h_s is the clear height of the wall.

In the lower part of the wall, where a plastic hinge would be expected to form, there are particularly great demands on the outer edges of the wall, which are known as boundary elements. These need to accommodate the flexural tension steel, and also confinement steel to sustain the concrete compressive strains. EC8 requires a minimum thickness of 200 mm and between $(h_s/15)$ and $(h_s/10)$ in these boundary elements, depending on their length. Evidence from shear wall failures in New Zealand presented by the report of the Canterbury Earthquakes Royal Commission (2012) suggests these thickness ratios may be too low to prevent buckling reliably.

8.5.3.2   Flexural and shear strength of slender shear walls

As with the design of beam–column joints, there is a clear distinction between simpler and more empirical US practice, and the more complex and rigorous procedures of both EC8 (particularly for DCH structures) and also the New Zealand concrete code.

US practice takes design bending and shear forces directly from analysis, without any capacity design considerations. Flexural strength is determined exactly as for beams or columns, while shear strength is based on a simple formula depending on the amount of web reinforcement, the concrete strength and the aspect ratio of the wall. Often, a shear wall is designed as contained within a beam–column frame; the frame is sized to support the tributary gravity loads without any support from the wall. The structure can then be treated in ASCE 7 as having a separate frame, rather than as being a 'bearing wall system', and attracts a more favourable structural or R factor.

There are two important differences from this US approach in EC8. First, design values of both bending moments above the base and shear forces are not taken directly from the seismic analysis, but on a capacity design approach. This is intended to ensure that the flexural hinges form only in the lower part of the wall and also that the shear strength everywhere exceeds the value needed to develop the wall's flexural strength. Figure 8.39 shows how the bending

Figure 8.39 Bending and shear force distributions from EC8 for design of shear walls

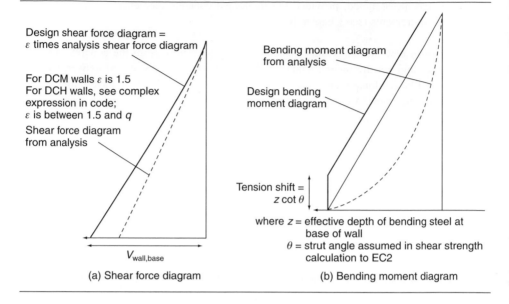

Design shear force diagram =
ε times analysis shear force diagram

For DCM walls ε is 1.5
For DCH walls, see complex
expression in code;
ε is between 1.5 and q

Shear force diagram
from analysis

$V_{wall,base}$

(a) Shear force diagram

Bending moment diagram
from analysis

Design bending
moment diagram

Tension shift =
$z \cot \theta$

where z = effective depth of bending steel at
base of wall
θ = strut angle assumed in shear strength
calculation to EC2

(b) Bending moment diagram

moment and shear force distributions obtained from analysis relate to the distributions used for checking design strength. The design shear force diagram exceeds the distribution obtained from analysis both because of the (uncertain) influence of higher mode effects (which are relatively much more important, compared with bending moments). It also allows for the more classic capacity design consideration, that the flexural strength provided at the base of the wall will in general exceed the analysis value, hence allowing correspondingly large shear forces to develop. The 'tension shift' in the bending moment diagram arises purely from static considerations. As shown in Figure 8.40, the force in the flexural steel at the top of a diagonal crack is more closely related to the bending moment at the bottom of the same crack, giving rise to the tension shift in Figure 8.39. The straightening of the bending moment diagram allows (roughly) for higher mode effects. In the New Zealand code (but not the Eurocode, except for squat DCH shear walls), the bending strength in upper sections of the wall must be further increased to allow for the flexural strength actually provided at the base, which will in general exceed the value from the analysis.

Different requirements apply to frame–wall or dual systems (where shear walls combine with frames to provide lateral resistance – see Figure 8.44) because of the complex interaction between frames and walls. A discussion of these factors is given in section 2.2.3 of Fardis (2006).

8.5.3.3 Boundary elements

Boundary elements are needed to sustain the highly stressed edges of shear walls in plastic hinge regions. When the concrete strains exceed around 0.35% under the design seismic loads, confinement steel is required, similar to that specified for the critical regions of columns and taking the form of closed horizontal loops or horizontal cross-ties. Both US

Figure 8.40 Tension shift mechanism in shear walls

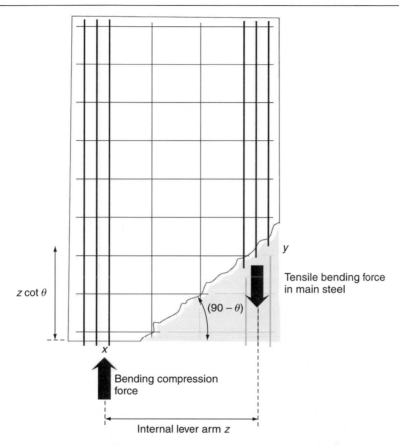

The figure shows the free body diagram above a main shear crack at the base of a shear wall. Neglecting the contribution of the web steel crossing the crack to the bending moment at the base and also of the concrete interfaces in the crack (both of which are relatively small), it can be seen that the bending force in the main steel at the top of the crack at level y is most closely related to the bending moment at the bottom of the wall level x. The difference between levels x and y is effectively the 'tension shift' shown in Figure 8.39.

and Eurocode practice are clearly based on this principle, but once again the execution is rather different. For DCH structures, EC8 requires a 'first principles' approach in which the concrete strains are determined from a 'plain sections remain plane' analysis for the local ductility demand determined from the behaviour factor q used in design, and the ratio of flexural strength provided to that obtained from analysis. Appropriate confinement steel is then specified in the region of the wall where the concrete strain exceeds 0.35%. The same rules apply to DCM, but no confinement steel is required where the axial load due to gravity is not excessive, and where a slightly reduced q factor has been assumed to lower the ductility demand.

ACI 318 also allows a 'first principles' approach but in addition gives a single closed form equation to determine whether or not confinement steel is needed, resulting in a much simpler calculation.

8.5.3.4 Squat shear walls

As discussed in Section 8.2.9, squat shear walls with a height to length ratio of less than approximately 2 behave in a rather different way to slender walls. Ductility is limited because failure is more likely in shear than flexure, and this is reflected in lower q factors in EC8 and R factors in ACI 18. Diagonal reinforcement may be needed to resist high levels of shear; for DCH structures, EC8 requires 50% of the shear at the base of a squat shear wall to be resisted by diagonal bars, but this is not specified for DCM structures.

8.5.3.5 Openings in shear walls

Functional reasons often necessitate openings in shear walls. These will affect the flow of forces through the walls, and hence their strength. If they are large enough and placed in critical locations, they can form potential failure triggers (Figure 8.41).

In the case of regular vertical arrays of openings, this can be turned to the designer's advantage by the formation of coupled shear walls (Figure 5.11). The most detailed rules for the design of coupled shear walls, including capacity design requirements, are provided in the New Zealand code NZS 3101 (NZS, 2006), although they are referred to in both EC8 and ACI 318. The coupling beams will usually require diagonal reinforcement, and the resulting eight layers of reinforcement result in a minimum practical thickness of 300 mm.

Figure 8.41 Damage around opening in a shear wall, 1985 Chile earthquake (© E Booth)

Code advice for the design of openings in other cases is limited. In a ductile shear wall, the objective is to ensure that the weakening associated with the introduction of the shear wall does not reduce shear and bending strength locally below the level corresponding to plastic hinge formation at the base of the wall. Local flexural strength can be determined on a plane sections remain plane assumption, while ACI 318 advises that the overall shear strength of the wall at a cross-section with openings can be taken as the sum of contributions from individual 'wall piers' between openings. More satisfactorily, a 'strut and tie' approach may be used, as described in section 5.7.7 of Paulay and Priestley (1992), and the method is discussed more generally in appendix A of ACI 318 and appendix B of NZS 3101 (NZS, 2006). It is unlikely that significant openings can be accommodated near the outside edges of shear walls in the highly stressed compressive region of the plastic hinge without reducing the available ductility.

8.5.3.6 Large panel precast buildings

The assembly of medium to high-rise buildings from precast storey-height concrete wall panels offers the advantage of rapid site construction and casting of concrete under well controlled factory conditions. It was extensively used in western Europe in the 1960s and 1970s and more recently in eastern Europe. A rather different system, called 'tilt-up' construction is found in the USA and elsewhere, primarily for warehouses, where the walls to low-rise buildings are cast in a horizontal position on site, and are then tilted up once cured to form the sides of the building.

The engineering problem with such construction lies in making adequate connections between the precast units; the potential for loss of a panel leading to progressive collapse has been recognised at least since the collapse of the Ronan Point building in England after a gas explosion (Griffiths *et al.*, 1968). However, the record of large panel buildings in the 1978 Bucharest earthquake (Bouwkamp, 1985) and 1988 Armenian earthquake (Wyllie and Filson, 1989) is good, despite some very poor construction. Tilt-up buildings in California have proved more vulnerable, often failing at the connection between roof and wall (Figure 8.42).

Four main approaches may be used in the design of precast wall systems, as described by Mattock (1981).

1 Ductile cantilever shear walls, which dissipate energy by plastic hinge formation at the wall bases, exactly as for a monolithic structure. In this case, horizontal joints must be strong enough to transmit moments and shears, determined on a capacity design basis, and vertical joints must be capable of transmitting a similar level of shear.

2 Energy dissipating connections, in which some of the connections are treated as ductile or semi-ductile fuses, limiting the forces on the other connections and the rest of the structure. Design rules for self-centering precast walls systems, (see Figure 5.15 in Section 5.5.8.4) are given in an appendix to the New Zealand concrete code NZS 3101: 2006, and in two ACI documents (ACI, 2007b, 2009) but are not currently referred to in EC8.

3 Elastically responding structures, in which sufficient strength is provided to prevent inelastic response in either structure or connections under the most severe anticipated

Figure 8.42 Failure of tilt-up warehouse building, Loma Prieta earthquake, California, 1989 (courtesy ABS Consulting, photograph from Peter Yanev, EQE, San Francisco)

ground motions. The use of base isolation may be an alternative means of limiting structural response to within elastic limits.

4   Energy dissipating panels, in which special panels are introduced that absorb energy under severe seismic loading. This type of system has been developed and used in Japan.

EC8 provides advice on the first three of these approaches. Hamburger *et al.* (1988) gives guidance on the design of tilt-up buildings.

8.5.4   Concrete floor and roof diaphragm

8.5.4.1   Structural functions of diaphragms

The floors and roof of a building, in addition to resisting gravity loads, are also generally designed to act as diaphragms. In this respect, they are required both to distribute seismic forces to the main elements of horizontal resistance, such as frames and shear walls, and also to tie the structure together so that it acts as a single entity during an earthquake. The robustness and redundancy of a structure is highly dependent on the performance of the diaphragms. Wyllie and Filson (1989) have suggested that the inadequacy of the floor diaphragms played an important part in the catastrophic performance of precast concrete buildings in the 1988 Armenian earthquake.

The seismic forces in a diaphragm may arise from two distinct causes, namely 'local' or 'transfer', and it is important to distinguish between the two, as follows.

Figure 8.43 Transfer diaphragm in a tower and podium building

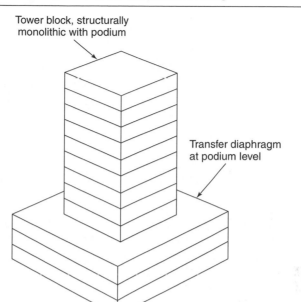

Tower block, structurally
monolithic with podium

Transfer diaphragm
at podium level

1 Local forces are those arising from the transfer of inertial loads arising at the level of
the diaphragm, which need to be taken back to the main horizontal load-resisting
structure.

2 Diaphragms may also be required to resist transfer forces that arise where there is an
offset in the horizontal load-resisting structure, for example, at the transition level
between a tower and podium (Figure 8.43). Transfer forces may also arise in the case
of a building in which the lateral resistance is provided by both shear walls and frames.
The transfer forces in the diaphragms arise from tying the walls and frames to describe
the same vertical deflected shape, which separately would be different (Figure 8.44).

As noted in Section 8.2.5.4, beam elongation may impose significant tensile strains in floor
diaphragms, particularly in the case of in situ toppings to precast floors, which may severely
reduce their shear strength, unless adequate reinforcement is provided. The New Zealand
standard NZS 3101: 2006 provides advice, but US and European codes do not.

8.5.4.2 Preliminary sizing of diaphragms

In-situ diaphragms are unlikely to be governed in size by seismic forces. However, in precast
floor construction, seismic forces are usually transferred back to the lateral load-resisting
structure solely through the in-situ topping on the precast units, which may therefore be
highly stressed. ACI 318 requires the topping to be at least 50 mm thick, with 75 mm thickness required at the edges in some cases. EC8 specifies a minimum thickness of 70 mm and
minimum reinforcement in two directions. Precast floors without an in-situ topping are not
recommended in seismic areas.

Figure 8.44 Transfer diaphragms in a frame wall or dual system building

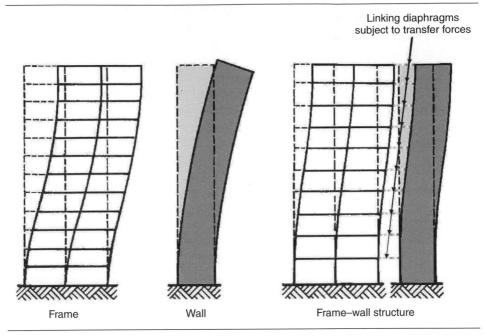

Linking diaphragms
subject to transfer forces

Frame   Wall   Frame–wall structure

8.5.4.3 Capacity design considerations

In a ductile structure, diaphragms will almost always be required to remain elastic, so that they can sustain their function of transferring forces to the main lateral resisting structure, and tying the building together. Diaphragms should in principle therefore have the strength to sustain the maximum forces that may be induced in them by the chosen yielding mechanism within the rest of the structure.

EC8 deals with this rather simply by specifying that diaphragms should be designed for 1.3 times the shear forces obtained directly from the analysis. ASCE 7 (ASCE, 2010) also specifies diaphragm forces that are generally greater than the analysis forces.

8.5.4.4 Diaphragm flexibility

Usually, the seismic analysis of buildings is carried out on the assumption that deflections in the diaphragms are so small compared with those in the main lateral load-resisting structure that the diaphragms can be treated as rigid. In most cases, this is quite satisfactory, because usually diaphragm flexibility affects neither overall structural stiffness (and hence natural period) nor the distribution of forces within a structure. Moreover, during a major earthquake, in ductile structures in which the diaphragms are designed to remain essentially elastic, the superstructure deflections are likely to include large plastic deformations, increasing the disparity still further. However, diaphragm flexibility can be important in two cases, as follows.

1 In structures with 'transfer' diaphragms, where the flexibility can significantly affect the distribution of load between lateral resisting systems.

2   When considering serviceability limit states in buildings with relatively flexible diaphragms, because the lateral load-resisting system would be expected to remain elastic, and so remain with a comparable stiffness to the diaphragms.

8.5.4.5   Local and transfer forces

Local diaphragm forces are likely to be appreciably greater than the code-described equivalent static forces at each level, because the latter reflect the change in peak shear force at each level, whereas higher-mode effects may give rise to accelerations causing greater local forces. The special procedures for assessing diaphragm forces given in ASCE 7 recognise this effect, although it is not considered in EC8.

Transfer forces should in principle be based directly on capacity design considerations, based on the as-built strength of the potential yielding zones of the structure, with due allowance for strain hardening in steel. However, neither ASCE 7 nor EC8 requires this, although the New Zealand concrete code NZS 3101(NZS, 2006) does.

ASCE 7 recognises the difference between local and transfer forces and requires them to be added together for the purposes of design. This is likely to be conservative, because they arise from different modes of vibration within the structure, and a square root sum square combination is more likely to be appropriate. EC8 provides no advice in this respect.

8.5.4.6   Strength of diaphragms

Both ACI 318 and EC8 allow diaphragm strength to be assessed as a deep beam or by 'strut-and-tie' methods, whereby the tie forces are taken in the reinforcement and the concrete provides compression struts (see Section 5.11 of Fardis *et al.*, 2005 and also appendix A of ACI 318). However, the requirements of EC8 for diaphragms in DCM structures are not entirely clear and it appears providing a minimum thickness and reinforcement ratio without an explicit strength check may satisfy the rules, although those for DCH structures are more stringent. The EC8 manual (ISE/AFPS, 2010) recommends the following

> 'Although not required by EC8 for DCM, it is recommended that a formal verification of concrete floors should be performed under the same circumstances specified by EC8 for DCH structures, as follows
> - irregular geometries or divided shapes in plan, diaphragms with recesses and re-entrances
> - irregular and large openings in the diaphragm
> - irregular distribution of masses and/or stiffnesses (as e.g. in the case of set-backs or off-sets)
> - basements with walls located only in part of the perimeter or only in part of the ground floor area.'

Other conditions in which an explicit strength check on diaphragms is advised are

- Diaphragms formed of precast slabs with a thin in-situ concrete topping. ACI 318 advises a minimum topping thickness of 50 mm, or 65 mm if the topping is not designed compositely with the precast elements.

- Precast diaphragms supported on beams expected to experience significant elongation during the design earthquake (see Section 8.2.5.3).

REFERENCES

ACI (2003) ITG-1.2-03 and ITG-1.2R-03: Special hybrid moment frames composed of discretely jointed precast and post-tensioned concrete members and commentary. American Concrete Institute, Detroit, MI, USA.

ACI (2007a) ITG-4.3R-07: report on structural design and detailing for high strength concrete in moderate to high seismic applications. American Concrete Institute, Detroit, MI, USA.

ACI (2007b) ITG-5.1-07: Acceptance criteria for special unbonded post-tensioned precast structural walls based on validation testing. American Concrete Institute, Detroit, MI, USA.

ACI (2009) ITG-5.2-09: Requirements for design of a special unbonded post-tensioned precast shear wall satisfying ACI ITG-5.1. American Concrete Institute, Detroit, MI, USA.

ACI (2011) ACI 318-11/318R-11: Building code requirements for structural concrete and commentary. American Concrete Institute, Detroit, MI, USA.

ASCE (1998) ASCE 4-98: Seismic analysis for safety-related nuclear structures. American Society of Civil Engineers, Reston, VA, USA. NB. A revised version is due to be published in 2013.

ASCE (2006) ASCE/SEI 41-06: Seismic rehabilitation of buildings. American Society of Civil Engineers, Reston. VA, USA. New edition expected 2014.

ASCE (2010) ASCE/SEI 7-10: Minimum design loads for buildings and other structures. American Society of Civil Engineers, Reston, VA, USA.

Bouwkamp J (1985) *Building construction under seismic conditions in the Balkan regions.* Vol 2: Design and construction of prefabricated reinforced concrete building systems. UNDP/UNIDO project RER/79/1015, United Nations Industrial Development Organization, Vienna, Austria.

Canterbury Earthquakes Royal Commission (2012) Final Report Volume 2: The performance of Christchurch CBD buildings. See http://canterbury.royalcommission.govt.nz/Final-Report – Volumes-1-2-and-3.

CEN (2004) EN 1998-1: 2004: Design of structures for earthquake resistance. Part 1: General rules, seismic actions and rules for buildings. European Committee for Standardisation, Brussels, Belgium.

Derecho AT and Kianoush MR (2001) Seismic design of reinforced concrete structures. In *The seismic design handbook* (Naeim F (ed.)). Kluwer Academic Publishers, Boston, USA.

Fardis M (2006) *Design and retrofit of reinforced concrete buildings to Eurocode 8.* Springer.

Fardis M, Carvalho E, Elnashai A *et al.* (2005) *Designers' guide to EN 1998-1 and EN 1998-5.* Thomas Telford, London, UK.

Fenwick RC (1994) The behaviour of reinforced concrete under cyclic loading. In *Concrete structures in earthquake regions.* (Booth E (ed.)). Longman Scientific and Technical, Harlow, UK.

Fenwick RC and Megget LM (1993) Elongation and load deflection characteristics of reinforced concrete members containing plastic hinges. *Bulletin of the New Zealand National Society for Earthquake Engineering* **26(1)**: 28–41.

Fenwick RC, Tankat AT and Thom CW (1981) *Deformation of reinforced concrete beams subjected to inelastic cyclic loading – experimental results.* University of Auckland School of Engineering, Report no. 374.

Goodsir WJ (1985) *The design of coupled frame-wall structures for seismic actions.* Department of Civil Engineering, University of Canterbury, Research report no. 85-8.

Griffiths H, Pugsley A and Saunders O (1968) *Report of the inquiry into the collapse of flats at Ronan Point, Canning Town.* HMSO, London, UK.

Hamburger RO, McCormick DL and Hom S (1988) Performance of tilt-up buildings in the 1987 Whittier Narrows earthquake. *Earthquake Spectra* **4(2)**: 219–254.

ISE/AFPS (Institution of Structural Engineers/Association Française du Génie Parasismique) (2010) *Manual for the design of steel and concrete buildings to Eurocode 8*. Institution of Structural Engineers, London, UK.

Koutromanos I, Stavridis A, Shing PB and Williams K (2011) Numeric modeling of masonry-infilled RC frames subjected to seismic loads. *Computer and Structures* **89**: 1026–1037.

Kowalsky MJ and Dwairi HM (2003) SP218-03: Review of parameters influencing the seismic design of lightweight concrete structures. American Concrete Institute, Detroit, MI, USA.

Lowes LN, Mitra N and Altoontash A (2004) *A beam–column joint model for simulating the earthquake response of reinforced concrete frames*. PEER 2003/10 Pacific Earthquake Engineering Research Center. See http://peer.berkeley.edu/publications/peer_reports/reports_2003/0310.pdf.

Mander JB, Priestley MJN and Park R (1988) Theoretical stress–strain model for confined concrete. *ASCE Journal of Structural Engineering* **114(8)**: 1804–1826.

Mattock AH (1981) A survey of precast wall systems. In *ATC8: Proceedings of a workshop on design of prefabricated buildings for earthquake loads*. Mayes R (principal investigator). Applied Technology Council, Redwood City, CA, USA.

Mo Y (2003) Reinforced concrete structures. In *Earthquake Engineering Handbook* (Chen W-F and Scawthorn C (eds)). CRC Press, Boca Raton, FL, USA.

Neville AM (1995) *Properties of concrete* (4th edn). Longman, Harlow, UK.

New Zealand Concrete Society and Society for Earthquake Engineering (1991) Guidance for the use of structural precast frames concrete in buildings. The Societies, Wellington, New Zealand.

NZS (New Zealand Standards) (2006) NZS 3101: Part 1: The design of concrete structures. NZS, Wellington, New Zealand.

Paulay T (1994) *The fourth Mallet–Milne lecture: Simplicity and confidence in seismic design*. Wiley, Chichester, UK.

Paulay T and Priestley MJN (1992) *Seismic design of reinforced concrete and masonry buildings*. Wiley-Interscience, New York, USA.

Peng B, Dhakal RP, Fenwick RC, Carr AJ and Bull DK (2007) Analytical model on beam elongation with the reinforced concrete plastic hinges. New Zealand Society for Earthquake Engineering conference paper. See http://hdl.handle.net/10092/4217.

Priestley MJN (1998) Brief comments on elastic flexibility of reinforced concrete and significance to seismic design. *Bulletin of the New Zealand National Society for Earthquake Engineering* **31(2)**: 73–85.

Priestley MJN, Calvi GM and Kowalsky MJ (2007) *Displacement-based seismic design of structures*. IUSS Press, Pavia, Italy.

Sivaselvan M and Reinhorn AM (1999) *Hysteretic Models For Cyclic Behavior of Deteriorating Inelastic Structures*, Report MCEER-99-0018, MCEER/SUNY/Buffalo, USA.

Tjokrodimuljo K (1985) *Behaviour of reinforced concrete under cyclic loading*. University of Auckland School of Engineering, Report no. 374.

Westenenk B, de la Llera JC, Jünemann R et al. (2013) Analysis and interpretation of the seismic response of RC buildings in Concepción during the February 27, 2010, Chile Earthquake. *Bulletin of Earthquake Engineering* **11(1)**: 69–91.

Wyllie L and Filson J (eds) (1989) Armenian earthquake report. *Earthquake Spectra* special supplement.

Earthquake Design Practice for Buildings
ISBN 978-0-7277-5794-4

ICE Publishing: All rights reserved
http://dx.doi.org/10.1680/edpb.57944.227

Chapter 9
Steelwork design

Many practising engineers have believed for years, albeit incorrectly, that steel structures were immune to earthquake-induced damage as a consequence of the material's inherent ductile properties. … However (recent earthquakes have) confirmed research findings that material ductility alone is not a guarantee of ductile structural behaviour.

Bruneau, Uang and Whittaker (1988)

This chapter covers the following topics.

- The lessons from earthquake damage.
- The behaviour of steel members under cyclic loading.
- Ductility in steelwork, and how to achieve it.
- Material specification.
- Special considerations for analysis.
- Design and detailing of braced frames.
- Design and detailing of moment-resisting frames.

Steel appears to be a much more straightforward material than reinforced concrete to design for earthquake resistance; it consists of only one component and is highly ductile. However, it is much more prone to buckling than concrete; moreover, welding it may induce brittle fracture. Consequently, current seismic design rules for steel are no less complex than those for concrete. This chapter does not set out to reproduce these rules but to make the designer aware of the principal issues which lie behind them.

9.1. Introduction

Structural steel is in many ways an ideal material for earthquake resistance, possessing high material ductility and high strength to mass ratio. However, considerable care is needed in the design and detailing of steel structures in order to ensure that a ductile end result is achieved under the conditions of extreme cyclic loading experienced during an earthquake. Special attention is needed in the design of connections (particularly welded connections) joining members intended to yield, and to compression struts intended to buckle during the design earthquake. Also, in general, steel structures tend to be more flexible than equivalent concrete structures and, unless controlled, the resulting larger displacements may lead to higher levels of damage to non-structural components and to more significant P–delta effects.

9.2.   Lessons learned from earthquake damage

The collapse of the 21-storey Pinot Suarez building in the 1985 Mexican earthquake (Figure 9.1) was the first example of failure in a major modern welded steel building in an earthquake (EEFIT, 1986). Collapse appears to have been triggered by compression failure of the welded steel plate box columns of the braced central core. Despite this high-profile failure of a seismically designed building dating from 1971, Yanev *et al.* (1991) were able to conclude in 1991, on the basis of studying steel building performance in 11 earthquakes between 1964 and 1990, that

> Buildings of structural steel have performed excellently and better than any other type of substantial construction in protecting life safety and minimizing business interruption due to earthquake induced damage. The superior performance of steel buildings, as compared to buildings of other construction, is evident even in structures that have not been specifically designed for seismic resistance.

However, subsequent events have forced this optimistic view to be modified. Although there are undoubted intrinsic advantages in using steel in earthquake country, the widespread failures in steel buildings that occurred in the 1994 Northridge (California) and 1995 Kobe (Japan) earthquakes showed that steel is far from immune, seismically. It became evident that at least some of the apparently good performance of steel buildings was due to the fact that very few had been severely tested in an earthquake before 1994. Elnashai *et al.* (1995) concluded after the Kobe earthquake

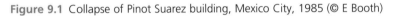

Figure 9.1 Collapse of Pinot Suarez building, Mexico City, 1985 (© E Booth)

Figure 9.2 Pre-Northridge beam to column moment connection

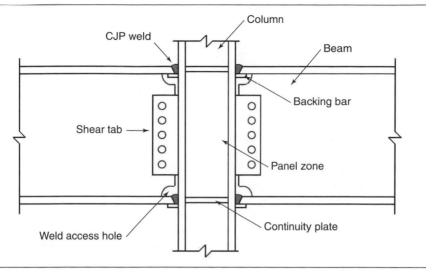

'The behaviour of steel structures was on the whole disappointing. It confirmed the serious doubts raised in the Northridge earthquake regarding the adequacy of existing design guidance. It will take very considerable efforts to establish the causes of the observed damage patterns. It will take even longer to regain confidence in steel as the primary seismic resistance material, if at all.'

No steel buildings collapsed in the 1994 Northridge California earthquake, and it at first appeared that the many tall steel buildings in the epicentral area had been undamaged. However, after cladding and other architectural finishes were removed to reveal the structural steel underneath, approximately 100 modern steel moment frame buildings were found to have suffered severe cracks. The problem occurred in the moment-resisting connections between beams and columns; the detail that proved vulnerable consisted of bolting the beam webs to shear tabs and then welding the beam flanges directly to the column flanges (Figure 9.2). This detail was recommended in the then current US seismic code, and was widely used; it was subject to brittle fracture in the connections between beam and column flanges, initiating in the welds and associated heat-affected zones. Predominantly, it was the bottom beam flange connections that were affected (Figure 9.3). Subsequently, the removal of finishes in some buildings in the San Francisco area affected by the earlier 1989 Loma Prieta earthquake revealed they had also suffered similar damage.

Engelhardt and Sabol (1996) concluded

Based on the available evidence, no single factor has been isolated as the sole cause of the damage. Rather, it appears that a number of interrelated factors combined to cause the non-ductile failures of steel moment connections in the Northridge earthquake. Both welding related factors and a poor connection design appear to be the foremost among contributing factors. Problems with the welds included the use of low

Figure 9.3 Brittle failure of steel structure in the Northridge California earthquake of 1994 (courtesy of ABS Consulting, California)

toughness filler metals combined with the presence of notches caused by welding defects and left-in-place backing bars. The basic connection design also contributed to the failures by generating excessively high stresses in the region of the beam flange groove welds. In addition to welding and design deficiencies, several other factors have been conjectured as playing some role in the failures. These include base metal factors, scale effects, composite floor slab effects and strain rate effects.

The failures in Kobe the following year were also associated with brittle fracture around welds, but the failure modes were more diverse and the construction details were different. Unlike US practice, columns (often tubular rather than flanged sections, as in the USA) were prefabricated with stub-beams to form 'column trees' and the frames were then assembled on site by bolting the remaining parts of the beams to the stub-beams. Thus, the welded beam–column connections were shop welded under conditions of high quality control, and the bolted site connections occurred away from the area of highest stress and without the potential for brittle fracture introduced by welding. Bruneau *et al.* (2011) observe that, unlike the case of Northridge, the most recently constructed steel buildings in Kobe generally performed relatively well; however, they report that in a survey of 998 modern steel buildings, 332 were found to be severely damaged, with 90 collapses and 113 buildings in which the beam–column connection was damaged.

Elnashai *et al.* (1995) quote a report by the Architectural Institute of Japan, which specifies the main damage patterns observed in Kobe as follows.

- Cracking at beam-to-column connections (very high incidence rate, up to 70%).
- Complete severance of members near the weld access hole.
- Severe damage or failure of column bases (101 out of 218 buildings inspected!).

Figure 9.4 Fracture in eccentrically braced frame link in Christchurch, New Zealand, 2011 2011 (reproduced from Clifton *et al.* (2011); New Zealand Society for Earthquake Engineering Inc.)

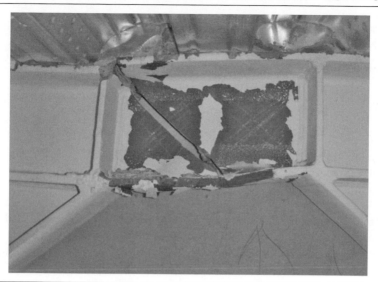

- In a few cases, beam hinging was observed.
- Fracture at the location of internal stiffeners.
- Buckling of members and collapse at connections of tubular steel frames.
- Fracture and overall buckling of slender bracing members.

Since the Northridge and Kobe earthquakes, there has been greater emphasis on improved weld filler material and weld detailing to remove stress concentrations, and also on designs that move the plastic hinging regions away from the immediate vicinity of welds; for example, by reinforcement at the connections. Design solutions are discussed in greater detail later in this chapter.

Subsequent earthquakes had not, at the time of writing, shaken large collections of modern steel buildings to their design level or beyond, as they had been in Loma Prieta, Northridge and Kobe. In the 2011 Tohoku, Japan earthquake, the ground shaking caused little structural damage to steel buildings designed after the 1981 Japanese code revisions following Kobe, although there was extensive non-structural damage, and the massive tsunami following the earthquake also cause a lot of damage (Midorikawa *et al.*, 2012). Only a few steel buildings were strongly shaken by the 2011 Christchurch, New Zealand earthquake and none collapsed, but links in two eccentrically braced frame (EBF) (see Section 9.6.3) buildings were found to be fractured (Figure 9.4).

9.3. The behaviour of steelwork members under cyclic loading
9.3.1 Introduction
Steel is a highly ductile material, and can achieve tensile and compressive strains of many times the yield strain (typically around 0.2% for high yield steel) before fracturing. Two

phenomena, however, may drastically limit the ductility that can be achieved in practical structures subjected to reversing load cycles well into the plastic range.

First, fractures may develop from points of stress concentration, predominantly (but not exclusively) in the heat-affected zone next to welds, where the material ductility may have been reduced. This is the phenomenon of low cycle fatigue, and led to many of the failures at Northridge and Kobe referred to above.

Second, buckling under compressive stress may reduce failure strength to well below that corresponding to yield. The effect is particularly marked for reversing loads because of the Bauschinger effect. This describes a fundamental property of steel, whereby its stiffness is reduced under loading in one direction if it has previously yielded due to loading in the opposite direction. As buckling is governed by member stiffness, the buckling strength of a member is reduced if it has previously yielded in tension, and each successive cycle reduces the stiffness, and so buckling strength, still further. This affects not only the overall buckling of a bracing member acting alternately as a tie and then a strut in successive loading cycles, but also local flange buckling of a plastic hinge under reversing moments.

These two phenomena therefore have the potential both to reduce the ductility of a steel structure, and to degrade its stiffness under successive load cycles.

9.3.2 Cyclic loading of struts

The behaviour of struts under reversing loads depends greatly on the slenderness ratio of the strut – that is, the ratio l/r_y of its effective length to the radius of gyration along its weak axis.

A stocky strut is one in which yielding and local buckling dominate response. The maximum compressive load a stocky brace can sustain occurs when it yields in compression over its entire cross-section or suffers a local buckling failure. Typically, a stocky strut in grade 50 steel has a slenderness ratio of 50 or less.

Figure 9.5 shows the response of a stocky strut to reversing loads. The compressive strength is somewhat less than the tensile strength, and reduces to some extent with successive cycles (by about a third in four cycles for the strut of Figure 9.5), but there is very little reduction in stiffness.

A slender strut under compression is dominated by elastic buckling and will fail at a compressive load much less than its tensile strength. Typically, a slender strut has a slenderness ratio of 120 or more. Figure 9.6 shows the response of a slender strut to reversing loads. In contrast to the stocky strut, there is a large loss in compressive strength with successive cycles (over half in four cycles for the strut of Figure 9.5) and a huge reduction in stiffness. The loss of both strength and stiffness on the compression cycle are very significant in the seismic response of braced frames, especially for V-braced and other frames that rely on both compression and tension members; this aspect is discussed in more detail in Section 9.6.2.

The loss of strength and stiffness in slender struts is due both to the Bauschinger effect, described in Section 9.3.1, and to the strut becoming increasingly bent, even in the tension

Figure 9.5 Response of a stocky strut under cyclic loading (reproduced, with permission, from Jain *et al.*, 1978)

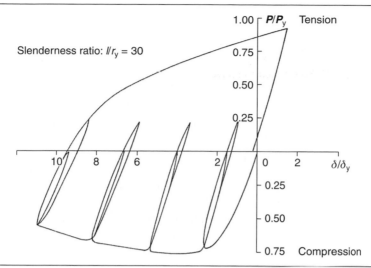

part of the loading cycle. This arises because of the following. Consider a slender strut subjected to cyclic loading sufficient to buckle the strut. In the compression cycle the strut will form a plastic hinge near mid-span, and plastic hinge rotation will remain when the compressive force is removed (Figure 9.7). Applying a tensile axial force can never completely remove this plastic rotation, even if the strut yields in tension. This is because a moment equal to the plastic hinge strength must be applied at the middle of the strut in order to reverse the hinge's plastic rotation. The residual deformation cannot therefore become less than the plastic hinge strength divided by the tensile yield strength for a pin-ended strut

Figure 9.6 Response of a slender strut under cyclic loading (reproduced, with permission, from Jain *et al.*, 1978)

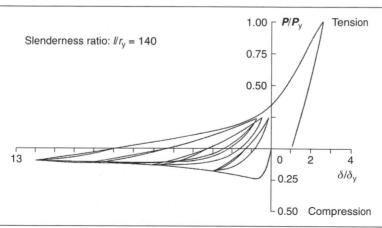

Figure 9.7 Residual deformations forming on cyclic loading of a slender strut

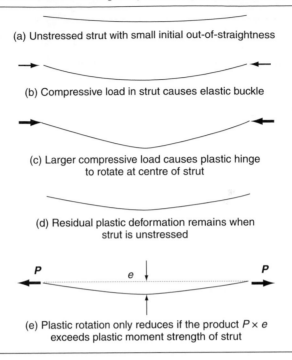

(a) Unstressed strut with small initial out-of-straightness

(b) Compressive load in strut causes elastic buckle

(c) Larger compressive load causes plastic hinge
to rotate at centre of strut

(d) Residual plastic deformation remains when
strut is unstressed

(e) Plastic rotation only reduces if the product $P \times e$
exceeds plastic moment strength of strut

(Figure 9.7). In many cases, the maximum tensile force on the strut will be much less than yield, and the residual plastic rotation of the hinge in the strut will increase with each loading cycle. This will successively reduce both compression stiffness and strength.

Struts of intermediate slenderness (slenderness ratio between about 50 and 120 for grade 50 steel) respond in compression primarily by plastic buckling; that is, a plastic hinge forms soon after elastic buckling starts. The initial compressive strength is significantly less than the tensile strength, and reduces with successive cycles, as does the stiffness, but to a much lesser extent than is the case for slender struts. Both the Bauschinger effect and residual deformations apply, but there is a smaller tendency for non-recoverable plastic residual deformations to develop, because the elastic stage of buckling produces relatively smaller lateral deformations (Figure 9.8).

9.3.3 Cyclic loading in flexure
Assuming that premature weld failure does not occur, the cyclic behaviour under flexural loading is controlled either by local flange or web buckling or by lateral torsional buckling.

The onset of local flange buckling (Figure 9.9) is governed by the local slenderness of the flange, which is one of the determinants of the 'compactness' of the section. (The other determinant of compactness is the web slenderness). Eurocode 3 recognises four ranges of compactness, which depend on the flange breadth to thickness ratio (b_f/t_f) and steel yield stress f_y; AISC 341-10 (AISC, 2010a) in the USA has similar classifications.

Figure 9.8 Response of a strut with intermediate slenderness under cyclic loading (reproduced, with permission, from Jain *et al.*, 1978)

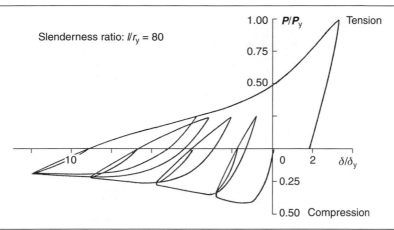

1 Class 1: Plastic cross-sections that can form a plastic hinge with significant rotation capacity.
2 Class 2: Compact cross-sections that can develop their plastic capacity but with limited rotation capacity.
3 Class 3: Semi-compact sections that can develop the yield moment but not the plastic moment capacity of the cross-section.
4 Class 4: Slender cross-sections that are unable to develop the yield moment due to the early occurrence of local buckling.

Figure 9.10 shows the hysteretic response of a class 1 section cycled to a ductility ratio of 7.2. The beam was laterally braced to prevent lateral torsional buckling. Local flange buckling was first observed during the second half-cycle, but it can be seen that the beam survived to 12 cycles with a fairly gradual loss in strength and stiffness. Essentially the same mechanisms are responsible for this loss as was described above for cyclically loaded struts, namely the Bauschinger effect and increasing residual deformation, but they act on a local rather than member level. In dissipative structures, in which plastic yielding is intended to occur to dissipate energy in the design earthquake, Eurocode 8 (EC8) (CEN, 2004) requires class 1 or 2 sections, with class 1 required to achieve a behaviour factor q greater than 4. These requirements apply to all members intended to achieve plasticity in the design earthquake ('dissipative members' in EC8's parlance), including flexural members in moment frames, compression struts in concentrically braced frames (CBF) and ductile links in EBF.

Lateral torsional buckling is the phenomenon caused by overall instability of the beam's compression flange buckling between lateral restraint points. Onset is determined by a number of factors, principally the flange width to thickness ratio (b_f/t_f) and the slenderness ratio of the beam l/r_y (where l is the effective unrestrained length of the beam and r_y is the radius of gyration about the minor axis). Vann *et al.* (1973) concluded that loss of stiffness is much more significant when lateral torsional buckling dominates response, rather than

Figure 9.9 (a) Local flange buckling and (b) fracture in a cyclically loaded flexural member (photographs courtesy of SAC Project 7.03 – Georgia Tech, USA (R. Leon and J. Swanson)

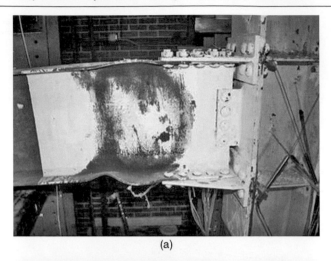

(a)

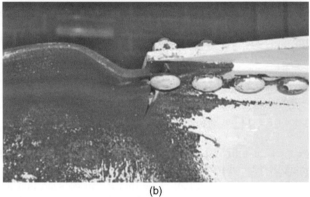

(b)

local web or flange buckling. They also found that deterioration of strength is only severe when local flange buckling is combined with either local web buckling, or lateral torsional buckling. In addition to rules affecting compactness (and hence the b_f/t_f ratio), AISC 341-10 restricts the maximum slenderness ratio in special ductility structures to $(17\,238/f_y)$. This gives $l/r_y \leq 73$ and 48 for $f_y = 235$ and 355 MPa, respectively. EC8 places no further limitations on l/r_y for flexural members other than those given by Eurocode 3 for non-seismic situations; however, the effective length must be calculated assuming a plastic hinge forms at one end of the beam.

This discussion concerns the flexural behaviour of beams. Columns are subject to axial loading, which increases the tendency to buckle and hence reduces available ductility. Strong column/weak beam design will reduce the tendency of columns to yield in flexure, but does not eliminate it; therefore, compact sections and low slenderness ratios are particularly important in the columns of moment frames.

Figure 9.10 Compact steel member ($b_f/2t_f = 7.8$) loaded cyclically in flexure (Vann *et al.*, 1973)

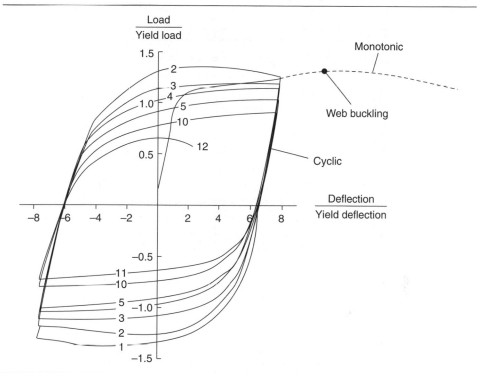

9.3.4 Cyclic loading of welds

Figure 9.3 showed the low cycle fatigue failure of a weld. A number of factors tend to promote such failure, most importantly the following.

- The flexural strength of the connection of which the weld is a part is less than that of the beam it connects. Note that if the beam supports a floor slab, this will act as a flange, strengthening the flexural strength of the beam but most likely not the connection.
- The weld profile contains a number of stress raisers, including the effect of stopping and starting the bottom flange weld at the web, and the effect of backing strips, if not removed and ground out.
- The weld metal is of low ductility, and the welding procedure tends to promote brittleness in the heat-affected zone of the connected member.

Measures to avoid such factors applying are central to current US code recommendations for the seismic design of welds, discussed in Section 9.6.5.6.

9.3.5 Panel zones

The junctions between beams and columns, called 'beam–column joints' in concrete structures, are known as 'panel zones' in steel moment frames. They are subject to very high shear stresses (see Figure 8.37 in the chapter on concrete) and are just as vital to the functioning of steel moment frames as they are to concrete ones. However, unlike the case of concrete, steel panel zones, represented by the column web adjacent to the beam, exhibit a ductile behaviour under cyclic loading. The zone is generally well restrained by the column flanges and (often) beam flange continuity plates (see below), which allow large cyclic shear distortions to be developed without loss of strength. There is a penalty for plastic deformation; moment frames depend for their stiffness on joint rigidity, which is greatly reduced when the panel zone yields. The implications for design are discussed in Section 9.6.5.5.

The point at which the beam flanges attach to the column flange is another potential point of concern in panel zones, particularly where flanged (I-section) beams connect to flanged columns. The beam flange force must eventually be transmitted into the column web, but if it must first flow through the column flange, high stresses will be induced around the root of the column flanges. This is because the free edges of the column flanges are being loaded out-of-plane and are much more flexible than the in-plane stiffness of the beam flanges. Continuity plates, which effectively continue the beam flanges across the column web, mitigate this problem; they also serve to restrain the column web in the panel zone, as noted above.

9.3.6 Scale effects

Two factors linked to scale effects have been suggested as contributing to the connection failures observed in the Northridge earthquake (Engelhardt and Sabol, 1996). Both are connected with the Californian practice of using very large member sizes in the seismic resisting elements, to minimise their number.

First, the use of very deep members in relation to the span means that, compared to more conventional sizes, relatively large plastic strains need to occur at the extreme fibres before the full plastic moment can develop. This will tend to limit the maximum rotational capacity of the plastic hinge.

Second, very large member and weld thicknesses result in the development of significant triaxial states of stresses, which have been shown to limit material ductility, even in highly ductile parent material.

Steel moment frames formed from very deep and wide members formed from high plate thickness are therefore particularly prone to brittle fracture.

9.4. Materials specification

Modern ductile steels produced for non-seismic environments are generally suitable for earthquake resistance, and EC8 places no requirements on basic steel material specification additional to those given in Eurocode 3 for non-seismic situations. In US practice, steel with a yield strength exceeding 380 MPa is not generally permitted by AISC 341-10, and minimum Charpy notch toughness values are also specified where the steel thickness exceeds 38 mm.

The major difference in specification for seismic applications involves the specification of upper bounds on the yield strength. This is to ensure that, during a severe earthquake, parts of the structure designed to yield (the 'dissipative' members, in EC8's parlance) do not have strengths much greater than their nominal design strength. If they did, brittle elements with yields closer to nominal values might reach their failure strength before the intended yield mechanism formed, thus greatly reducing the available ductility. EC8 treats this issue by offering the designer three options.

1   By requiring an upper bound on the yield strength of the dissipative members to be less than a given factor (recommended as 1.375) times the nominal yield strength.
2   By using a lower grade of steel in the dissipative members than the brittle, or non-dissipative, ones, which are, however, designed assuming they are also formed from the lower grade steel.
3   For existing structures, by basing capacity design checks (i.e. the checks that brittle elements do not reach their failure strength when the dissipative elements yield) on measurements of the actual yield strength of the dissipative elements, as measured in situ.

Since the connection failures in Northridge and Kobe, there has been greater recognition of the importance of specifying ductile welding materials. Low hydrogen weld metals with good notch ductility are needed, and it is important to ensure that the welding sticks are kept dry before use. AISC 341-10, unlike EC8, has a section on special requirements for weld material in seismic situations.

9.5.   Analysis of steelwork structures

Steelwork structures can be analysed by any of the methods described in Chapter 3. Some particular aspects that apply to steel are now discussed.

9.5.1   Ductility reduction factors

In linear elastic methods of analysis, relatively large ductility reduction factors are permitted by codes for the most ductile configurations. In EC8, a behaviour factor q of typically 6 to 8 may be applied to specially detailed moment-resisting frames and EBF, and the maximum value of the equivalent R factor in the US loading code ASCE 7-10 (ASCE, 2010) is 8. Lower values apply to CBF, typically $q = 4$ and $R = 6$ for the most ductile arrangements. The recommended minimum factors for steel structures without seismic detailing are $q = 1.5$ to 2 in EC8 (compared to 1.5 for reinforced concrete and masonry) and $R = 3.5$ in ASCE 7-10 (compared to 3 for reinforced concrete and 1.5 for plain masonry), although both codes advise that non-seismically detailed structures should not be used except in regions of low seismicity, unless they are seismically isolated.

9.5.2   Rotational demand and capacity of steel flexural hinges

If a non-linear static or dynamic analysis is performed, direct information is obtained on the plastic demands on the yielding regions. In principle, this allows a much more rigorous assessment of the ultimate performance of a structure than by the use of linear analysis with crude ductility factors such as q in EC8 or R in ASCE 7-10. However, assessing the rotational capacity of steel plastic hinges from first principles is much more complex than is the case for reinforced concrete members, because of the effect of local and global buckling on response. In practice, empirical expressions for plastic hinge rotation are needed; data are given in ASCE 41-06 (ASCE, 2006) and in annexe B of EC8 part 3. For design, both EC8 and AISC 341-10 set minimum limits for rotational capacity, which must be demonstrated directly by testing, as discussed in Section 9.6.5.6.

9.5.3   Allowing for flexibility in unbraced steel frames

Two points are worth noting in this context. First, because moment-resisting steel frames are relatively flexible, P–delta effects (Figure 3.23) may be significant in tall unbraced structures.

Second, the flexibility of the panel zones, where beams and columns intersect, may contribute significantly to the deflection of a moment-resisting steel frame, and should be allowed for. Bruneau *et al.* (2011) provide an extensive discussion on how to allow for this effect in an analysis.

9.5.4   Analysis of CBF

When using linear analysis methods, EC8 requires that X-braced frames (Figure 9.11(a)) are analysed neglecting the stiffness contribution of the compression members, and that the tension members are sized on the same assumption. By contrast, US practice requires that two cases in each direction of loading are examined; compression members should be assumed as first included and then neglected, with all members designed for the worst case. It is advisable to neglect the contribution of the diagonal braces to resisting gravity loads in any bracing configuration, and also to supporting the horizontal beams in the V-braced configurations of Figure 9.11(e and f).

9.6.   Design of steel building structures

The remaining sections of this chapter discuss design considerations for steel structures, with particular reference to the provisions of the European code EC8 and the US code AISC 341-10. For a much more detailed discussion of the AISC provisions, refer to Bruneau *et al.* (2011), which provides numerous worked examples. The steel provisions of the Eurocode are discussed in the EC8 manual (ISE/AFPS, 2010), and they have been reviewed by the European Convention for Constructional Steelwork (ECCS) (ECCS, 2013a). ECCS has also published a freely downloadable design manual for steel structures (ECCS, 2013b).

9.6.1   Design levels of ductility

In both EC8 and AISC 341-10, different levels of ductility are recognised and the discussion in Section 8.5.1 for concrete applies equally to steel structures. The ductility classes in EC8 are high (DCH), medium (DCM) and low (DCL) while the (non-equivalent) ductility levels in AISC 341-10 are special, intermediate and ordinary. However, in EC8, the additional design effort for steel DCH buildings, compared to DCM buildings, is much less than is the case for concrete. Also, the lower design lateral forces resulting from the higher q values allowed by DCH design are less attractive if deflection governs design rather than strength, because design deflection (unlike design strength) does not depend on q – see Section 6.7. This is more likely to apply to steel than concrete buildings, because they are generally more flexible.

9.6.2   Concentrically braced frames

General planning considerations for CBF were given in Section 5.5.4 of Chapter 5. Some of the more important design requirements given in EC8 are as follows. Somewhat different rules apply in AISC 341-10, as discussed by Elghazouli (2003).

Figure 9.11 Examples of bracing arrangements for concentrically braced frames

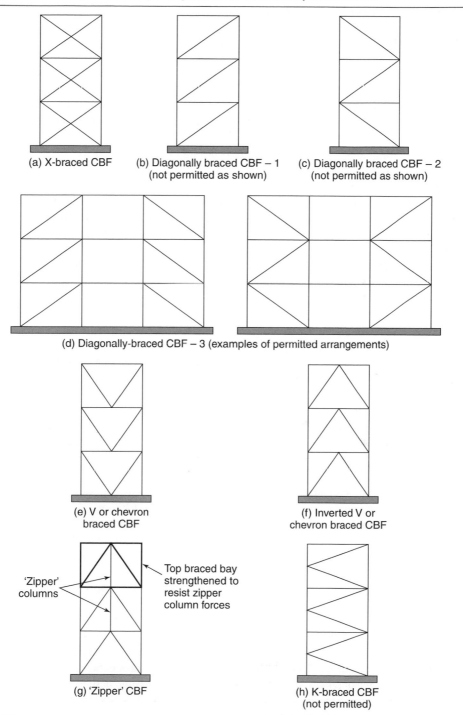

(a) X-braced CBF

(b) Diagonally braced CBF – 1
(not permitted as shown)

(c) Diagonally braced CBF – 2
(not permitted as shown)

(d) Diagonally-braced CBF – 3 (examples of permitted arrangements)

(e) V or chevron
braced CBF

(f) Inverted V or
chevron braced CBF

'Zipper'
columns

Top braced bay
strengthened to
resist zipper
column forces

(g) 'Zipper' CBF

(h) K-braced CBF
(not permitted)

9.6.2.1 General

- Generally speaking, tension diagonals are designed as the 'dissipative' or yielding elements, and other elements, including beams, columns and connections, must be designed on capacity design principles to resist the yield force from these diagonals with sufficient overstrength. In particular, connections must be designed to resist a load equal to 1.375 times the yield strength of the members they connect. This applies to bolted or fillet welded connections, but full strength butt welds are considered to satisfy the overstrength condition without the need for further analysis. Connections may alternatively be designed as dissipative elements, but in this case special design and analysis procedures are required.
- In DCH structures, bracing members must belong to compactness class 1 (see Section 9.3.3); both class 1 and 2 members are permitted in DCM structures; class 3 is only permitted if q is 2 or less.
- Bracing systems that rely on both compressive and tensile braces to resist lateral load (e.g. the V-braced arrangements of Figure 9.11(e, f and g) qualify for a much less favourable q factor (2.5 and 2 for DCH and DCM, respectively) than X or diagonally braced systems (Figure 9.11(a and d)). The latter have $q = 4$ for both DCH and DCM.
- Within any plane of bracing, the compression diagonal braces should balance the tension diagonal braces at each bracing level, in order to avoid tension braces contributing most to lateral resistance in one direction and compression braces in the other – compare Figure 9.11(b and c) with Figure 9.11(a and d). This is to satisfy the general principle that the diagonal elements of bracings should be placed in such a way that the load deflection characteristics of the structure are the same for both positive and negative phases of the loading cycle. Failure to satisfy this will result in drift of the structure in the more flexible direction during the course of an earthquake, leading to premature failure. The similar provision in AISC 341-10 for balancing tension and compression resistance is less restrictive than EC8.
- In principle, capacity design procedures are required to ensure that the columns can sustain the yield strength of the braces without buckling. In practice, this raises a difficulty, because a strict interpretation would then require the lower columns to be designed for the simultaneous yielding of all bracing higher in the structure, a condition that is unlikely to occur. The implementation in EC8 (and AISC 341-10) therefore requires the columns to be designed for the force obtained from the seismic analysis (i.e. not directly from capacity design principles) but increased by a suitable factor to ensure the columns remain essentially elastic. The factor in EC8 is approximately 1.375 times the minimum ratio of actual to required strength in the bracing; the use of this factor represents a direct application of capacity design principles.
- It is important to ensure a reasonably uniform distribution of ductility demand in the braces over the height of the structure. If this is not achieved, and the braces at one level yield well before the others, a weak storey might form, concentrating most of the ductility demand at that level. To avoid this, EC8 places a restriction on the ratio of bracing member strength to strength required from the seismic design. The ratio between maximum and minimum values of this ratio must not exceed 125%. There is no similar requirement in AISC 341-10.

9.6.2.2   X and diagonally braced systems

These are generally designed assuming that the compression braces do not contribute stiffness or strength (see Figure 9.11(a and d)). EC8 places upper and lower limits on the slenderness of diagonal braces in X-braced systems. The upper limit corresponds to a slenderness l/r_y of around 180 (depending on yield strength), and is designed to prevent the strength and stiffness degradation shown for a slender strut in Figure 9.6. The lower limit of around 110 is intended to prevent column overloading; columns to which the diagonal braces are connected will be sized to resist the full yield strength of the tension brace assuming no force in the compression brace, but higher axial forces might occur in the columns before very stocky braces have buckled. In AISC 341-10, there is a similar limit on upper bound slenderness, but no lower limit.

9.6.2.3   Diagonal and V-braced systems

These systems rely on both compression and tension braces for stability, and so the stiffness and strength of the compression braces must be explicitly accounted for (see Figure 9.11 (e and f)). The same upper bound limits on slenderness apply, but there is no lower bound limit in EC8, because the concern about neglecting the compression brace force does not apply.

In V-braced systems, the horizontal brace is subjected to an out-of-balance force when the compression brace begins to buckle, and in EC8 this must be designed for. Also, the horizontal brace must be designed to carry any gravity loads without support from the diagonal braces. The AISC 341-10 rules are similar.

9.6.2.4   Zipper braced systems

Zipper braced systems are formed from V braces with an additional vertical member ('zipper column') linking the connection point of the diagonal members (see Figure 9.11(g)). The intention is to distribute the out-of-balance force that develops when the compression brace buckles, thus increasing redundancy and ductility. The bracing in the top storey can be provided with additional strength to transfer the zipper column forces back to the main columns. EC8 does not refer explicitly to zipper braced systems; AISC 341-10 refers to them in the commentary on special CBF, implying that they can be designed as such. Further information is given in Yang *et al.* (2008).

9.6.2.5   K-braced systems

K-braced systems are not permitted, because buckling of the compression brace imposes an out-of-balance force not on the horizontal beam (as in the case of V-braced systems) but the column, and this is clearly unacceptable (see Figure 9.11(h)).

9.6.3   Eccentrically braced frames

In EBFs, the joints of diagonal bracing members are deliberately separated from those of the vertical and horizontal members to form a link element that can act as a ductile fuse under extreme lateral loads. General planning considerations for EBFs were given in Section 5.5.5 of Chapter 5. Design rules based on the research effort of the past 30 years (mainly conducted in the USA) are contained in codes such as EC8 and AISC 341-10. They are intended to ensure that the link elements have sufficient ductility to sustain the inelastic cyclic deflections to which they would be subjected in a design earthquake, and to ensure that the surrounding members always remain within the elastic range. The latter objective is met by the use of capacity design procedures – that is, designing the surrounding members for the yielding actions developed in the links. EBFs are assigned ductility factors similar to those of ductile moment frames; that is, a q of up to 8 in EC8 and an R of 8 in AISC 341-10.

Replaceable EBF links have recently been researched (Mansour *et al.*, 2008) so that links damaged after a severe earthquake can be easily replaced, to enable the building's original seismic resistance to be restored relatively cheaply and quickly.

9.6.4   Buckling restrained braced frames
Buckling restrained braced frames (BRBFs) are CBFs with one vital difference; the conventional diagonal bracing members of CBFs are replaced by proprietary members with the same strength and ductility in compression as in tension. Section 5.5.6 of Chapter 5 explained the principles involved.

The avoidance of buckling means that the causes of strength and stiffness degradation shown in Figure 9.7 are not present, and available ductility is improved. Moreover, the uncertainty in predicting the load corresponding to the onset of buckling is also removed; hence capacity design protection of the columns and connections is also more reliable. BRBFs are assigned the same force reduction (R) factor as EBFs in ASCE 7-06 and AISC 341-10 (AISC, 2010) gives design rules. EC8 does not currently treat BRBFs.

9.6.5   Moment-resisting frames
The general characteristics of moment-resisting (unbraced) frames were discussed in Section 5.5.3 of Chapter 5.

9.6.5.1 General considerations
The design intent is normally to limit yielding to flexural hinges forming in the beams, and to ensure that columns and connections remain elastic by the use of capacity design procedures. To this end, EC8 requires that the flexural strength of columns at a joint exceeds the flexural strength of the beams at the joint by 30% (20% in AISC 341-10), except in single-storey frames and in the top storey of multi-storey frames, where the requirement does not apply. However, if the columns are fixed against rotation at their base, plastic hinges must also form there if a sway mechanism is to develop (see Figure 3.19 in Chapter 3) and the columns must be able to sustain the plastic rotations involved. Alternatively, the column bases may be pinned; however, this substantially increases deflections to an extent that may be unacceptable. Given the poor performance of column bases in the Northridge and Kobe earthquakes noted by Smith (1996), attention to the design of column bases is required whatever solution is adopted.

Note that, unlike the case of reinforced concrete frames, a capacity design check for shear strength is not required, because yielding in shear (at any rate for reasonably compact sections without excessively thin webs) is a relatively ductile mechanism. However, column buckling under axial loading is a highly undesirable mechanism, and should be avoided by methods discussed in Section 9.6.5.4.

9.6.5.2 Preliminary sizing

A rough preliminary indication of required member sizes may be obtained for a typical building by assuming a total seismic weight of say 10 kN per metre of floor area (to include the structure, permanent finishes and a proportion of the live load). Equivalent lateral forces can then be obtained, using standard code procedures, and the bending forces in the beams estimated by assuming points of contraflexure exist at the mid-span of beams and mid-height of columns and that inner columns take twice the shear of external columns. The required beam flexural strengths can then be checked. The column must be sized on capacity design principles; their bending strength should be 30% greater than that of the beams, while simultaneously resisting axial loads due to gravity, plus those induced by flexural yielding of the beams, which, for a preliminary design, can conservatively be assumed to take place simultaneously in all beams.

However, deflections rather than strength may well govern. A preliminary estimate of storey drift can be obtained from the equation

$$d = x\left(\frac{k_b + k_c}{k_b k_c}\right)\left(\frac{h^2}{12E}\right)V_c \tag{9.1}$$

where x is the 'ductility factor', d is storey drift (m), k_b is (I_b/L) for a representative beam (m^3), k_c is (I_c/h) for a representative internal column (m^3), $I_b I_c$ are moments of inertia of beam, column (m^4), L is centre-to-centre spacing of columns (m), h is storey height (m), E is Young's modulus of steel (kPa) and V_c is shear in the representative column.

The 'ductility factor' x is the factor by which the deflections obtained from an elastic analysis must be multiplied to allow for plastic deformations; in EC8, x is taken as the behaviour factor q, and in ASCE 7-10 it is the factor C_d given in Table 1617.6.2 in that code.

The storey drift must then be compared with the maximum permitted in the governing code. In EC8, this would generally be 1% of the storey height under the ultimate design earthquake, but up to twice this deflection is allowed where the cladding and partitions are not brittle, or are suitably isolated from the frame. ASCE 7-10 typically requires a drift limit of 2% of the storey height, but this may vary between 0.7% and 2.5% depending on the type of structure and importance category.

A procedure such as this can form the basis for a more rigorous design, perhaps using a computer program, by selecting member sizes that will allow a satisfactory solution to be found without too many trial iterations. A very different procedure, which starts from setting an acceptable drift limit and then derives the member strength needed to achieve it, is the displacement-based design procedure, outlined in Section 3.8 of Chapter 3.

9.6.5.3 Beams

As described in Section 9.3.3, the beam's flange and web thickness must be sufficient to limit local flange and web buckling, and EC8 requires sections that conform to Eurocode 3 class 1 (or class 2 for DCM structures) specifications. Lateral torsional buckling must also be controlled by adequate lateral restraint.

Beam flexural strength is assessed on the basis of the seismic analysis, although as noted earlier this may not provide sufficient stiffness to limit deflections to within code limits in tall buildings.

Flexural hinges forming in beams must be capable of sustaining an adequate plastic rotation. A minimum rotation capacity of 0.025 radians for DCM and 0.035 radians for DCH is specified by EC8, and in addition the loss of stiffness and strength under an unspecified number of cyclic loads should not exceed 20%. In AISC 341-10, the plastic rotation capacity of special moment frames must be capable of sustaining at least 0.040 radians, reducing to 0.020 radians for intermediate ductility frames. Achievement of these rotational capacities is highly dependent on the connection design, discussed in Section 9.6.5.6.

9.6.5.4 Columns

Columns are generally designed to be protected against yielding and therefore do not have to conform to the same compactness requirements referred to above for beams. The exception (noted in Section 9.6.5.1) is fixed column bases; here, the seismic compactness rules must apply because the columns are 'dissipative'.

Capacity design procedures are used to ensure adequate flexural strength (see Section 9.6.5.2). It might be thought they would also be suitable to set design axial loads in columns, to prevent highly undesirable axial buckling failure. In principle, this can be done by adding the axial loads generated by simultaneous yielding of all beams to the gravity loads. However, it is unlikely that all the beams will yield simultaneously, and this simple addition would result in excessive conservatism for tall buildings. Generally, codes specify that the columns are designed for the forces derived from the seismic analysis, increased by a simple factor. In EC8, the factor equals 1.375Ω. Here, Ω is the minimum ratio of resistance moment to design moment at plastic hinge positions in the beams. As the resistance moment must not be less than the design moment, Ω is always at least one. When the beams are all sized considerably in excess of the minimum requirement, the structure will start to yield at a lateral force considerably greater than that effectively assumed in the analysis. Therefore, the column forces will also be greater than predicted by the analysis. Factoring the column forces by 1.375Ω therefore allows for this. The factor is set at 1.375Ω rather than Ω to allow for strain hardening in the beam plastic hinges, and to provide some degree of additional reserve of strength. US practice is similar, but the details are different.

Column splices need to be designed to transmit safely the design axial force in the column, calculated as above, together with the column bending moment at the splice position. An elastic analysis is a potentially unsafe way of predicting the latter, both because the beams may be stronger than required by the analysis, and so yield at higher loads, and also because the point of contraflexure in the column under inelastic dynamic loading tends to be poorly predicted by an elastic analysis. At the extreme, if the splice were placed at the point of contraflexure (zero moment) predicted by an elastic analysis, and were designed only for axial load, it could fail if the point of contraflexure shifts. EC8 has no specific requirement but AISC 341-10, unlike EC8, has a number of requirements intended to ensure the safe design of column splices, including special rules for splices subjected to net tension forces.

9.6.5.5 Panel zones

The panel zones of columns are the parts to which the beams connect. Although yielding of the panel zone in shear is a ductile failure mode, it significantly reduces the stiffness of the frame and is generally discouraged. Panel zones therefore need to be designed to withstand the shear forces induced in them when the beams connected to them yield, by capacity design principles; both plastic yielding and buckling need to be considered. Both EC8 and AISC 341 give detailed rules for the strength design of panel zones.

Panel zones also need detailing. Continuity plates, discussed in Section 9.3.5, may be needed across the top and bottom of the zone to transmit the beam flange forces into the zone, because these are what give rise to the high shears in the zone. A minimum plate thickness of the panel zone is also advisable to control inelastic web buckling, even if the zone has been designed not to yield using capacity design principles. ASCE 341 requires a minimum panel zone thickness t_z given by

$$t_z = (d_z + w_z)/90 \qquad (9.2)$$

where d_z is the panel zone depth, corresponding to the depth between beam flanges, and w_z is the width between column flanges. If the column web is less than t_z, a 'doubler plate' needs to be welded to the column flange in the panel zone.

9.6.5.6 Connections

Beam to column connections are generally designed on capacity design principles to withstand the yielding forces in the beams that they connect; EC8 requires design for moments and shears generated by 1.375 times the yield moments in the beams.

Following the failures in Northridge and Kobe, both AISC 341-10 and the Eurocode specify a minimum plastic rotational capacity for the beams of a ductile moment frame, as explained in Section 9.6.5.2. This capacity, which is highly dependent on the beam–column connection design, must be demonstrated by testing, and not calculation, but EC8 currently gives no further information. By contrast, extensive codified requirements have been developed in the USA. Two possibilities are offered. The most commonly used is to specify the standard pre-qualified designs defined in AISC 358-10 (AISC, 2010b). Alternatively, for non-standard designs, special testing of representative beam–column sub-assemblies must be carried out; detailed specifications for the tests are given in AISC 341-10, which requires that the testing procedures (including the number of specimens to be tested) be agreed by a special review panel.

AISC 358-10 describes two types of predominantly welded joint. The first is shown in Figure 9.12. Here, the beam is bolted on site to a single plate welded to the column flange; this serves as a temporary connection while the beam flanges and webs are joined with full penetration welds to the column flange. The single plate is then welded to the beam web with a fillet weld. Detailed rules are given for welding procedures, including the removal of all potential stress raisers from the weld, and for non-destructive testing. The beam flange thickness is restricted to 25 mm, to avoid the scale effects referred to in Section 9.3.6. In principle, this is like the pre-Northridge connection (Figure 9.2), but with many strength and detailing improvements.

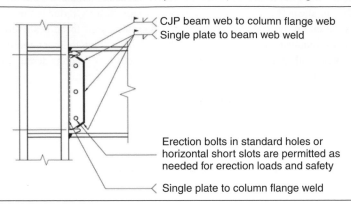

In the second welded arrangement (Figure 9.13), the beam has been deliberately weakened at a point slightly away from the connection. Therefore the beam yields at this weakened section, which reduces the maximum moment to which the connection can be subjected – an application of capacity design principles. In special (high ductility) structures, full penetration welds are still required to connect beam flanges and web to the column flange, but larger section sizes can be used than for the detail of Figure 9.12. This 'reduced beam section' arrangement is currently one of the most commonly used in the USA for new structures. It has also been used to improve the ductility of existing structures with inadequate, pre-Northridge connections, because the reduction in moment at the beam–column junction caused by the beam weakening makes potential stress-raisers around the connection less critical. By weakening the beams, the overall lateral strength of the frame is of course reduced, but the stiffness (which may govern design) is much less affected.

A number of bolted configurations are also offered by AISC 358-10. They include arrangements with extended endplates and beam flanges that are either unstiffened (Figure 9.14(a)) or stiffened (Figure 9.14(b and c)). Alternatively, bolted flange plates can be used, which force the plastic hinge to form away from the column face (Figure 9.15). There is also a proprietary system – the Kaiser bolted bracket – in which the flange stiffeners of Figure 9.14(b and c) are replaced by a cast steel bracket bolted to the column flange and beam web, thus forming the principal means of connection. Another proprietary arrangement, the ConXtech ConXL moment connection (Figure 9.16) involves attaching the beams to the column by means of a bolted collar; the column comprises a concrete-filled box section. Unlike the other details, this arrangement allows the column to be part of frames in two orthogonal directions.

Other types of joint not specified in AISC 358-10 are described in an earlier document, FEMA 350 (FEMA, 2000). They are not prohibited by US seismic codes, but would generally require additional testing before being accepted for a specific project.

9.6.5.7 Energy-dissipating connections
Another solution to the problems of connection design in steel moment frames involves special connections, which (unlike the prequalified joints of Figures 9.12 to 9.16) are

Figure 9.13 Prequalified reduced beam section joint in AISC 358-10 (reproduced from AISC, 2010b © American Institute of Steel Construction. Reprinted with permission. All rights reserved)

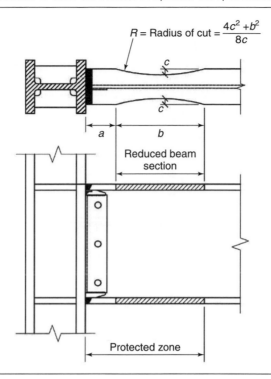

$$R = \text{Radius of cut} = \frac{4c^2 + b^2}{8c}$$

intended to prevent any plastic hinges from forming in the beams. Instead, post-elastic deformation, and energy dissipation, occurs in the connection itself. Figure 9.17 shows one such connection, based on a clamping mechanism, which slips at a load determined by friction. The relative movement between the clamped plates after slipping dissipates energy but is set to occur at a load at which the beams, column and their connections remain elastic.

Figure 9.14 Prequalified bolted joints with extended end plates in AISC 358-10 (reproduced from AISC, 2010b © American Institute of Steel Construction. Reprinted with permission. All rights reserved)

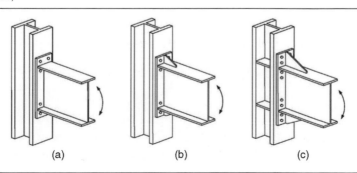

Figure 9.15 Prequalified bolted flange plate joints in AISC 358-10 (reproduced from AISC, 2010b © American Institute of Steel Construction. Reprinted with permission. All rights reserved)

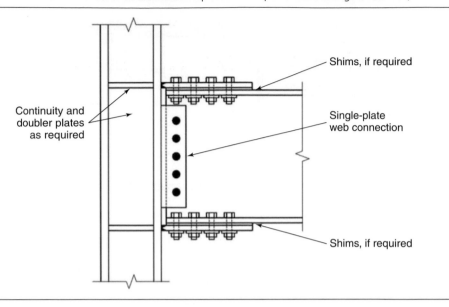

Shims, if required

Continuity and doubler plates as required

Single-plate web connection

Shims, if required

Figure 9.16 Prequalified ConXtech ConXL moment connection in AISC 358-10 (reproduced from AISC, 2010b © American Institute of Steel Construction. Reprinted with permission. All rights reserved)

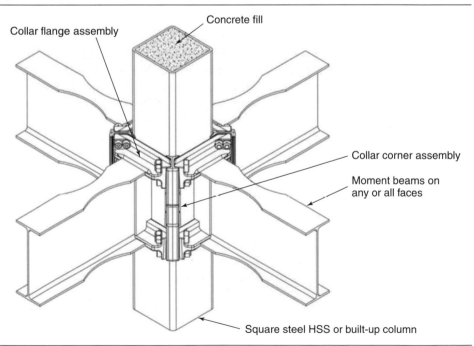

Concrete fill

Collar flange assembly

Collar corner assembly

Moment beams on any or all faces

Square steel HSS or built-up column

Figure 9.17 Dissipative, bolted connection (reproduced from MacRae *et al.*, 2010; New Zealand Society for Earthquake Engineering Inc.)

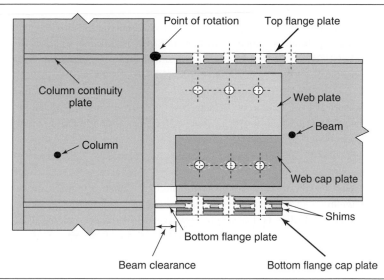

Note that the relative movement occurs in the bottom beam flange, to minimise the damage in floor slabs supported by the top beam flange.

9.6.5.8 Frames not proportioned to resist lateral loads

A common framing plan involves a moment-resisting perimeter frame designed to resist all seismic loads, and internal frames designed only to resist gravity loads. It is of course essential that the latter do not collapse under the deflections to which they are subjected during the design earthquake, and this condition needs to be checked. EC8 exempts such elements from seismic detailing rules, provided their contribution to lateral resistance is neglected in the analysis, and also that their contribution to lateral stiffness is not more than 15% of the total. Beyond checking for gravity load carrying capacity under the design seismic deflections, EC8 has no further rules. AISC 341-10 gives no specific guidance.

9.6.5.9 Moment-resisting frames with masonry infill panels

EC8 allows masonry infill panels either to be designed as structurally separated from the steel frame, or to interact with it, in which case the effects of interaction must be considered. Section 8.5.2.8 in Chapter 8 for infills to concrete frames is also relevant to steel frames, because similar considerations apply. Further discussion is given by Ravichandran and Klinger (2012).

9.7. Steel concrete composite structures

Steel sections filled with reinforced concrete have certain advantages. The compressive strength added by the concrete increases member resistance in buckling, with the possibility of improving cyclic performance. The concrete, if properly detailed, can also control the onset of local flange or web buckling. EC8 devotes a chapter to steel concrete composite structures, as does AISC 341-10.

REFERENCES

AISC (2010a) ANSI/AISC 341-10: Seismic provisions for structural steel buildings. American Institute of Steel Construction, Chicago, IL, USA.

AISC (2010b) ANSI/AISC 358-10: Prequalified connections for special and intermediate steel moment frames for seismic applications. American Institute of Steel Construction, Chicago, IL, USA.

ASCE (2006) ASCE/SEI 41-06: Seismic rehabilitation of buildings. American Society of Civil Engineers, Reston, VA, USA. New edition expected 2016.

ASCE (2010) ASCE/SEI 7-10: Minimum design loads for buildings and other structures. American Society of Civil Engineers, Reston, VA, USA. New edition expected in 2014.

Bruneau M, Uang CM and Sabelli R (2011) *Ductile Design of Steel Structures*, 2nd edn. McGraw Hill, New York, USA.

CEN (2004) EN 1998-1: 2004: Design of structures for earthquake resistance. Part 1. General rules, seismic actions and rules for buildings. Section 6: Special rules for steel buildings. Section 7: Special rules for composite steel-concrete buildings. Comité Européen de Normalisation, Brussels, Belgium.

Clifton G, Bruneau M, MacRae G, Leon R and Fussell A (2010) Steel building damage from the Christchurch earthquake series of 2010 and 2011. *Bulletin of the New Zealand Society for Earthquake Engineering* **44(4)**: 297–318.

ECCS (European Convention for Structural Steelwork) (2013a) *Assessment of EC8 Provisions for Seismic Design of Steel Structures*. ECCS, Brussels, Belgium.

ECCS (2013b) *ECCS manual on the design of steel structure in seismic zones*. ECCS, Brussels, Belgium.

EEFIT (Earthquake Engineering Field Investigation Team) (1986) *The Mexican Earthquake of 19th September 1985*. EEFIT, Institution of Structural Engineers, London, UK.

Elghazouli A (2003) Seismic design procedures for concentrically braced frames. *ICE – Structures and Buildings* **156(4)**: 381–394.

Elnashai A, Bommer J, Baron C, Lee D and Salama A (1995) *Selected engineering seismology and structural engineering studies of the Hyogo-ken Nanbu (Great Hanshin) earthquake of 17th January 1995*. Civil Engineering Department, Imperial College, London, UK. ESEE research report no. 95-2.

Engelhardt M and Sabol T (1996) Lessons learnt on steel moment frame performance from the 1994 Northridge earthquake. In *Seismic Design of Steel Buildings after Kobe and Northridge* (Burdekin M (ed.)). Institution of Structural Engineers, London, UK.

FEMA (Federal Emergency Management Agency) (2000) FEMA 350: Recommended design criteria for new steel moment-frame buildings. FEMA, Washington, DC, USA.

ISE/AFPS (Institution of Structural Engineers/Association Française du Génie Parasismique) (2010) *Manual for the design of steel and concrete buildings to Eurocode 8*. Institution of Structural Engineers, London, UK.

Jain AK, Goel SC and Hanson RD (1978) Hysteresis behaviour of bracing members and seismic response of braced frames with different proportions. Department of Civil Engineering, The University of Michigan, Ann Arbor, MI, USA. Report no. UMEE 78R3. July.

MacRae G, Clifton G, Mackinven H *et al.* (2010) The sliding hinge joint moment connection. *Bulletin of the New Zealand Society for Earthquake Engineering* **43(3)**: 202–212.

Mansour N, Yunlu Shen Y, Christopoulos C and Tremblay R (2008) Experimental evaluation of nonlinear replaceable links in eccentrically braced frames and moment resisting frames. In *14th World Conference on Earthquake Engineering, Beijing*. See www.nicee.org/wcee/.

Midorikawa M, Nishiyama I, Tada M and Terada T (2012) Earthquake and tsunami damage on steel buildings caused by the 2011 Tohoku Japan earthquake. *Proceedings of the International Symposium on Engineering Lessons Learned from the 2011 Great East Japan Earthquake, Tokyo, Japan.*

Ravichandran S and Klinger R (2012) Seismic design factors for steel moment frames with masonry infills. Parts 1 and 2. *Earthquake Spectra* **28(3)**: 1189–1222.

Smith DGE (1996) An analysis of the performance of steel structures in earthquakes since 1957. In *Seismic Design of Steel Buildings after Kobe and Northridge* (Burdekin M (ed.)). Institution of Structural Engineers, London, UK.

Vann WP, Thompson LE, Whalley LE and Ozier LD (1973) Cyclic behaviour of rolled steel members. *Proceedings of the 5th World Conference on Earthquake Engineering, Rome*. Vol 1: 1187–1193. International Association for Earthquake Engineering, Tokyo, Japan.

Yanev P, Gillengerten JD and Hamburger RO (1991) *The Performance of Steel Buildings in Past Earthquakes*. American Iron and Steel Institute, Washington, DC, USA.

Yang CS, Leon RT and DesRoches R (2008) Design and behavior of zipper-braced frames. *Engineering Structures* **30(4)**: 1092–1100.

Earthquake Design Practice for Buildings
ISBN 978-0-7277-5794-4

ICE Publishing: All rights reserved
http://dx.doi.org/10.1680/edpb.57944.255

Chapter 10
Masonry

Masonry materials – mortar and stones or bricks – are stiff and brittle, with low tensile strength, and are thus intrinsically not resistant to seismic forces. However, the earthquake resistance of masonry as a composite material can vary between good and poor, depending on the materials used [and] the quality of workmanship.

Sir Bernard Feilden.
Between Two Earthquakes – Cultural Property in Seismic Zones.
ICCROM, Rome/Getty Conservation Institute, Marina del Rey, CA, 1987

This chapter covers the following topics.

- Forms of masonry construction.
- The lessons from earthquake damage.
- Analysis of masonry buildings.
- Design of engineered masonry buildings.
- Rules for simple 'non-engineered' masonry buildings.

Masonry buildings are heavy, usually very brittle and often have low lateral resistance; their collapse in earthquakes has therefore killed many people and will continue to do so. A significant complicating (but not always adverse) additional factor is that they are often built by owners or local builders without involving any formally trained engineers. This chapter reviews the various types of masonry building and how they respond to strong ground shaking, and discusses the design and analysis of both engineered and non-engineered construction.

10.1. Introduction

Brick and stone masonry is a widely available, low-energy material with good thermal and acoustic insulating properties, and the skills are found all over the world to use them for creating highly practical and often very beautiful buildings. However, its low tensile strength limits the available ductility and places reliance on its ability to sustain high compressive stresses during an earthquake. If the compressive strength is low (as is the case, for example, with the earthen bricks known as 'adobe') then the consequences in an earthquake can be disastrous, and often have been (Figure 10.1). However, well-designed buildings made from good quality brick or stone can perform well. If the following three fundamental points (Bothara and Brzev, 2011) are addressed, then simple, low-rise unreinforced masonry buildings can achieve safety, even in the absence of engineering design.

Figure 10.1 Failure of weak masonry buildings in Gujarat, 2001 (© E Booth)

- Good quality materials and workmanship.
- Compact layout, with well distributed walls in both directions firmly attached to each other, to the foundations and to the floors and roofs they support.
- Attention to good seismic detailing, such as the provision of horizontal bands wrapping round the entire periphery of the building and firm connection of inner and outer wythes of brickwork.

Further improvements to the performance of unreinforced masonry can be achieved by various strengthening measures intended to increase the tensile strength and ductility of the masonry, as described in the next section.

10.2. Forms of masonry construction
10.2.1 Unreinforced masonry
Masonry consists of blocks or bricks, usually bonded with mortar. A wide variety of forms exist. The weakest is when cohesive soil is placed in a mould and sun-dried to form a building block. This type of construction (called adobe in Latin America and elsewhere) is cheap, widely available and requires only basic skills to form, but cannot usually be relied on to resist strong ground motion. Stabilising the soil with lime or other cementitious material improves matters.

Random rubble masonry consists of rough cut or natural stones held in a matrix of soil or mortar. It may form the core of a wall with a cladding of dressed (i.e. cut) stone, called ashlar. The seismic resistance depends on the matrix holding the stones together; if this is weak, the seismic performance will be poor or very poor. Stone shape is also an important factor – walls with round-shaped or wedge-shaped (semi-dressed) stones show poor seismic performance compared to the semi-dressed stones. Without horizontal steel or other ties to connect them, the outer layers of the wall have a tendency to delaminate and separate from the core during an earthquake (Figure 10.8), with a consequent loss of strength.

Carefully cut rectangular blocks of stone (dressed stone) of good quality arranged to resist lateral resistance without developing tensile stresses can possess surprisingly good earthquake resistance. Here, the presence of vertical prestress, usually coming from the weight of masonry above, is important for two reasons. First, seismically induced tensile stresses may not develop if the prestress is great enough, thus stopping stones from falling out and preserving the integrity of the structure. Second, the shear strength of dressed stone relies primarily on friction; the higher the contact forces between stones, the higher the shear strength. As compressive gravity loads are higher at the base of a building, often the seismic resistance is also greater. Therefore, unlike concrete and steel buildings, the damage observed in dressed stone masonry is often less at the bottom of a building than at the top (Figure 10.2) where the earthquake-induced accelerations are also greatest. Inducing compressive stresses by introducing vertical or inclined steel prestressing cables is thus a powerful way to improve the seismic resistance of good quality stone masonry buildings (Beckmann and Bowles, 2004: section 4.5.10; Collins and Jordan, 1997; Ma *et al.*, 2012).

Manufactured bricks or blocks can approach the compressive strength of natural stone without requiring the special skills and equipment needed to dress natural stone. The blocks can be made of natural clays fired to a high temperature (bricks) or from concrete (concrete masonry units or CMU). Solid blocks have the greatest compressive strength but hollow blocks are lighter, and so easier to handle, and can be reinforced with vertical steel bars (Figure 10.3).

Proprietary brick systems have also been developed that provide a mechanical interlock between bricks, which improves the shear strength of the completed wall – see, for example, www.habiterrabuildingsolutions.com.

Eurocode 8 (EC8) (CEN, 2005) part 1 recommends restricting the use of unreinforced masonry to areas where the 475-year return peak ground acceleration is less than 2 m/s^2. In order to conform to EC8, even 'unreinforced' masonry is required to have continuous horizontal reinforcing bands of reinforced concrete or steel to be placed peripherally around the building at every floor level. In North America, unreinforced masonry is similarly restricted to areas of low seismicity.

10.2.2 Reinforced masonry

Horizontal and vertical steel bars within the masonry units forming a wall increase its flexural and shear strength and also its ductility, resulting in greatly improved seismic performance. EC8 requires a minimum of 0.5% reinforcement horizontally and 0.8% vertically, based on cross-section area, in order to classify masonry as 'reinforced'.

Figure 10.2 Increase in seismic damage with height in a stone masonry building, Gujarat, India, 2001 (© E Booth)

Practice varies greatly on how the reinforcement is placed. The most common approach is to use hollow concrete blocks (Figure 10.3), typically with two voids in each block. In the main body of the wall, some or all of the voids have single vertical bars in them, and the voids are then filled with a concrete grout; in Figure 10.3, every third void is shown as having a vertical bar. Horizontal bars can be laid in the 'bed joints' between the side walls of successive blocks. This bed joint reinforcement often takes the form of 'ladder reinforcement' – pairs of bars separated by spacers; it is primarily intended to restrain cracking in the walls and is not counted towards the shear resistance under seismic loading. Alternatively, the horizontal bars can be placed centrally in grooves formed in the blocks, which is the more common European practice.

To qualify as 'reinforced masonry', additional vertical reinforcing bars should be placed at the junctions between walls and to either side of openings; EC8 requires this reinforcement to be

Figure 10.3 Typical reinforced hollow concrete blocks

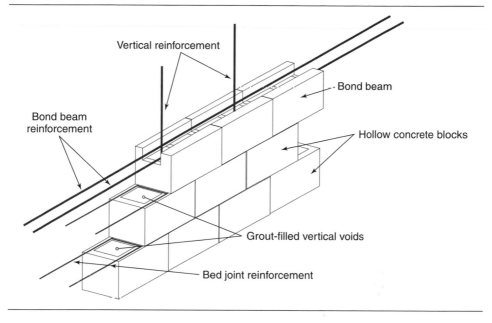

confined by horizontal hoop steel. The reinforcement cage so formed will usually be too large to fit into the void space of a standard block so it can either be placed in special blocks with large voids or separate 'columns' cast against the ends of the blocks, as in confined masonry (Figure 10.4). Additional horizontal steel may also be required in continuous peripheral 'bond

Figure 10.4 Forming a corner column in confined masonry: (a) during construction; (b) completed after concreting

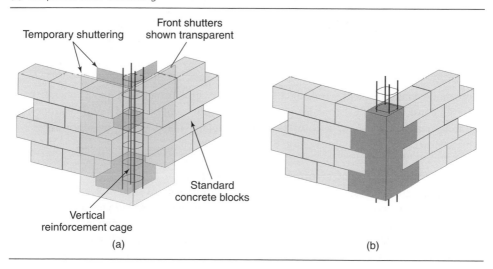

beams' (Figure 10.3), although EC8 does not specify them; they are placed at floor levels and possibly also at intermediate levels when shear demands are high. Lintel beams over door and window openings (a universal requirement) may be formed in a similar way.

Figure 10.3 shows a wall with full grouting, in which all the voids are filled with grout, whether or not they have vertical reinforcing bars; this is required to achieve higher shear strength in the plastic hinge regions of ductile shear walls. For designs in which a limited ductility demand is envisaged, partially grouted walls in which the unreinforced voids are not grouted are commonly used; however, this practice results in a lower shear strength, and it may be difficult to prevent some of the grout leaking into the unreinforced voids. North American codes restrict the use of partial grouting to low-rise buildings, and it is not permitted in Colombia in areas of high seismicity. Where horizontal reinforcement is placed centrally rather than in the bed joints, full grouting is also needed to protect and anchor the horizontal bars.

The reinforcement must of course be appropriately detailed. In EC8, the additional bands of vertical steel must be confined with horizontal hoops, as explained previously, and all lap splices of vertical and horizontal bars must be at least 60 bar diameters in length. However, there is no requirement to confine horizontal bond beams. North American practice is more complex, and depends on the ductility level that the designer has assumed. A comprehensive guide to the seismic design of reinforced masonry to Canadian practice and standards is provided by Anderson and Brzev (2009); the details they give are typical of North American practice, although as noted above practice elsewhere may be rather different.

10.2.3 Confined masonry

Improving the seismic performance of masonry by the introduction of tensile ties in the form of timber beams is a practice many centuries old (Ali *et al.*, 2012), long predating the emergence of the profession of engineer. Given the scarcity of timber in many places, the more common current practice is to use reinforced concrete instead to form 'confined masonry'. Here, the walls are surrounded by vertical and horizontal reinforced concrete elements, which must be cast into the walls after their construction, thus ensuring a good bond between the concrete confining elements and the masonry. Vertical elements are needed at minimum at the corners of the building (Figure 10.4), at the free ends of walls and around openings, while the horizontal elements are placed at floor levels.

Note that 'confined masonry' is distinct from masonry infill panels built into a concrete frame after its construction. EC8 contains provisions for both 'confined' and 'infill' masonry; the latter is treated by EC8 as a concrete frame structure, rather than a masonry shear wall structure and was discussed in Section 8.5.2.8.

10.2.4 Other forms of strengthening masonry

The use of vertical prestressing to improve the performance of unreinforced masonry has already been mentioned. Many other methods of strengthening techniques also exist, such as providing steel ties between pairs of external walls at floor levels, applying strengthening layers such as reinforced mortar to wall surfaces and wrapping strengthening bands around the outside of the structure.

10.3. Lessons from earthquake damage

The different forms of masonry construction described in the previous section perform in different ways. Weak masonry walls formed from adobe or from random rubble perform the worst, and 100% destruction of such buildings is quite common in a severe earthquake (Figure 10.1); this type of construction is commonly built by the occupiers in poorer areas. Buildings formed from regular, dressed blocks of good quality granite or other rock are less likely to suffer total collapse, although there may be damage (Figure 10.5); this form of construction is often found in large public or official buildings. The ability of good quality masonry to remain (just) stable after experiencing major cracking and deformation is frequently amazing (Figure 10.5). Adding confinement or reinforcement to masonry is found to improve performance still further.

Ground shaking has two distinct effects on masonry walls depending on whether it is acting along the plane of the walls or is perpendicular to them. Out-of-plane shaking may make walls unstable, and cause toppling of parapets (Figure 10.2), gable ends (Figure 10.6) or whole walls (Figure 10.7). As the swaying of a building in an earthquake means that accelerations increase up its height, damage tends to be greater towards the top, as these figures show. Experience from earthquakes shows that walls that are well connected to floor and roofs are far less prone to the destabilising effect of out-of-plane shaking. These effects are also reduced by horizontal and vertical ties around the building, which serve to hold the structure together. The concrete elements in confined masonry perform both functions, and confined

Figure 10.5 Damage to a palace after the Gujarat India earthquake of 2001 (© E Booth)

Figure 10.6 Gable end failure in the Erzincan, Turkey earthquake of 1992. Note that the sloping (hipped) gable on the farthest building on the right has not failed (© E Booth)

Figure 10.7 Out-of-plane failure of walls: (a) two-storey building, Padang, Indonesia earthquake of 2009; and (b) top floor of six-storey building, Boumerdes, Algeria earthquake of 2003 ((a) Reproduced from Bothara *et al.*, 2010; New Zealand Society for Earthquake Engineering Inc.; (b) photograph courtesy of M. Farsi)

(a)   (b)

Figure 10.8 Delamination of the outer surface of a wall in the Maharashtra, India earthquake of 1993 (photograph courtesy of S. Brzev)

masonry buildings have generally performed significantly better than their unreinforced equivalents; see, for example, Astroza *et al.* (2012) for evidence from the 2010 Chile earthquake. There are also local effects of out-of-plane shaking; outer layers of a semi-dressed masonry wall with a rubble core may delaminate (Figure 10.8).

Walls are much stiffer in their plane, and shaking in this direction means they attract much larger forces, which may lead to diagonal cracking and shear failure (Figure 10.9). Here, it is the shear strength of the masonry that is critical, and measures to improve it, such as the addition of surface strengthening layers or internal steel bars to form reinforced masonry, improve performance. Overall, however, it is found that out-of-plane failures are more common than in-plane ones; therefore, computing in-plane shear stress and comparing it with the available strength, although useful, is only a partial guide to the earthquake resistance of a masonry building.

Many earthquake damage surveys have revealed two other expected but crucial features affecting performance. First, box-like structures that are well provided with substantial

Figure 10.9 In-plane failure of walls in the Kashmir earthquake of 2005 (reproduced from Bothara and Hiçyilmaz, 2008; New Zealand Society for Earthquake Engineering Inc.)

walls perform much better than buildings with complex plan shapes, irregular distributed walls, and door or window openings that reduce in-plane strength. Second, and crucially, poor quality construction greatly reduces performance. A classic example is that of the Haiti earthquake of 2010 in which at least 100 000 people (perhaps many more) died, mainly as a result of the poor workmanship and sub-standard materials used in Port-au-Prince and its surroundings (Marshall *et al.*, 2011).

10.4. Designing masonry buildings for seismic resistance
10.4.1 Minimum material strength
EC8 recommends a minimum compressive strength of block or brick of 5 N/mm^2 normal to the bed face and 2 N/mm^2 parallel to the bed face in the plane of the wall. A minimum mortar strength of 5 N/mm^2 is recommended for unreinforced and confined masonry and 10 N/mm^2 for reinforced masonry.

These limits are appropriate for new construction. Existing buildings with lime mortars are unlikely to comply, however, although it may still be possible to demonstrate satisfactory seismic performance. Generally, it is desirable for shear failure to occur in the mortar before failure in the masonry unit, because a more ductile response is then likely, mobilising frictional resistance.

10.4.2 In-plane shear strength of masonry walls
In simple masonry wall buildings, practically all the lateral resistance is provided by the in-plane stiffness of the walls. The designer's task is then to ensure that all the seismic forces can be safely transmitted back to the walls by the floors and roof, and to ensure that the walls have sufficient in-plane shear strength to resist them. In most cases, the in-plane shear strength will be governed by the mortar, but with weak bricks or blocks and strong mortar, the masonry units may fail first (a less ductile mode, as noted previously). When the mortar

shear strength governs, in-plane shear strength is determined by adding the shear resistance of the masonry under zero compression to the shear resistance provided by friction between the masonry blocks. EC8 specifies that the design seismic in-plane shear strength should be determined from Eurocode 6 (CEN, 2005), which provides the following equations. Equation 10.1 applies to masonry in which the vertical (header) joints are completely filled with mortar. The limit in Equation 10.2 represents failure of the masonry units before slip develops in the mortar.

$$v_d = \qquad (v \qquad + \qquad 0.4\sigma_v)/\gamma_m \qquad\qquad (10.1)$$

$$\text{Intrinsic shear strength} \qquad \text{Frictional component}$$

or

$$0.065f_b/\gamma_m \quad \text{if less} \qquad\qquad (10.2)$$

where v_d is the design in-plane shear strength, v is the masonry shear strength under zero compressive load, σ_v is the vertical stress due to permanent loads, and f_b is the compressive strength of the masonry unit. γ_m is the material factor, which is at least 1.5.

Note that v_d is related to the average shear stress over the wall, and allows for the fact that the shear stresses in the centre of the wall are greater than those at its ends. Typical values of v for unreinforced masonry are given in Table 10.1. In existing buildings, v may be determined from in-situ tests, such as UBC 21-6 (ICBO, 1997). In this test, both head joints of a selected brick are cleared and a flat jack is introduced into one of them to stress the brick horizontally in shear. The shear force and deflection at first slip and at failure are noted. Ideally, the force-deflection characteristics should be measured at various levels of vertical load by introducing a second jack to vary the vertical stress in the brick being tested. This enables the frictional component of resistance and coefficient of friction to be estimated.

Reinforced masonry can achieve much higher shear strength. Not only does the reinforcement increase the shear strength, but it also provides a measure of ductility, as reflected in the q factors shown in Table 10.3.

Door and window openings will of course reduce in-plane shear strength. The shear strength of a wall should be based on its net area, after allowing for openings, and timber, reinforced concrete or steel lintels should be placed over the openings. Preferably, the vertical sides and bottom should be similarly reinforced, and this is a requirement in EC8 for openings greater

Table 10.1 Indicative values for intrinsic shear strength of unreinforced masonry

Masonry quality	v: MPa
Poor	0.15
Average	0.30
Good	0.45

Figure 10.10 Guidelines for placing doors and windows in stone masonry buildings: (a) walls in mud mortar; (b) walls in cement mortar (Bothara and Brzev, 2011; IAEE, 1986)

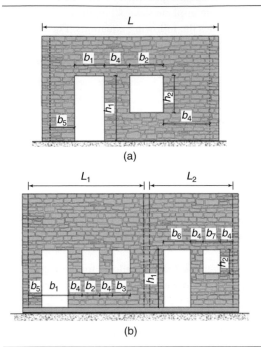

$$b_1 + b_2 \leq 0.33L$$
$$b_4 \geq 0.5h_2 \text{ and } b_4 \geq 600 \text{ mm}$$
$$b_5 \geq 0.25h_1 \text{ and } b_5 \geq 600 \text{ mm}$$

(a)

One storey buildings: $b_1 + b_2 + b_3 \leq 0.5L_1$
$$b_6 + b_7 \leq 0.5L_1$$
Two storey buildings: $b_1 + b_2 + b_3 \leq 0.42L_1$
$$b_6 + b_7 \leq 0.42L_2$$
$$b_4 \geq 0.5h_2 \text{ and } b_4 \geq 600 \text{ mm}$$

(b)

than $1.5\,\mathrm{m}^2$ in confined masonry. Door and window openings are of course essential! – but their spacing and distance from building corners should be positioned so as not to comprise the wall's shear strength too much. Figure 10.10 suggests empirical guidelines related to size and location of openings in a stone masonry wall.

Unreinforced masonry walls are likely to suffer a considerable loss of in-plane shear strength and stiffness once their ultimate strength is reached. If there are adjacent, less highly stressed walls, these may then be able to relieve some of the loads, provided the floor or roof diaphragms are strong and stiff enough for the redistribution of forces involved. EC8 allows for up to 25% of seismic loads to be redistributed from more highly to less highly stressed walls, provided the diaphragms can make the necessary transfers. However, walls that are heavily damaged by in-plane shear forces are also likely to have their ability to carry gravity loads compromised; hence the likelihood of collapse is increased. For this reason, EC8 recommends that unreinforced masonry walls should be designed for a q factor of 1.5, reflecting an essentially elastic response (Table 10.3). The improved ability of confined and reinforced masonry to maintain vertical resistance after sustaining significant in-plane shears is reflected in the higher q factors shown in Table 10.3.

Slender masonry walls with a low ratio of base length l to height h may start to uplift at one edge before their in-plane shear strength is reached. This constitutes rocking; although a wall will maintain some shear resistance after rocking has started, the shear stiffness will drop

Table 10.2 Geometric limits on masonry walls from EC8

Masonry type	Min. thickness, $T_{ef,min}$: mm	Max. out-of-plane slenderness, $(h_{ef}/T_{ef})_{max}$	Min. stockiness, $(l/h)_{min}$
Unreinforced, with natural stone units	350	9	0.5
Unreinforced, with any other type of units	240	12	0.4
Unreinforced, with any other type of units, in cases of low seismicity	170	15	0.35
Confined masonry	240	15	0.3
Reinforced masonry	240	15	No restriction

h, greater clear height of the openings adjacent to the wall; h_{ef}, effective height of the wall (see Eurocode 6 (CEN, 2005)); l, length of the wall; T_{ef}, thickness of the wall (see EN 1996-1-1 (CEN, 2005))

considerably, and it is likely that most of its shear load will shed to other walls. EC8 specifies that slender walls not satisfying the minimum (l/h) ratios shown in Table 10.2 should be taken as secondary elements that are not counted as contributing to lateral resistance.

10.4.3   Out-of-plane forces on masonry walls

The main task for masonry walls is to resist the overall seismic shear forces developed in the direction in which they run. However, the walls will also be subjected to seismic accelerations perpendicular to their plane, and these will give rise to out-of-plane forces, which may be at least as damaging as the in-plane forces. They arise only from the self-mass of the walls and any finishes applied to them, which form only a small proportion of the building's total mass, but the shears must still be transferred back to the points of lateral restraint to the wall.

Where the lateral restraint from cross-walls and floors is ineffective, out-of-plane failure may occur by an overall overturning of the wall (Figures 10.6 and 10.7); D'Ayala and Speranza (2003) provide a detailed discussion. This type of failure mechanism, commonly found in historic masonry buildings, points to the importance of ensuring that adequate lateral restraint is provided; for example, by providing horizontal tying bands around the building and ensuring strong connections between floor and walls. Delamination is a common out-of-plane failure mechanism in low-rise stone masonry walls built using rounded or semi-dressed stone boulders with small stones and pieces of rubble; it takes place when outer wall layers bulge and collapse outward under out-of-plane inertia forces (Figure 10.8).

The out-of-plane forces also result in the development of significant out-of-plane bending moments. This is unlikely to be a problem for reinforced masonry due to its tensile strength. However, in unreinforced (and to a lesser extent confined) masonry, the very low tensile strength implies that out-of-plane bending strength relies on the compressive prestress due to the gravity loads the walls support. Clearly, out-of-plane bending is likely to be most significant if the wall thickness is small in relation to the distance to lateral restraint. The problem is complex, but several factors make it less severe than it appears to be at first. Tests at Bristol University (Zarnic *et al.*, 1998) showed that compressive membrane action can considerably improve the out-of-plane resistance of wall panels. Moreover, once the wall has cracked in out-of-plane bending, it loses stiffness and this is likely to decouple it from the input motions. Unlike in-plane effects, which involve tributary inertia forces from the entire building, out-of-plane response is driven only by self-mass, and hence is likely to be displacement limited. However, excessively small ratios of thickness to storey height or wall length should be avoided. If the out-of-plane deflection of a load-bearing wall takes the line of action of the gravity loads much beyond the central third of the wall thickness, its collapse becomes much more likely. Geometric limits proposed by EC8 are given in Table 10.2.

10.5. Analysis of masonry structures

The analysis of masonry buildings poses at least as many inherent problems as that of concrete ones; the response is likely to be highly non-linear and issues of stability as well as strength will be important. For regular, well-engineered masonry buildings, a linear elastic response spectrum analysis should yield useful (and perhaps sufficient) information on the magnitude and distribution of in-plane shears. Table 10.3 shows the q factors recommended by EC8 for such an analysis. However, the analysis can give no information on instability effects or on redistribution after cracking. Non-linear static analysis takes account of the redistribution of shears, and guidance on this sort of analysis is given in ASCE 41 (ASCE, 2006). However, it does not address out-plane effects and in particular wall instability.

As a comment on the effectiveness of conventional linear or non-linear finite element analysis for masonry, such analysis would have been unlikely to predict that the ancient pavilion shown in Figure 10.11 could have survived the 2001 Gujarat earthquake, which toppled many nearby modern reinforced concrete buildings in the Indian city of Ahmadabad. The pavilion appears to disobey conventional rules of seismic design with its top-heavy roof and proneness to a column sway mechanism, yet it survived, as did a number of other similar ancient structures. Clearly, other analysis tools are needed for such cases, particularly when historic or non-engineered (vernacular) buildings are involved.

Table 10.3 Seismic reduction factors for masonry walls in EC8 (CEN 2004)

Type of construction	Behaviour factor q in EC8
Unreinforced masonry in accordance with EN 1996 (CEN, 2005) alone (recommended only for low seismicity cases)	1.5
Unreinforced masonry in accordance with EN 1998-1 (CEN, 2004)	1.5
Confined masonry	2.0
Reinforced masonry	2.5

Figure 10.11 Unreinforced masonry structure which survived the Gujarat, India earthquake of 2001 (© E Booth)

One possibility is the use of the program FaMIVE (D'Ayala and Speranza, 2003), which considers potential collapse mechanisms for a given wall geometry, and calculates a value of ground shaking sufficient to trigger each mechanism, based on energy considerations. The lowest level of ground shaking is then taken as an estimate of seismic resistance. FaMIVE can only consider one single wall or plane façade at a time, so that three-dimensional interactions between the walls in a building are not considered. However, it can be a useful adjunct to a three-dimensional finite element analysis – for example, a response spectrum or non-linear static analysis, which does allow for such effects.

Alternative approaches may be used in dressed stone structures, in which the individual blocks are taken as rigid, infinitely strong bodies that are held together by frictional forces, and that can rock relative to each other. Failure is judged to occur when all contact is lost between the blocks. De Jong and Ochsendorf (2010) describe analytical methods for applying such an analysis to masonry arches. Computer based 'discrete element' analyses (Jean, 1998) can deal with much more complex geometries than is possible with analytical approaches, but are not currently considered to provide a complete solution. Small geometric perturbations that are hard to capture in a computer model can cause local failures, such as bricks dropping out of vaults, which can in turn lead to a more general collapse. An application of the method is given by Brookes and Mehrkar-Asl (1998), who describe its use to investigate various options for strengthening a two-storey unreinforced masonry building.

10.6. Simple rules for masonry buildings

In conventional masonry wall buildings, the key features to consider are limiting in-plane shear stresses, ensuring a reasonable distribution of walls in both horizontal directions and an absence of torsional eccentricity, providing efficient floor and roof diaphragms to tie the building together and distribute seismic loads back to the walls, and ensuring that door and window openings do not introduce local points of weakness. Rules of thumb for masonry design that account for these factors in a mainly qualitative way may often be just as valuable as a detailed, complex analysis; in many cases, the rules of thumb may actually be more useful.

EC8, section 9.7 provides some rules for 'simple' low-rise masonry buildings, which avoid the need for a detailed seismic analysis. The main quantitative requirement concerns the minimum area of wall that should be provided in each direction as a percentage of the total floor plan area; this is known as the wall density ratio. The required wall density ratio depends on the level of seismicity, the number of storeys and the type of construction, identified as either unreinforced, confined or reinforced masonry. The rules given by EC8 are thought to be rather conservative, and are likely to be substantially changed in the next revision of the Eurocode. Meli *et al.* (2011) provide recommendations for the minimum wall density requirements in confined masonry buildings. Wall density has been used as an indicator of seismic performance for masonry buildings in Latin America, especially in Mexico and Chile (Moroni *et al.*, 2003).

EC8 also sets a number of conditions that have to be met for masonry buildings to qualify as 'simple', which can be regarded as a statement of good practice, as follows.

- The plan shape must be approximately rectangular, with a recommended minimum ratio of shortest to longest side of 0.25, and with projections or recesses from the rectangular plan area not exceeding 15%.
- The building should be stiffened by shear walls, arranged almost symmetrically in plan in two orthogonal directions.
- A minimum of two parallel walls should be placed in two orthogonal directions, the length of each wall being greater than 30% of the length of the building in the direction of the wall under consideration.
- At least for the walls in one direction, the distance between these walls should be greater than 75% of the length of the building in the other direction.
- At least 75% of the vertical loads should be supported by the shear walls.
- Shear walls should be continuous from the top to the bottom of the building.
- Differences in mass and shear wall area between any two adjacent storeys should not exceed 20%.
- For unreinforced masonry buildings, walls in one direction should be connected with walls in the orthogonal direction at a maximum spacing of 7 m.

Two excellent guides on the design of buildings in stone masonry (Bothara and Brzev, 2011) and confined masonry (Meli *et al.*, 2011) have been prepared by the US-based Earthquake Engineering Research Institute and the International Association for Earthquake Engineering through the World Housing Encyclopedia project. These publications contain simple, very well illustrated advice on the design, construction and strengthening of such buildings, and are freely available online: follow links from www.world-housing. net/tutorials.

REFERENCES

Ali Q, Sacher T, Ashraf M *et al.* (2012) In-plane behaviour of the Dhajji–Dewari structural system (wooden braced frame with masonry infill). *Earthquake Spectra* **28(3)**: 835–858.

Anderson D and Brzev S (2009) Seismic design guide for masonry buildings. Canadian Concrete Masonry Producers Association. Free download from http://www.ccmpa.ca.

ASCE (2006) ASCE/SEI 41-06: Seismic rehabilitation of buildings. American Society of Civil Engineers, Reston, VA, USA. New edition expected in 2014.

Astroza M, Moroni O, Brzev S and Tanner J (2012) Seismic performance of engineered masonry buildings in the 2010 Maule earthquake. Special issue on the 2010 Chile earthquake. *Earthquake Spectra* **28(S1)**: S385–S406.

Beckmann P and Bowles R (2004) *Structural Aspects of Building Conservation*. Elsevier, Amsterdam, The Netherlands.

Bothara J and Brzev S (2011) *A tutorial: Improving the Seismic Performance of Stone Masonry Buildings*. Earthquake Engineering Research Institute, California, USA, and International Association of Earthquake Engineering, Tokyo, Japan.

Bothara JK and Hiçyilmaz K (2008) General observations of behaviour during the 8 October 2005 Pakistan earthquake. *Bulletin of the New Zealand Society for Earthquake Engineering* **41(4)**: 209–223.

Bothara J, Beetham R, Brunsdon D *et al.* (2010) General observations of effects of the 30 September 2009 Padang earthquake, Indonesia. *Bulletin of the New Zealand Society for Earthquake Engineering* **43(3)**: 143–173.

Brookes C and Mehrkar-Asl S (1998) Numerical modelling of masonry using discrete elements. In *Seismic Design Practice into the Next Century* (Booth E (ed.)). Sixth SECED Conference. Balkema, Rotterdam, The Netherlands.

CEN (2004) EN 1998-1: 2004: Design of structures for earthquake resistance. Part 1: General rules, seismic actions and rules for buildings. European Committee for Standardisation, Brussels, Belgium.

CEN (2005) Eurocode 6. EN 1996-1-1: Design of masonry structures. Common rules for reinforced and unreinforced masonry structures. European Committee for Standardisation, Brussels, Belgium.

Collins BJ and Jordan JW (1997) Earthquake strengthening and repair of Christ Church cathedral, Newcastle NSW. In *7th International Conference on Structural Faults and Repair, Edinburgh, Scotland*, July 1997. See www.newcastlecathedral.org.au/common/download/Earthquakerepairs.pdf.

D'Ayala D and Speranza E (2003) Definitions of collapse mechanisms and seismic vulnerability of historic masonry buildings. *Earthquake Spectra* **19(3)**: 479–510.

De Jong MJ and Ochsendorf JA (2010) Dynamics of in-plane arch rocking: an energy approach. *ICE – Engineering and Computational Mechanics* **163(3)**: 179–186.

IAEE (International Association for Earthquake Engineering) (1986) *Guidelines for Earthquake-Resistant Non-Engineered Construction*. IAEE, Tokyo, Japan.

ICBO (International Conference of Building Officials) (1997) UBC 21-6: In-place masonry shear tests. ICBO, Whittier, CA, USA.

Jean M (1998) The non-smooth contact dynamics method. *Computer Methods in Applied Mechanics and Engineering* **177**: 235–257.

Ma R, Jiang L, He M, Fang C and Feng L (2012) Experimental investigations on masonry structures using external prestressing techniques for improving seismic performance. *Engineering Structures* **42(September)**: 297–307.

271

Marshall JD, Lang AF, Baldridge SM and Popp DR (2011) Recipe for disaster: construction methods, materials and identification of needs for rebuilding after the January 2010 earthquake. *Earthquake Spectra* **27(51)**: 323–344.

Meli R, Brzev S, Astroza M *et al.* (2011) Seismic design guide for low-rise confined masonry buildings. Earthquake Engineering Research Institute, California, USA, and International Association of Earthquake Engineering, Tokyo, Japan. Downloadable from www.world-housing.net/tutorials.

Moroni MO, Astroza M and Acevedo C (2003) Performance and seismic vulnerability of masonry housing types used in Chile. *ASCE – Journal of Performance of Constructed Facilities* **18(3)**: 173–179.

Zarnic R, Gostic S, Severn RT and Taylor CA (1998) Testing of masonry infilled frame reduced scale building models. In *Seismic Design Practice into the Next Century* (Booth E (ed.)). Balkema, Rotterdam, The Netherlands.

Earthquake Design Practice for Buildings
ISBN 978-0-7277-5794-4

ICE Publishing: All rights reserved
http://dx.doi.org/10.1680/edpb.57944.273

Institution of Civil Engineers

publishing

Chapter 11
Timber

Wood construction is light, and while there have been some horrible failures, there have been very few casualties.

Henry Degenkolb. *Connections*.
Earthquake Engineering Research Institute, Oakland, CA, 1994

This chapter covers the following topics.

- The lessons from earthquake damage.
- Characteristics of timber as a seismic-resisting material.
- Codes, standards and design recommendations.

Timber is lightweight, easy to construct and can have excellent seismic-resisting properties. This chapter surveys the design advice, currently much less extensive than for other materials, for both traditional and also more recent forms of this attractive building material.

11.1. Introduction

Timber is widely used for building construction in highly seismic areas, both in the developed world and also the developing world, particularly Central and South America where houses built from timber and bamboo are common. This use is increasing; timber is a renewable, potentially carbon-neutral construction material, particularly when supplied from sustainable sources and its low weight relative to its strength make it easy to transport. Moreover, it is straightforward to connect and form. Current technology can assemble small components into strong, reliable composites, avoiding the dependence on natural growth to achieve long spans or special shapes. Architects – and their clients – like the natural beauty of exposed timber. These aspects combine to make it an attractive building material.

However, until recently timber was perhaps the least researched and written about of seismic-resisting materials. That situation is changing. The relatively poor performance of timber houses in the Northridge California earthquake of 1994, and the need to establish the seismic performance of recent technologies such as cross-laminated panels (Section 11.4.3) has prompted extensive recent research (Figure 11.1).

11.2. Lessons learnt from earthquake damage
In the ancient world, wooden pagodas and temples have successfully survived earthquakes in Japan and China, and the timber buildings of Anatolia, Turkey have a good record. Timber

Figure 11.1 Testing a full scale three-storey timber building on the Tsukuba shaking table, Japan (reproduced from Ceccotti *et al.*, 2007; photograph courtesy of Maurizio Follesa, taken as part of the research project, Progetto Sofie (Sofie Project), funded by the Province of Trento, Italy, and carried out through CNR-IVALSA (Trees and Timber Institute), Italy)

also has a long and successful history of use as a tying element within masonry buildings to improve earthquake resistance.

In recent Californian earthquakes, low-rise timber houses have suffered where they have been inadequately anchored to their foundations, and have shifted. Unbraced 'cripple walls' (short walls lifting the lowest floor of ground level to allow the passage of services) have also commonly failed. A number of buildings with garages at ground level failed due to 'soft storey' formation, particularly 'tuck-under buildings' with the lowest level open on three sides and closed on the fourth. Generally, however, the seismic performance of low-rise timber-frame buildings in California is assessed as far superior to that of their unreinforced masonry equivalents, and comparable to that of low-rise buildings with reinforced concrete or reinforced masonry shear walls (Anagnos *et al.*, 1995).

Traditional single-family Japanese houses are one and two-storey wood post and lintel, with bamboo reinforced mud infill and heavy fired clay tiling. They perform very poorly in earthquakes, being prone to pancake collapse, as seen in the 1995 Kobe earthquake. More recent construction has lighter, shingle roofs, with a more substantial timber frame often on a reinforced concrete base, and this has performed much better (Scawthorn *et al.*, 2005).

In developing countries, single or two-storey housing using bamboo or more conventional timber has generally fared better than masonry construction. Often, however, the housing is occupied by the poorest families forced onto marginal and unsuitable land, and the failures that occur are of unstable slopes on which the houses are built, or are due to weakening of the frames by insect or fungal attack.

Devastating fires have broken out after earthquakes in dense areas of wooden buildings causing extensive damage and loss of life, notably in Tokyo in 1923, but also more recently, for example, in the Marina district of San Francisco in 1989 and Kobe, Japan in 1995.

A detailed survey of the performance of timber buildings in earthquakes is given by Karacabeyli and Popovski (2003).

11.3. Characteristics of timber as a seismic-resisting building material

Timber has a high ratio of tensile and compression strength to weight, which is a favourable seismic-resisting feature. Timber joints can also dissipate significant amounts of energy when they are stressed in an earthquake. When yielding of steel elements (nails, screws, bolts) within the joint is involved, damping ratios of as much as 45% can occur, and in most timber structures the damping will be at least 15%, which is two or three times the typical level in concrete or steel structures (Dolan, 2003). The failure of the parent timber, however, tends to be rather brittle in most failure modes, and so overstrength members and under-strength connections are usually indicated. However, compression failure perpendicular to the grain is ductile, involving collapse of the wood's cellular structure, and some of the ductility in nailed and bolted joints arises from this mechanism. Glued joints are not able to dissipate much energy, and nor are joints made with large steel bolts in which failure occurs by crushing, shearing or splitting of the timber. 'Carpenter joints' (such as a tenon or halving joint, in which the forces are transferred directly through the wood without mechanical fasteners) may be dissipative, provided shear failure or tension failure perpendicular to the grain does not occur.

Three other favourable seismic features of timber housing should be mentioned. First, it is easy to achieve a good tensile strength in the connections between timber elements, and so a timber-frame building with timber floors is well tied together in a way much harder to achieve in masonry. This greatly improves its earthquake resistance. Second, timber frames tend to be quite highly redundant, which also improve resistance. Third, it is straightforward to nail plywood wall panels to a timber frame, which provides lateral strength and stiffness together with excellent energy dissipation through the screwed or nailed joints. This form of construction has a good performance record in Californian earthquakes.

Two unfavourable features should also be noted. First, the strength of timber reduces when its moisture content increases. It is also susceptible to insect and fungal attack, and this can effectively destroy its resistance. Timber treatment, combined with suitable detailing to dissipate moisture and deter insects, is needed to counter this. The second feature, noted in the previous section, is the proneness of large groups of timber buildings to major conflagrations.

11.4. Design of timber structures
11.4.1 Provisions of Eurocode 8
11.4.1.1 Introduction
The chapter in Eurocode 8 (EC8) on the seismic design of timber structures, which is contained in section 8 of EC8 part 1 (CEN, 2004a), is relatively short, running to six pages compared to the 58 for concrete and 23 for steel. It is mainly concerned with setting out broad principles for successful design, rather than giving detailed design rules. It is based on research predating 2000, and although the basic principles all remain sound, the rules need updating in parts particularly to cover more recent forms of construction (see Section 11.4.3 below). Follesa et al. (2011a) propose revisions to EC8 to address these issues.

The next sections of this chapter summarise the main current provisions of EC8.

11.4.1.2 q Factors and ductility classes in timber structures
EC8 recognises three classes of timber structure, depending on the ability to dissipate energy (Table 11.1). Low ductility (DCL) structures must be designed as elastically responding, with a q factor of 1.5, and the partial factors for materials γ_m given in Eurocode 5 (CEN, 2004b) for fundamental load combinations apply. Individual countries (in their national annexe) may prohibit the use of DCL structures in areas of high seismicity, although the main part of EC8 gives no advice on this. Medium and high ductility (DCM and DCH) structures can be designed for q factors as high as 5, and the more favourable γ_m factors for accidental

Table 11.1 Upper limits of q factors for timber buildings from EC8 (data taken from CEN, 2004a)

Design concept and ductility class	q	Examples of structures
Low capacity to dissipate energy (DCL)	1.5	Cantilevers; beams; arches with two or three pinned joints; trusses joined with connectors
Medium capacity to dissipate energy (DCM)	2	Glued wall panels with glued diaphragms, connected with nails and bolts; trusses with dowelled and bolted joints; mixed structures consisting of timber framing (resisting the horizontal forces) and non-load-bearing infill
	2.5	Hyperstatic portal frames with dowelled and bolted joints
High capacity to dissipate energy (DCH)	3	Nailed wall panels with glued diaphragms, connected with nails and bolts; trusses with nailed joints
	4	Hyperstatic portal frames with dowelled and bolted joints
	5	Nailed wall panels with nailed diaphragms, connected with nails and bolts

load combinations apply. For all ductility classes, load combinations including earthquake are regarded as instantaneous loading, for which the appropriate strength increase may be applied.

At the most fundamental level, the distinction between 'medium' and 'high' ductility may be established by test. DCM structures must thus have dissipative regions able to survive three fully reversed cycles to a displacement ductility of 4 with a loss of not more than 20% in resistance. The same test applies for DCH, except that the displacement ductility demand rises to 6. In practice, rules are given to allow design to proceed for the majority of straightforward cases without testing, as summarised in the next section.

11.4.1.3 Summary of EC8 rules for dissipative (DCM and DCL) timber structures

The main rules for construction to qualify as 'dissipative' without recourse to testing are as follows.

- Sheathing for shear walls or floor diaphragms should consist of particle board with a minimum density of $650 \, kg/m^3$ and a minimum thickness of 13 mm, or of plywood at least 9 mm thick.
- Blocking (backing timbers) is required in sheathing for shear walls or floor diaphragms at free edges and over supporting walls.
- Glued connections must be regarded as non-dissipative, although, as can be seen from Table 11.1, where they are combined with ductile connections, medium or high ductility can be achieved.
- 'Carpenter joints' (defined in Section 11.3 above) can be regarded as dissipative provided they have sufficient overstrength in shear (a factor of 1.3 on required resistance compared with demand is recommended) and do not fail in tension perpendicular to the grain.
- When smooth nails, dowels or staples are used, there must be additional provision (e.g. retaining straps) to prevent their withdrawal. As a matter of general principle, screws are always preferred to nails or staples, although this is not explicitly stated in the code.
- In nailed, dowelled or bolted timber-to-timber or timber-to-steel connections, the minimum thickness of timber must be $10d$ and the maximum fastener diameter d must be 12 mm.
- When nailing wood-based sheathing materials to a timber backing to form shear walls or floor diaphragms, the minimum sheathing thickness must be $4d$ and maximum fastener diameter d should not exceed 3.1 mm.
- Some relaxations are permitted on the previous two rules, but lower q values then apply.

11.4.1.4 Capacity design rules for dissipative timber structures

Medium and high-ductility structures must be checked on capacity design principles to ensure that yielding occurs in the connections intended to yield, and that other parts of the structure remain elastic. These elastic parts include the timber members themselves, although EC8 refers particularly to

- anchor-ties and any connections to massive sub-elements
- connections between horizontal diaphragms and lateral load-resisting vertical elements.

The overstrength factor to be used in these capacity design checks is not stated in EC8; it is merely required to be 'sufficient'. Jorissen and Fragiacomo (2011) propose an overstrength factor of 1.6 for multiple steel nailed, screwed or bolted connections loaded parallel to the grain. For metal connectors (anchoring elements and angle brackets) used in CrossLam construction (see Section 11.4.3), a value of 1.3 is suggested (Follesa et al., 2011a).

The seismic resistance of shear walls should be greatest in the lower storeys to ensure that development of plasticity in the connections is well spread throughout the height of the building. To achieve this, Follesa et al. (2011a) recommend that the seismic resistance of shear walls should decrease proportionally to the decrease of the storey seismic shear. However, this is not an explicit requirement of EC8.

11.4.2 Practice in the USA

By contrast with the Eurocode, US codes for seismic-resisting timber design primarily consist of application rules rather than a statement of broad principles. An extensive summary is provided by Dolan (2003). The American Forest and Paper Association (2008) gives seismic design rules for timber structure in code format, with a commentary; this document is described by the National Earthquake Hazards Reduction Program recommended seismic provisions (BSSC, 2009) as the 'national consensus standard for seismic design of engineered wood structures'.

11.4.3 Construction using cross-laminated timber and other forms of manufactured panel

New forms of precisely manufactured wooden panels, tailor made to suit a specific design, offer an attractive building material for low to medium-rise houses in many parts of the world, and the high lateral strength and stiffness, low mass and moderate ductility make them highly suitable in earthquake country. CrossLam is one such type of proprietary cross-laminated wall and floor panel; developed in Germany and Austria during the 1990s, it consists of three or five layers of rectangular timber elements laid in alternate directions glued to form a composite (Figure 11.2). Wall panel sizes up to 3 m wide by 12 m long by 0.3 m thick, with preformed door and window openings, are shop manufactured and then connected on site to similar floor panels.

The 2005 edition of EC8 part 1 contained no advice on the seismic design of CrossLam panels. Follesa et al. (2011a) proposes design values for the q (behaviour) factors, which are summarised in Table 11.2. The same source proposes q factors and other design parameters, including overstrength factors for connections and interstorey drift limits for other recent forms of manufactured timber construction material, such as log house buildings.

11.4.4 Bamboo construction

Bamboo (strictly a grass, not a timber) has long been part of traditional construction in seismic regions of South and Central America. Known locally as bahareque construction,

Figure 11.2 Typical construction details of a CrossLam building (reproduced from Follesa and colleagues, 2011a, 2011b; courtesy of Maurizio Follesa)

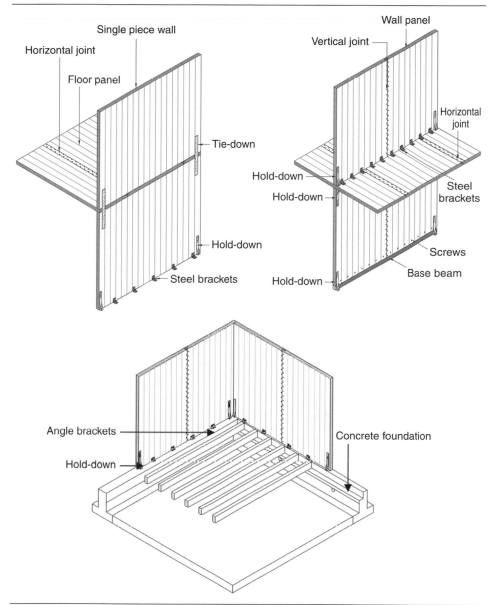

this consists of a bamboo lattice supporting clay earth plaster. Bahareque has a mixed, but generally quite favourable, record of resisting earthquakes. More recently, more reliable forms of bamboo frame have been developed for earthquake-resistant construction. The research has centred on methods of selecting bamboo of sufficient age and therefore strength

Table 11.2 *q* (Behaviour) factors for cross-laminated buildings proposed by Follesa *et al.* (2011a) for use with EC8

Proposed *q* factors for cross-laminated buildings		
Design concept and ductility class	*q*	Examples of structures
Low capacity to dissipate energy (DCL)	1.5	Walls formed from a single panel (i.e. without vertical joints), not conforming to note 1
	2	Walls composed of several panels connected with vertical joints, not conforming to note 1
Medium capacity to dissipate energy (DCM)	2	Walls formed from a single panel (i.e. without vertical joints), conforming to note 1
		Walls composed of a unique element without vertical joints, conforming to note 1
High capacity to dissipate energy (DCH)	3	Walls composed of several panels connected with vertical joints made with mechanical fasteners (nails or screws)

Note 1: To conform with this note, EC8 rules for dissipative structures, summarised in sections 11.4.1.3 and 11.4.1.4 above, should be followed and ductile connections should be provided in certain locations (described below) with sufficient overstrength in the remaining connections to develop the design strength of the ductile connections. The ductile connections should be provided in the following locations

■ vertical connection between wall panels in case of walls composed of more than one element
■ shear connection between upper and lower walls, and between walls and foundation
■ anchoring connections against uplift placed at wall ends and at wall openings.

The overstrength connections designed not to yield should be provided in the following locations

■ connections between adjacent floor panels in order to limit at the greater possible extent the relative slip and to assure a rigid in-plane behaviour
■ connections between floors and walls underneath thus assuring that at each storey there is a rigid floor to which the walls are rigidly connected
■ connections between perpendicular walls, particularly at the building corners, so that the stability of the walls themselves and of the structural box is always assured
■ wall panels under in-plane vertical action due to the earthquake and floor panels under diaphragm action due to the earthquake.

for construction, cost-effective ways of preventing fungal and insect attack, and methods of connection between bamboo members. The connection methods have been based on the use of steel bolts, with the connection area strengthened by filling the central hollow of the bamboo with a cement mortar (Figure 11.3). Cement-based renders have been used to enhance the in-plane shear strength of the walls. The Columbian Association of Earthquake Engineering, in conjunction with a number of other organisations, has published a well-illustrated manual in Spanish, which provides design advice (AIS, 2001). The UK-based organisation TRADA International, in partnership with the Indian Plywood Industries Research and Training Institute, has also developed a bamboo building system for earthquake resistance (Jayanetti and Follett, 2004).

Figure 11.3 Bamboo beam-to-column connection (r)

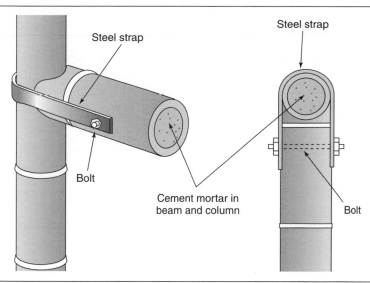

REFERENCES

AIS (Asociación Colombiana de Ingeniería Sísmica) (2001) Manual de Construccion Sismo Resistente de Viviendas en Bahareque Encementado (Manual for the seismic resistant construction of dwellings in cemented bahareque). Columbian Association for Seismic Engineering. See www.desenredando.org/public/libros/2001/csrvbe/guadua_lared.pdf (in Spanish).

American Forest and Paper Association (2008) ANSI/AF&P SDPWS 2008: Special design provisions for wind and seismic, with commentary. American Wood Council, Washington, DC, USA. See www.awc.org/pdf/2008WindSeismic.pdf.

Anagnos T, Rojahn C and Kiremidjian A (1995) NCEER-ATC Study on Fragility of Buildings. Technical Report NCEER-95-0003. National Center for Earthquake Engineering Research, State University of New York, Buffalo, NY, USA.

BSSC (Building Seismic Safety Council) (2009) FEMA 750: Recommended seismic provisions for new buildings and other structures. BSSC, Washington, DC, USA. See www.fema.gov/library/viewRecord.do?id=4103.

Ceccotti A (2008) New technologies for construction of medium rise buildings in seismic regions – the XLAM case. *Structural Engineering International* **18(2)**: 156–165.

CEN (2004a) EN 1998-1: 2004: Design of structures for earthquake resistance. Part 1: General rules, seismic actions and rules for buildings. European Committee for Standardisation, Brussels, Belgium.

CEN (2004b) Eurocode 5. EN 1995-1-1: Design of timber structures. General rules. European Committee for Standardisation, Brussels, Belgium.

Dolan JD (2003) Wood structures. In *Earthquake Engineering Manual* (Chen W-F and Scawthorn C (eds)). CRC Press, Florida, USA.

Follesa M, Fragiacomo M and Lauriola MP (2011a) A proposal for revision of the current timber part (section 8) of Eurocode 8. Part 1. In *Meeting 44 of the Working Commission W18 – Timber*

Structures, CIB, International Council for Research and Innovation, Alghero, Italy, August 29–September 1, Paper no. CIB-W18/44-15-1, 13 pp.

Follesa M *et al.* (2011b) *Edifici a struttura di legno: progettazione e realizzazione.* MADE Expo – Federlegno Arredo, Milan, Italy (in Italian).

Jayanetti L and Follett P (2004) Earthquake-proof house shakes bamboo world. *ICE – Civil Engineering* **157(3)**: 102.

Jorissen A and Fragiacomo M (2011) General notes on ductility in timber structures. *Engineering Structures* **33**: 2987–2997.

Karacabeyli E and Popovski M (2003) Design for earthquake engineering. In *Timber Engineering* (Thelandersson S and Larsen HJ (eds)). Wiley, Chichester, UK.

Scawthorn C (ed.) (2005) *Preliminary Observations on the Niigata Ken Chuetsu, Japan Earthquake of October 23, 2004.* Earthquake Engineering Research Institute, Oakland, CA, USA. See www.eeri.org/lfe/pdf/japan_niigata_eeri_preliminary_report.pdf.

Earthquake Design Practice for Buildings
ISBN 978-0-7277-5794-4

ICE Publishing: All rights reserved
http://dx.doi.org/10.1680/edpb.57944.283

Chapter 12
Building contents

The damage and/or loss potential of [building contents] so far has not received enough attention, particularly bearing in mind escalating values and value concentrations.

Herbert Tiedemann. *Earthquakes and Volcanic Eruptions – a Handbook on Risk Assessment.* Swiss Reinsurance Company, Zurich, 1992

This chapter covers the following topics.

- Acceleration and displacement sensitive items.
- Analysis of 'non-structural' elements.
- Testing and experience databases.
- Electrical and mechanical equipment.
- Vertical and horizontal services.
- Cladding elements.

Earthquake damage to the contents of a building, and to architectural features such as cladding elements, routinely causes more financial loss than structural damage and may also give rise to a severe threat to life. This chapter discusses the design and analysis tools that a structural engineer needs to minimise the threat of loss.

12.1. Introduction

Mechanical equipment, windows, ceilings and cladding may typically represent 70% of a building's value, and its contents can represent many times the value of the building. Failure of these 'non-structural' elements in earthquakes has given rise to both direct and indirect financial loss, the latter through the interruption of business caused by factory closures and uninhabitable offices, damage to bridges, airports and railways, loss of ports and harbours and so on. They have also posed a risk to life, either directly due to injury and suffocation from the collapse of false ceilings, cladding elements, etc., or indirectly due to blocking of escape routes. Another source of hazard has arisen from the failure of fire-fighting equipment and of vessels containing flammable, noxious or radioactive materials.

Damage surveys of earthquakes have shown that, in many cases, buildings that have only suffered minor structural damage have been rendered uninhabitable due to the failures of mechanical, electrical and plumbing systems, and damage to architectural elements. For all these reasons, setting performance goals for the non-structural elements may be of equal importance to setting them for the building structure – see Table 5.1 in Chapter 5. This is of particular importance for buildings such as casualty hospitals required to remain functional immediately after an earthquake, or those such as nuclear power plants whose failure could cause substantial risks to society.

Usually, mechanical and electrical engineers or architects are responsible for specifying these 'non-structural' elements; the interface between the structural engineers and design professionals of other disciplines is therefore of particular importance.

12.2. Analysis and design of non-structural elements for seismic resistance

12.2.1 General principles of design and detailing

Non-structural elements may be damaged during an earthquake due to two distinct mechanisms, namely relative displacement and acceleration.

'Displacement-sensitive' elements become damaged by distortions imposed on them by the structure; cladding elements attached to the façade are an example (Figure 12.1). There are two design strategies that can be employed here. The first option is to make the structure so stiff that the imposed displacements are sufficiently small not to cause damage; limits on building storey drifts are intended to help achieve this. The second option is to make items sufficiently flexible to accommodate the imposed deflections, either by flexibility within the items themselves, or at their points of attachment to the structure. The displacement-based design techniques discussed in Section 3.8 of Chapter 3 are of course particularly well suited to checking whether performance objectives are met for displacement-sensitive elements.

'Acceleration-sensitive' elements are items for which the relative movements between the points of support to the structure are not a significant cause of distress, but which become damaged due to the accelerations (and hence inertia forces) imposed on them by the structure. Usually, the damage takes the form of the item becoming detached from its support. The design strategy is then to make the anchorage of the items strong enough to develop the shear and overturning forces needed to prevent failure.

Design for both types of element may proceed by means of an analysis of the displacements and/or accelerations that the non-structural element has to accommodate. For essential plant items, such as standby generators in critical facilities, this may need to be supplemented by direct testing on a shaking table, or by reference to databases recording experience of plant performance in earthquakes. These types of approach are discussed in the following sections.

Alternatively, for standard items, a more qualitative approach is generally the most valuable; much damage to items such as false ceilings, storage shelves and cabinets can be protected by simple and inexpensive attention to detail. FEMA 74 (ATC, 2011) provides extensive design information, including many standard details and illustrations of damage, for a very wide range of architectural, mechanical, electrical and plumbing components, and provides the most comprehensive practical guidance on non-structural elements available at the time of writing. Like so much guidance from the US Federal Emergency Management Agency

Figure 12.1 Failure of a 'displacement-sensitive' non-structural element: façade cladding Mexico City, 1985 (© E Booth)

(FEMA), it is freely downloadable from the FEMA website (www.fema.gov/earthquake-publications).

12.2.2 Analysis for displacement-sensitive items

In principle, the requirement is to establish the maximum relative displacement d_r between points of attachment. Equation 12.1 assumes the use of a response spectrum analysis to Eurocode 8 (EC8) (CEN, 2004); similar procedures would apply to other codes.

$$d_r = qv\sqrt{(d_{r1})^2 + (d_{r2})^2 + (d_{r3})^2 + \cdots} \qquad (12.1)$$

In Equation 12.1, the relative displacement d_r is calculated from the square root sum of the squares combination of contributions to the relative displacement in each of the building's modes of vibration, d_{r1}, d_{r2}, d_{r3}, etc. calculated for the ultimate limit state event. However, the calculated deformation from the response spectrum analysis must be multiplied by the behaviour factor q, to allow for the post-yield deformation of the structure (for explanation, see Chapter 3) and then reduced by a factor v, which EC8 gives as 0.5 for standard buildings. v allows for the fact that limiting damage to non-structural elements is considered a serviceability limit state, rather than an ultimate limit state consideration, and a shorter return period and hence lower design accelerations are appropriate.

Equation 12.1 may seem unnecessarily cumbersome; direct calculation of the difference in maximum displacement between points of support would seem to be a more straightforward procedure. However, this would be unconservative; such a calculation would underestimate the contribution of higher modes of vibration to relative displacement. Although they usually contribute little to overall (and hence maximum) displacement, higher modes may be very significant in the relative displacement of the upper levels of tall buildings.

Extended items such as pipes, which have multiple supports to the structure, may require more sophisticated analysis; this is briefly discussed in Section 12.4.

In practice, protection to many standard architectural elements such as cladding and partitions is often checked solely by complying with code-specified limits on storey drift (the relative displacement between the top and bottom of a storey). In EC8 part 1, where cladding elements are rigidly attached to the structure, the serviceability limit state storey drift is limited to 0.5% of storey height, but this rises to 0.75% for rigidly attached ductile cladding. When the cladding fixings is designed to accommodate the structural deformations, EC8 increases the drift limit to 1% (a limit based on structural rather than non-structural considerations). This drift-based approach is increasingly being recognised as rather crude; ensuring satisfactory earthquake performance of non-structural elements may need more precise analysis, and will certainly require the sort of detailing recommended in FEMA 74 (ATC, 2011).

12.2.3 Analysis of simple acceleration-sensitive items
During an earthquake, an item of equipment inside a building, such as a pump, will generally experience different motions to those of a similar piece of equipment attached to the ground outside. As the building sways in the earthquake, the accelerations at ground-floor level will generally be similar to those of the ground outside, but will change up the height of the building, generally becoming greater, except in tall, flexible buildings or those with base isolation. Not only is the amplitude of motion affected, but so too is its frequency content.

Codes of practice give simple formulae allowing for the variation in amplitude of motion with floor level and the modification in frequency content. These formulae are adequate for most practical cases found in standard buildings, for the purposes of designing floor fixing for simple items such as pumps or cabinets. They can also be used for the inertia term for cladding fixings; however, the fixings will also experience forces due to imposed deformation.

The equation in EC8 for the inertia force on non-structural items is as follows

$$F_a = S_a \, (m_a \gamma_a)/q_a \qquad (12.2)$$

Figure 12.2 Failure of an 'acceleration-sensitive' non-structural element: storage racks, Northridge earthquake, California, 1994 (reproduced from BSSC, 2006; image is in the public domain)

where F_a is the horizontal seismic force, acting at the centre of mass of the non-structural element in the most unfavourable direction (kN), m_a is the mass of the element (tonnes), γ_a is the importance factor of the element, equal to 1 for most items, but rising to 1.5 for critical items, q_a is the behaviour factor of the element, which varies between 1 and 2, and S_a is the acceleration that the element will experience in the design earthquake (m/s^2).

Figure 12.3 shows how S_a relates to the ground acceleration, which EC8 takes to be a_gS, the design peak ground acceleration (PGA) on rock times the soil factor S. How S_a relates to a_gS depends on the natural period of the item on its support, T_a, and on the dynamic response of the building, as it sways in the earthquake. Usually S_a will be amplified above a_gS, although EC8 specifies that it should never be taken as less than the PGA. For a rigid item such as a cabinet or plant item fixed rigidly to the structure without anti-vibration mounts ($T_a = 0$), the amplification factor is 1 at the ground level rising to 2.5 at the top of the building. In fact, this is a simplification; the true amplification at the top of the building depends on the damping in the building, how close the building is to resonance with the earthquake motions

287

Figure 12.3 Plant item amplification factors from EC8; PGA, peak ground acceleration

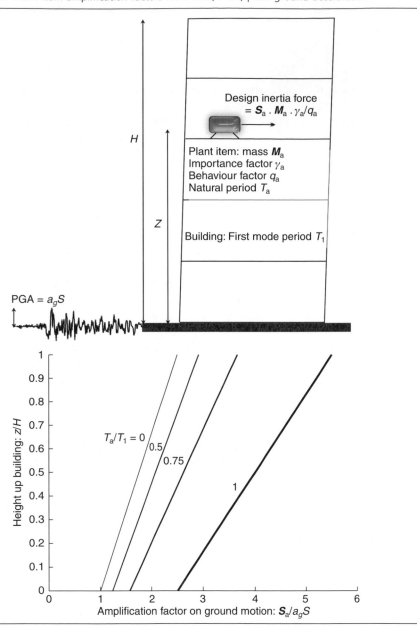

and the degree of yielding within the structure. In most cases, however, an amplification of 2.5 at the top of the building is reasonably conservative.

When the item has significant flexibility – a pump on anti-vibration mounts, for example – the amplification at the top may increase above 2.5. According to the EC8 formula, amplification reaches 5.5 when there is a perfect match between building period and non-structural item period ($T_a = T_1$) due to resonance between the building motions and the non-structural element. Once again, this is an approximation. For simple cases such as the design of fixing bolts for plant items, it should be sufficient; more sophisticated approaches are discussed in the next section.

12.2.4 Analysis of acceleration-sensitive items using 'floor response spectra'

The EC8 amplification factors in Figure 12.3 (and similar ones given in other seismic codes) assume that the non-structural element being considered is simple enough to be approximated by a single degree of freedom system. More complex cases may need more sophisticated methods of justification, if they are critical. Qualification by testing and using 'experience databases' are two such methods discussed in the following two sections. These are methods that dispense with the need for sophisticated analysis. However, analysis routes to qualification are also available. One obvious method would be to include the non-structural item directly in the model for the main structural analysis. There are difficulties here, however. One problem might be in ensuring that sufficient modes of vibration have been considered in the analysis to capture adequately the response of one relatively small part – that is, the non-structural item. A more practical (and insuperable) objection to this route is that the details of the non-structural element may well not have been finalised at the time of the structural analysis. The solution here is to use the structural analysis to produce 'floor response spectra' at the points where the non-structural items are expected to be attached. These floor response spectra are produced in exactly the same way as normal ground spectra, but they relate to the motions of the structure at the attachment points. Response spectra express both the amplitude and frequency content of motions, and so include the factors allowed for in a more simple way by the EC8 formula plotted in Figure 12.3.

The most direct way to derive floor response spectra is to carry out a time-history analysis on the main structure, which will then yield the time history of motions at the attachment points. This, however, requires at least one input time history at ground level. Usually, an input response spectrum will be specified and the appropriate choice of suitable time histories compatible with the design spectrum is not straightforward, as discussed in Section 2.8.3. As an alternative, ASCE 4-98 (ASCE, 1998) refers to a number of methods for producing a floor response spectrum directly from a ground spectrum and a structural model, which avoids having to select a time history. One such direct spectrum-to-spectrum method is given by Singh (1975, 1979).

The floor response spectra generated by either time history or direct spectrum-to-spectrum methods will show a strong peak at the first mode period of the building structure. This period, however, is likely to be subject to considerable uncertainty, and a broadening of the peak to allow for this uncertainty is advisable; the reader should refer to the relevant rules in ASCE 4-98.

Floor response spectra are an appropriate tool when the building response is expected to remain linear during the design earthquake, and they are often used for applications such as nuclear power plants where non-linear response is expected to be small or non-existent. When substantial non-linear response is expected, the direct spectrum-to-spectrum method of producing floor spectra is unlikely to be appropriate, and the time-history analysis would need to be non-linear.

It can be seen then that the use of floor response spectra is a way of decoupling the analysis of the main structure from that of the non-structural element, which is particularly useful when details of the former must proceed before those of the latter are available. The method implicitly assumes that the response of the non-structural item will not significantly affect that of the main structure. Usually, an estimate of the mass of the item can be made with sufficient accuracy at the time of the main structural analysis, and the implicit assumption is valid. However, when natural periods of the structure and non-structural item are very close, the assumption breaks down, even if the mass of the non-structural item is an order of magnitude less than that of the structure. The only way to treat this case of the 'tail wagging the dog' is to abandon floor response spectra, and analyse the system as a coupled whole.

12.2.5 Analysis of items that can rock

The previous two sections assume that the non-structural item responds linearly to earthquake-induced vibrations, and applies to rigid items or those with a well-defined period of vibration. Items that are unanchored to their base and can rock on their foundation, however, have no fixed period of vibration, and a different approach is needed; for information, see De Jong (2012) and ASCE 4 (ASCE, 1998).

12.2.6 Testing of acceleration-sensitive items

The analysis methods discussed in the two preceding sections are appropriate for designing fixings and structural members, but they are most unlikely to give an insight into whether the mechanical parts of a plant item will continue to function during and after an earthquake. It is easy (and inexpensive) to design the holding-down bolts for an emergency generator set that will ensure with high confidence that it remains fixed to the structure, but will it still produce electricity?

In fact, rotating machinery is robust, and generally continues to function if it does not become detached. A ship in a storm can experience accelerations of the same order as those in an earthquake, and yet its machinery generally continues to function, and hard drives of computers also continue to spin. However, some plant may be more sensitive; for example, mechanical relays found in older electrical switchgear may malfunction, and some electrical insulation is brittle and prone to fracture. The most direct way to ensure that a plant item performs adequately is by placing it on an earthquake shaking table, and subjecting it to suitable motions. For plant items within a building, these motions would need to correspond to the floor response spectra discussed in Section 12.2.4. This method of direct qualification has been much used for safety-critical plant items in nuclear power plants.

12.2.7 Qualifying acceleration-sensitive items from 'experience databases'

Testing of plant on a shaking table is expensive, and in any case most plant items are fairly robust against seismic motions. During the 1980s, the nuclear power industry in the USA started to develop a database that showed how plant items had fared in previous earthquakes. The idea was to qualify a standard piece of equipment in a nuclear power plant by showing that similar kit in conventional power stations had survived earthquakes that bounded the design motions. Of course, some plant in a nuclear station is highly specialised, but others, such as pumps and standby generators, are standard items that may still be required to perform a vital safety-related function that must be preserved during and after an earthquake.

It was for such items that the seismic qualification utilities group (SQUG) produced the 'generic implementation procedure' for seismic verification of nuclear power plant (Starck and Thomas, 1990). The procedure allows a plant item to be qualified from the database assembled, if the following conditions are met.

- The design spectrum is bounded by the estimated survival spectrum in the database.
- The plant item is similar in design to one of the items listed in the database.
- The plant item is adequate in terms of security and rigidity of fixing, and workmanship.
- The plant item does not interact adversely with other items. It must not be damaged by knocking against adjacent items, or by debris falling on to it.

Unfortunately, the SQUG database is not in public circulation, and is only available to subscribing members, for which a substantial subscription is necessary. However, a useful checklist in the public domain for non-structural items is given by ASCE/SEI 31-03 (ASCE, 2003); this is mainly based on observations of performance during past earthquakes.

12.3. Electrical, mechanical and other equipment

Plant items such as generators, pumps and computers have been referred to in previous sections; usually it is sufficient to ensure that they are adequately fixed to the structure. Small tanks holding liquid may be treated in the same way. However, larger tanks, particularly if they hold flammable liquids, need to be designed allowing for the interaction between the tank and its contents, which is complex. A report by the New Zealand Society for Earthquake Engineering (Whittaker, 2009) forms one of the best sources of advice on tank design, and there is also useful information in EC8 part 4 and ASCE 4 (ASCE, 1998).

12.4. Vertical and horizontal services

Extended non-structural elements within a building, such as pipes, ducts and lifts are attached to many points of the structure, and the simple type of analysis described in Section 12.2.2 is not applicable because the imposed deformations are more complex. Small pipes are likely to be sufficiently flexible to accommodate these imposed deformations without distress, although where buried services enter a building, it may be necessary to introduce additional flexibility.

Automatic shutdown valves are available at moderate cost which are intended to shut off gas supplies if the ground acceleration exceeds a threshold such as 10% g. These may be useful to reduce the risk of fire and explosion in the building.

Once again, useful qualitative design information is given in ASCE/SEI 31-03 (ASCE, 2003). Methods for the analysis of both buried and above-ground pipework are given in ASCE 4 (ASCE, 1998) and in EC8 part 4.

12.5. Cladding

Failure of the external cladding to buildings occurs frequently during earthquakes. It causes a serious falling hazard to people outside the building and may render the building itself temporarily uninhabitable.

Cladding elements are vulnerable to both imposed deformations in their plane and to accelerations normal to their plane. Cladding attached to the outside of the structural frame in general needs to be checked for both effects. Externally attached precast concrete panels may be particularly vulnerable; their in-plane rigidity can give rise to high deformation forces, while their large mass causes high out-of-plane inertia forces. One solution is to provide slotted connections at the base of the panels, which allow in-plane movement, releasing the associated forces, but providing out-of-plane restraint. For glass curtain walling in a secondary metal restraint frame, special seismic gaskets have been developed that allow the glass to move relative to the restraint frame. External stone or brick cladding needs to be well tied back to the main structure for out-of-plane inertial restraint, while providing sufficient structural stiffness to limit storey drifts to the appropriate code-specified limits (Section 12.2.2) should limit damage due to imposed in-plane deformations.

Concrete or steel frames with infill masonry forming the cladding were discussed in Chapter 8, Section 8.5.2.8.

REFERENCES

ASCE (1998) ASCE 4-98: American Society of Civil Engineers Standard: Seismic analysis for safety-related nuclear structures and commentary. American Society of Civil Engineers, Reston, VA, USA. New edition scheduled for publication in 2014.

ASCE (2003) ASCE/SEI 31-03: Seismic evaluation of existing buildings. American Society of Civil Engineers, Reston, VA USA. New edition scheduled for publication with ASCE 41 in 2014.

ATC (Applied Technology Council) (2011) FEMA 74: *Reducing the risks of nonstructural earthquake damage – a practical guide*. Applied Technology Council for the Federal Emergency Management Agency, Washington, DC, USA. See www.fema.gov/earthquake-publications.

BSSC (Building Seismic Safety Council) (2006) *Seismic considerations for steel storage racks located in areas accessible to the public. FEMA 460*. Building Seismic Safety Council for the Federal Emergency Management Agency, Washington, DC, USA. See www.fema.gov/earthquake-publications.

CEN (2004) EN 1998-1: 2004: Design of structures for earthquake resistance. Part 1: General rules, seismic actions and rules for buildings. European Committee for Standardisation, Brussels, Belgium.

De Jong M (2012) Amplification of rocking due to horizontal motions. *Earthquake Spectra* **28(4)**: 1405–1422.

Singh MP (1975) Generation of seismic floor spectra. *ASCE – Journal of Engineering Mechanics Division* **101(5)**: 593–607.

Singh MP (1979) Seismic design input for secondary system. *Proceedings, ASCE Mini-Conference on Civil Engineering and Nuclear Power*, Boston, Vol. II, April.

Starck RG and Thomas GG (1990) Overview of the SQUG generic implementation procedure (GIP). *Nuclear Engineering and Design* **123(2–3)**: 225–231.

Whittaker D (ed.) (2009) *Seismic Design of Storage Tanks*. New Zealand Society for Earthquake Engineering, Wellington, New Zealand.

Earthquake Design Practice for Buildings
ISBN 978-0-7277-5794-4

ICE Publishing: All rights reserved
http://dx.doi.org/10.1680/edpb.57944.295

Chapter 13
Seismic isolation

Seismic isolation is a design strategy based on the premise that it is both possible and feasible to uncouple a structure from the ground, and thereby protect it from the damaging effects of earthquake motions.

Ian Buckle and Ronald Mayes. Seismic isolation: from idea to reality.
Theme issue of *Earthquake Spectra* Vol. 6, Issue 2, May 1990

This chapter covers the following topics.

- Principles of seismic isolation.
- Performance of isolated buildings in damaging earthquakes.
- Types of seismic isolation and their application.
- Analysis of seismically isolated structures.
- Standards for structural design and for testing of bearings.

Of all the special technologies listed in Section 5.5.8, seismic isolation – the strategy of protecting structures from violent ground movements by mounting them on flexible bearings – is the longest established, most used and still one of the most successful. This chapter describes the performance of seismically isolated buildings in real earthquakes, lists the types of bearings available, gives a method for preliminary analysis and points to building codes and other sources of advice for design.

13.1. Introduction
13.1.1 Seismic isolation: an idea whose time has come
The idea that a building could be protected from earthquakes by decoupling it from the ground with an isolation layer dates back to Roman times; Pliny the Elder wrote in his famous Natural History 'The temple of Diana in Ephesus has been built on a marshy soil to protect it from earthquake and fault effects. Between the soil and the foundations of the temple a layer of coal and wool fleeces has been interposed.' (Pliny was later killed while observing the eruption of Vesuvius in AD 79; the account he wrote of the eruption luckily survived him and has proved invaluable to succeeding generations). In the nineteenth century, John Milne, an English pioneer of engineering seismology working in Tokyo, constructed a building in which the isolation layer consisted of four cast iron balls, although it has to be admitted he later declared the attempt a 'failure' (Muir Wood, 1988). However, it was not until the development and use of elastomeric lead rubber bearings in New Zealand in the mid-1970s that seismic isolation became a practical reality; since then, thousands of buildings and bridges in many seismic

areas of the world have been built employing the principle. The extensive evidence from both testing laboratories and real earthquakes confirms that the principle lives up to its promise. It is the most used of all the special technologies for protecting buildings from earthquakes and is still rated as one of the most successful (Mayes *et al.*, 2013).

13.1.2 Basic principles of seismic isolation

The way in which seismic isolation works is, in concept, straightforward. In the same way that shock absorbers smooth out the ride of a car by absorbing the bumps in a rough road, seismic isolation works by decoupling a building from the violent ground motions caused by an earthquake (Figure 13.1). Perfect isolation – for example, a building on a frictionless surface – would leave the building completely unaffected by the earthquake, but would not

Figure 13.1 Basic principles of seismic isolation (reproduced from Mayes and Naeim, 2001, with kind permission from Springer Science + Business Media B.V.)

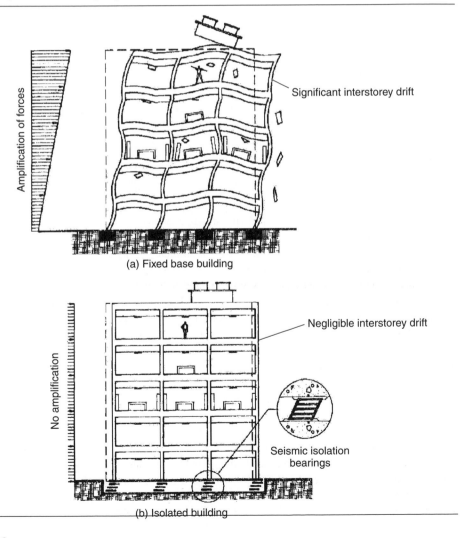

Figure 13.2 Effect of period lengthening and increased damping on response for earthquake motions with a predominant period of approximately 0.5 s

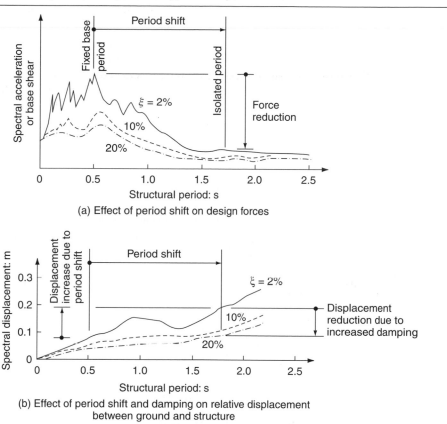

(a) Effect of period shift on design forces

(b) Effect of period shift and damping on relative displacement between ground and structure

of course be practical; for one thing, the building would then not be able to resist lateral forces such as wind loads. However, very substantial reductions are possible by mounting a building on bearings with high but finite horizontal flexibility. The bearings act to lengthen the natural period of the building, taking it away from the main periods of the ground motion (Figure 13.2). Usually, the isolation plane is at the base of the building – hence the common term 'base isolation' – but even in buildings this not always the solution adopted, and in bridges the isolation plane usually occurs near the top, at the deck support points.

A major advantage of seismic isolation is that it protects all the elements above the isolation plane. The reduction in accelerations above the isolation plane not only reduces the inertia forces that the structure must resist, but also the forces on attachments such as water tanks or plant items, and so these too are less prone to failure. Moreover, the reduction in structural forces reduces the shear deformations in the structure, and hence the damage to cladding, glazing, partitions and other non-structural elements. Thus, seismic isolation serves to protect both the building's structure and its contents, and an isolated building is much more likely to be able to function normally immediately after a strong earthquake.

Relative deflections within the superstructure are reduced at the expense of large deformations occurring across the isolation plane, and these must be considered in design. Services (for example, water and gas pipes) crossing into the building must be flexible enough to accommodate these deformations. An 'isolation gap' must be created and maintained between the building and the ground. It is also essential to ensure that the deformations do not become so large that they compromise the ability of the isolation bearings to carry the building's weight. In order to limit deformations across the isolation plane occurring during the design earthquake to a few hundred millimetres (a typical value for bearing capability), additional damping within the bearing is often provided (Figure 13.2). It is also desirable that the bearings should return to their original position after an earthquake. Practical means of achieving these objectives are discussed later.

So far, the discussion has been in terms of reducing the horizontal effects of earthquakes. What about the vertical motions that all earthquakes give rise to? In practice, it is usually much less important to provide protection against these vertical effects. All buildings, by necessity, are built with a substantial factor of safety against gravity loads, which are equivalent to 100% g vertically, but usually only need to resist horizontal wind forces equal to a few per cent of their weight. Moreover, buildings are much stiffer vertically than horizontally, and vertical deformations in an earthquake are unlikely to distress cladding and other non-structural elements. Therefore, protection against vertical seismic motions is not usually needed. In fact, it is desirable to make seismic isolation bearings very rigid in a vertical direction, because vertical flexibility in the bearings would cause the building to rock during an earthquake, negating some of the benefits of horizontal isolation.

13.1.3 Applications in practice

Seismic isolation has been widely used in new buildings in New Zealand, the USA, Japan, Indonesia, Italy (Figure 13.3) and elsewhere. However, the current perception is that additional costs of isolation mean that the total construction cost is larger than for conventional fixed-base buildings; as noted in Section 5.6, a premium of approximately 10% of structural cost is typical compared to a fixed-base building. The relative cost of the bearings to the total building cost (including contents and land) will be substantially less, however, and the extra cost of the bearings may be offset, at least to some extent, by savings in the foundations and from not having to provide seismic detailing in the superstructure.

These figures relate to initial cost, but neglect the improved performance of an isolated building. It is easy to see that the additional cost of seismic isolation would be justified if it prevented the design earthquake resulting in total loss, a possibility for a code-compliant fixed-base structure. As the technology becomes more familiar, seismic isolation is likely to become even more widely used for standard projects, particularly if design standards become less onerous relative to those for fixed-base buildings.

Seismic isolation has also been widely used to protect existing buildings with inadequate seismic resistance (Figure 13.4). The attraction is that structural intervention is concentrated at the isolation plane, and the need to strengthen elements elsewhere is much reduced or even eliminated. This concentration of effort reduces the disruption during the retrofitting works (a great advantage if the building has to remain in operation during this period) and

Figure 13.3 Telecom Administration Centre, Ancona, Italy: (a) external view; (b) interior detailing, showing bearing (photograph courtesy of Alga Spa)

(a)

(b)

Figure 13.4 Retrofitting of an existing reinforced concrete building, Antalya International Airport (reproduced from Booth *et al.*, 2006)

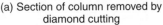

(a) Section of column removed by diamond cutting	(b) Isolator placed in removed section of column

also reduces the architectural impact, which is likely to be of crucial importance in historic buildings in which the original features must be preserved. Individual isolation of precious items in museums has also been used in California and elsewhere.

13.2. Performance of seismically isolated buildings in earthquakes

Since publication of the first edition of this book in 1987, many thousands of structures in seismic areas have been built with seismic isolation, and there is growing experience of how isolated buildings actually fare in strong earthquakes.

Five Californian buildings experienced moderate to high ground motions during the 1994 Northridge earthquake (Smith, 1996). Three performed very well, most notably the University of Southern California (USC) Hospital, where the peak ground acceleration (PGA) near the site was recorded at 0.49 g. Less successful were two three-storey steel frame houses, which were mounted on helical steel springs with viscous dampers. It appears that there was non-structural damage, although the structures survived during motions that were probably greater than those at the USC Hospital. The rocking introduced by the vertical flexibility of the springs has been suggested as a cause for the damage.

Two Japanese buildings were in the epicentral area of the 1995 Kobe earthquake and both performed well. The West Japan Postal Computer Centre is a six-storey, 47 000 m² building, mounted on 120 elastomeric bearings with steel and lead dampers. It was undamaged by the earthquake ground motions, recorded with a PGA of 0.4 g at the site, while a neighbouring fixed-base building reported some damage.

Several seismically isolated buildings and bridges were strongly shaken by the $M = 8.8$ Maule, Chile earthquake of 2010 and all performed well (Moroni *et al.*, 2012). One building and two bridges were instrumented with accelerometers so valuable response information was obtained.

Kasai *et al.* (2013) report on a survey of 327 seismic isolated buildings shaken by the $M = 9.2$ Tohoku Japan earthquake of 2011. Twenty of the buildings were instrumented, and were subjected to peak ground accelerations of between 5% and 40% g. The instrumented buildings studied were between two and 21 storeys high, and used a variety of bearing types, including high-damping rubber, lead rubber and sliding bearings. None of the buildings reported any structural damage; however, 28% of the 327 buildings studied recorded damage at expansion joints, 3% had some damage to finishes and furniture, and 1% recorded damage to equipment. Some damage was also reported in the steel and lead dampers (external to the bearings) in 30 of the buildings. Overall, however, the isolated buildings performed well.

Christchurch Women's Hospital was strongly shaken in the Canterbury, New Zealand earthquakes of 2010/2011, but experienced only minor damage and remained operational (Canterbury Earthquakes Royal Commission, 2012). It was fitted with lead rubber bearings.

13.3. Seismic isolation systems

13.3.1 Functional requirements of a seismic isolation system

An isolation system needs to provide the following.

1 Horizontal flexibility to lengthen the building period, while maintaining vertical stiffness.
2 Damping, to restrict the relative deformation at the plane of isolation and limit it to within the capacity of the bearings.
3 Sufficient stiffness to prevent damage under wind forces.
4 It is also desirable that residual horizontal deflections of the building relative to the ground are small after an earthquake.

These four aspects are now discussed in the following sections.

13.3.1.1 Providing horizontal flexibility

Horizontal flexibility can be provided by rubber or sliding bearings, as described in more detail in the next section. Less conventionally, slender structures can be allowed to lift off their bearings. This system has been used for a bridge (Beck and Skinner, 1974) and a chimney (Sharpe and Skinner, 1983) in New Zealand and forms the basis of the rocking wall system, shown in Figure 5.17 of Chapter 5. The Southern Cross Hospital in Christchurch New Zealand, completed in 2010, was fitted with this system and survived the Canterbury New Zealand earthquakes of 2010/2011; the structure performed 'extremely well', while there was 'minor cosmetic damage' to non-structural elements and some services needed repair (Canterbury Earthquakes Royal Commission, 2012).

13.3.1.2 Providing damping and initial stiffness

Various means of providing damping exist, which can be classified as hysteretic, Coulomb (frictional) or viscous. Hysteretic dampers consist of ductile metal elements designed to yield, and the lead rubber bearings described below fall into this category. Dampers based on the yielding of steel have also been quite widely used (Figure 13.5). The friction in sliding bearings provides Coulomb damping. The advantage of both hysteretic and Coulomb

Figure 13.5 Steel hysteretic damper at pilehead, Union House, Auckland (reproduced, with permission, from Boardman *et al.*, 1983 (photograph Les Megget))

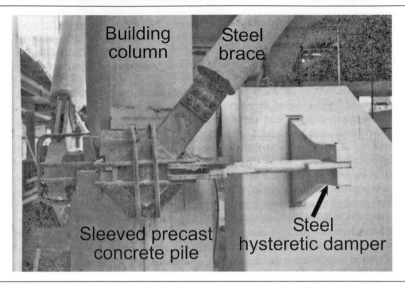

damping is that they provide the high initial stiffness (that is, before the metal yields, or the friction is overcome), which is needed for wind resistance. These two forms of damping are independent of velocity, at least to a first order; by contrast, viscous damping is zero without a relative velocity across the damper, but increases as that velocity becomes larger. A resulting advantage of viscous damping is that it is at maximum when the deformation and acceleration are at their minimum. Hence the damping and inertial forces are out of phase, and so the structure does not have to resist the maximum of the two effects simultaneously. Viscous dampers can take the form of conventional fluid-filled dampers (similar in principle to a car's shock absorber). The properties of elastomeric rubbers can also be modified to produce damping ratios of up to 20% of critical, and can be further modified to produce sufficient initial stiffness for wind loading, while softening considerably at larger deformation to provide the necessary period lengthening under severe earthquake loading.

13.3.1.3 Providing re-centring

Rubber bearings with viscous damping will always tend to return to their initial position after an earthquake. When combined with hysteretic dampers, the bearings must have enough horizontal stiffness to recover sufficient of the plastic deformation in the dampers. The horizontal stiffness of the system needs to be high enough to produce an acceptably low deformation when acted on by the yield force of the damper, because this will be the residual deformation when the system comes to rest. Similar considerations apply to systems with frictional damping. In planar sliding bearings, there is no tendency for a re-centring force to develop and supplementary elastic systems may be added to provide this, for example by means of elastomeric bearings combined in series with the sliding bearings. Alternatively, the sliding bearings may have a spherical shape, as described below for the friction pendulum system (Figure 13.8), which provides re-centring.

13.3.2 Types of seismic isolation bearing

Figure 13.6(a) shows a lead rubber bearing schematically. The layers of rubber provide the lateral flexibility, while the steel plates restrain the rubber from bulging outwards under vertical loading and so help maintain vertical stiffness. The lead plug provides hysteretic damping after it has yielded, and high initial stiffness before yield. The plug is stopped short of the top of the bearing to prevent it carrying vertical load. The base plate is fixed to the substructure and the top plate is usually dowelled to the superstructure in such a way as to achieve the necessary horizontal shear transfer, but to prevent uplift (vertical tension) forces developing. A horizontal deflection capacity of around 1 to 1.5 times the net rubber thickness (i.e. after deducting the thickness of the steel plates) can be achieved in lead rubber bearings, and an equivalent viscous damping ratio of up to approximately 30% of critical can be obtained.

Figure 13.6 Rubber seismic isolation bearings: (a) lead rubber bearing; (b) high damping rubber bearing

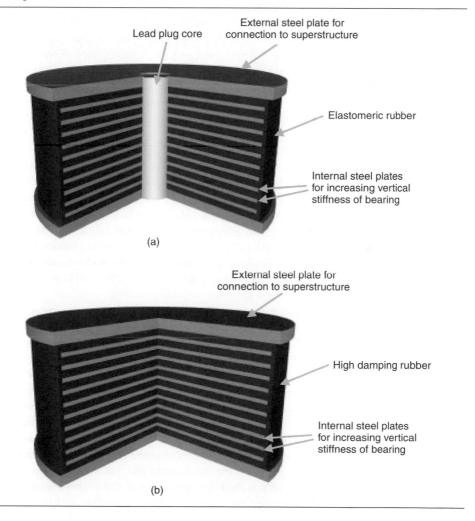

High damping rubber bearings are constructed in a similar way, but have no central lead plug. The elastomer forming the bearing is made with filler materials that modify the damping and also stiffness, providing high initial stiffness but lower horizontal stiffness at seismic deflections (Figure 13.7). The viscous damping ratio achievable, at approximately 20%, is less than for lead rubber bearings, but the deflection capacity is greater at up to twice the net rubber thickness. This larger deflection capacity arises not for reasons of overall stability, but from the lower heat dissipated per deflection cycle to a given limit; this leads to a lower temperature rise.

Friction pendulum bearings were originally produced under patent by Earthquake Protection Systems Inc. of California; following the expiry of the patent in 2005, a number of companies now manufacture them. The system consists of an articulated slider with a low friction coating moving on a spherical stainless steel surface supporting the structure (Figure 13.8). The bottom of the slider is also housed in a low friction spherical bearing, which allows it to rotate and maintain good contact with the upper plate. The restoring force arises because the spherical nature of the upper bearing plate causes it to rise when it moves relative to the slider, and the weight of the building then causes the system to return to the central position. In fact, the effective spring stiffness can easily be shown to depend on the radius of curvature of the upper plate and the vertical force it supports. When the vertical force arises only from gravity, again it is easy to show that the period of the system T depends only on this radius, just as the period of a pendulum depends only on its length and not the weight at its end. This property gives rise to the name 'friction pendulum'. T can be shown to be related to the radius R by

$$T = 2\pi\sqrt{R/g} \tag{13.1}$$

In fact, as the restoring and frictional forces are also proportional to the vertical force, an advantage claimed for the system is that the horizontal forces will always act through the centre of mass, thus minimising torsional effects. However, during an earthquake, vertical forces on the bearings may vary considerably due to seismic overturning moments, and so in practice response may be more complex. Also, local areas with high gravity loads will attract high lateral forces, which may not be desirable.

The deflection capacity of friction pendulum systems is limited by their size in plan. The equivalent viscous damping ratio ξ depends on the friction coefficient μ, the deflection and the radius of the bearing R; at maximum deflection u_{max}, ξ can be shown to equal approximately

$$\xi = \frac{2}{\pi}\left(\frac{\mu}{4\pi^2 u_{max}/(gT^2) + \mu}\right) \tag{13.2}$$

ξ increases with the period T and coefficient of friction μ, but if μ is too large, the bearing will tend to 'stick' before returning to its central position. In fact, a displacement of at least μR is needed to ensure that the system moves back towards the centre from rest and re-centring is never fully achieved after an earthquake. Taking into account practical limitations on T and μ, the maximum achievable damping ratio is therefore approximately 20%. There is likely to be more variation in μ than in the damping value of rubber bearings, resulting in a big variation

Figure 13.7 Hysteresis of a high damping rubber bearing at strains of up to 149% (photograph courtesy of Alga Spa); kips, kilo pounds

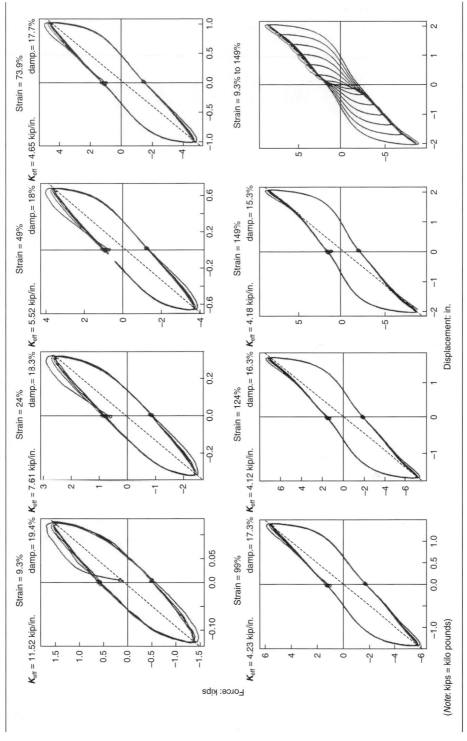

(*Note*: kips = kilo pounds)

Figure 13.8 Friction pendulum system

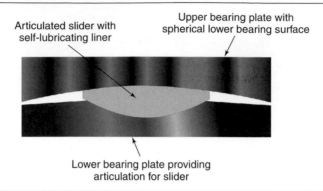

Articulated slider with
self-lubricating liner

Upper bearing plate with
spherical lower bearing surface

Lower bearing plate providing
articulation for slider

in effective structural damping ξ, although the period T is precisely controlled through the radius of the bearing.

13.4. Design considerations
13.4.1 Suitability of buildings for seismic isolation
13.4.1.1 Height limitations
Figure 13.2 shows that seismic isolation works by providing a large separation between the period of a building and the period of the ground motions to which it is subjected. Therefore, tall buildings with a period greater than 1 s are usually judged unsuitable, although taller buildings have been seismically isolated in Japan.

13.4.1.2 Soft soil conditions
Sites that are subject to particularly long period earthquake motions are also not suitable for isolated structures, and in this case lengthening the building period might even cause an increase in seismic demand. Therefore, buildings on very soft soil sites, which amplify the long period content of the earthquake motions, are more likely to benefit by being designed as stiff as possible.

13.4.1.3 Active faults
Sites near active faults may be affected by the long period 'seismic flings' described in Section 4.3.7 of Chapter 4, which give rise to very large displacement demands on bearings. Such sites may therefore be unsuitable for isolated buildings.

13.4.1.4 Wind loads
Where the design wind load exceeds 10% of the building weight, seismic isolation is unlikely to be attractive, because the building needs to respond relatively rigidly to wind loading, and therefore a flexible response would only be possible for earthquake excitations higher than the wind load, which would limit the effectiveness of the isolation. However, design wind loads exceeding 10% of weight are unusual in engineered buildings, particularly concrete ones.

13.4.1.5 Overall slenderness
Slender buildings may be unsuitable for isolation with conventional isolation bearings, because the bearings would need to be designed for large vertical forces due to rocking,

which might not be feasible, both in compression and tension. Rocking wall systems (Figure 5.17) might be a suitable alternative.

13.4.1.6 Building separations and structural joints

An isolated building has substantially larger horizontal deflections than its fixed-base equivalent (Figure 13.2), and therefore greater separations from adjacent structures are required than is the case for non-isolated buildings. A gap of at least 200 mm (and often more) is needed, and buildings on constricted sites where this is not possible may therefore not be suitable.

13.4.1.7 Building function

One of the benefits of isolation is that the building contents, as well as its structure, are protected. Therefore, isolation may be particularly advantageous when the building contents are valuable and its ability to function after an earthquake is crucial. In this context, it is significant that almost all hospitals built in Japan since the 1995 Kobe earthquake have been isolated (Takayama and Morita, 2012).

13.4.2 Suitability of isolation as a strengthening technique

The same considerations of suitability discussed above for new buildings apply when considering isolation, as opposed to member strengthening, as a technique for improving the seismic performance of an existing building. Isolation has the added advantage that operations are confined to one level (the isolation plane), which should reduce disruption to the normal functioning of the building, particularly if the isolation plane can be in the basement. It is also highly suitable for buildings in which the main deficiency is a lack of seismic detailing, because the ductility demands in the superstructure are greatly reduced by isolation. Isolation on its own, however, is unlikely to rectify situations in which the superstructure has a deficiency in lateral strength of more than 50%; the strength reduction in an isolated building is not as great as Figure 13.2 suggests at first sight, because the large reduction in spectral acceleration is offset by smaller ductility factors (see Section 13.5.4). When there is a large shortfall in lateral strength, additional lateral strengthening measures – for example, provisions of shear walls – are likely to be required to supplement seismic isolation, which may therefore become relatively less attractive.

13.4.3 Position of isolation level

Often, the isolation level will best be placed below ground level. In this way, none of the above-ground finishes within the building have to cross the isolation plane. Moreover, this position maximises the superstructure protected. Hence seismic isolation is often referred to as base isolation. However, other positions for the isolation plane may be indicated; for example, when retrofitting existing buildings without a basement. Note that in bridges, the isolation plane is generally between the deck and the pier; that is, at the top not the base of the structure.

13.5. Analysis of seismic isolation systems
13.5.1 Objectives of analysis

The analysis of a seismic isolation system will usually need to consider two levels of seismic loading: a design earthquake with a return period typically of 475 years and a rare, extreme earthquake with a return period of several thousand years. The objectives of the analysis will primarily consist of the following.

- Establish the maximum deformation of the isolator, to make sure that it does not exceed the isolator's limit in an extreme event.
- Establish the response of the structure above the isolation plane. Usually, the objective will be to ensure an essentially elastic response in the design event. The response of the superstructure and foundations below the isolation plane must also be considered, usually with the same objective.
- Check the performance of the non-structure – finishes, services – particularly where they cross the isolation plane.

13.5.2 Simplified analysis

As most of the deformation in a seismically isolated building occurs in the isolation bearings, an obvious approximation to its seismic behaviour is to assume that everything above the isolation layer responds rigidly. On this assumption, where the isolation layer is at or near the base of the structure, the building can be modelled as a rigid mass on a spring represented by the bearings. This type of analysis is only allowed by Eurocode 8 (EC8) (CEN, 2004) and ASCE 7 (ASCE, 2009) for final design when a range of conditions applies, although it will be useful for preliminary design more widely. The main conditions are

- a building height less than four storeys
- ground conditions where the soil stiffness is at least medium (EC8 soil type C or stiffer)
- the isolated period is not more than 3 s at the design displacement, but at least three times the fixed base period
- the structure is regular (no abrupt changes in stiffness with height and no excessive torsional response)
- the isolation system is independent of the rate of loading
- the isolators do not soften excessively with increasing deflection; as the displacement increases from 20% to 100% of the design displacement, the stiffness should not reduce by more than 2/3.

The model can then be further simplified by representing the isolation system as a linear spring with constant viscous damping. In practice, of course, isolation bearings are often far from linear – see Figure 13.7 – but for a first approximation, they can be represented by an equivalent linear spring whose stiffness is chosen to give the total lateral force in the bearings at the maximum deflection under the design earthquake, and a level of viscous damping reflecting the hysteretic energy dissipated (Figure 3.26). As the maximum deflection is not at first known, the process is iterative, as follows.

1. A specific type of isolator is selected.
2. A trial deflection of the isolators is chosen.
3. The lower bound $k_{eff,LB}$ and upper bound $k_{eff,UB}$ stiffness and ξ of the isolators corresponding to that deflection is found from test data published by the manufacturer as is the damping ξ; see Figure 3.26 for how this would be established from test data such as that shown in Figure 13.7.

4 The effective isolated period T_{eff} of the building is calculated as

$$T_e = 2\pi\sqrt{M/k_{eff,LB}} \tag{13.3}$$

where M is the total mass of the building above the isolation plane.

5 The deflection of the isolation system D_D under the design earthquake, neglecting for the moment any torsional response, can then be calculated as the dynamic force (product of mass times peak acceleration) divided by the isolator stiffness. This leads directly to

$$D_D = \eta S_e(T_e)M/k_{eff,LB} \tag{13.4}$$

where $S_e(T_e) = 5\%$ damped elastic spectral acceleration for the design earthquake at period T_e. $S_e(T_e)$ will usually be based on the design response spectrum in the governing seismic code. The implicit assumption is that the foundations are sufficiently rigid in both translation and rotation to transmit the ground motions essentially unmodified to the isolation system; this is an assumption that may need to be checked and if necessary refined at a later stage. Note also that $S_e(T_e)$ is based on the unmodified ground motion, calculated without structural reduction factors (e.g. q in EC8 or R in ASCE 7). η is the correction factor to allow for the level of damping provided by the isolation system, which will usually considerably exceed 5%. Priestley et al. (2007) propose η for standard sites not subject to 'scismic flings' (see Section 13.4.1.3 – Active faults)

$$\eta = \sqrt{7/(2 + \xi)} \tag{13.5}$$

for sites near active faults subject to 'seismic flings'

$$\eta = \{7/(2 + \xi)\}^{0.25} \tag{13.6}$$

where the damping ratio ξ is expressed as a percentage (e.g. for 10% damping, take $\xi = 10$). Similar, though slightly different, values are provided in ASCE 7. At this stage, the calculated deflection needs to be compared with the trial deflection of step 1, and the process iterated if necessary with revised values of spring stiffness and damping.

6 The deflection D_D in Equation 13.4 applies to the deflection at the centre of stiffness of the structure. Because of torsional response, deflections elsewhere will be different; ASCE 7 requires that D_D should be amplified to D_{TD} given by

$$D_{TD} + D_D\left[1 + y\left(\frac{12e}{b^2 + d^2}\right)\right] \tag{13.7}$$

where (see Figure 13.9) e is the distance between centres of mass and stiffness, to which 5% accidental eccentricity should be added, y is the maximum distance to an isolator from the centre of stiffness, measured perpendicular to the direction of seismic loading, b and d are the overall plan dimensions of the building.

7 The acceleration immediately above the isolating bearings in the design event is $\eta S_e(T_e)$. The assumption that the superstructure responds as a rigid block rigid body implies that this acceleration is constant up the height of the building. Hence, the superstructure forces can be calculated. Following EC8, this leads to

$$F_i = \eta m_i S_e(T_e)/(q = 1.5) \tag{13.8}$$

Figure 13.9 Allowing for torsional response in isolated structures: Equation 13.7

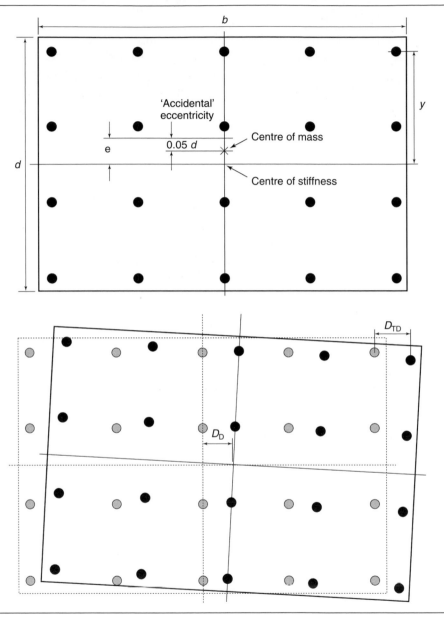

where F_j is the force to be applied at level j and m_j is the corresponding mass. Note that in this case, a structural reduction or 'behaviour' factor of $q = 1.5$ has been assumed; no special seismic detailing is then required.

In ASCE 7, a triangular rather than constant distribution of acceleration up the height of the building is assumed, which is more conservative. Equation 13.8 is then modified to

$$F_j = \eta m_j S_e(T_e)\left[h_j \sum_{i=1}^{N}(m_i) \bigg/ \sum_{i=1}^{N}(m_i h_i)\right]\bigg/ R_I \tag{13.9}$$

where h_i is the height of the ith floor above the top of the bearings, N is the total number of floors. R_I is the behaviour modification factor for isolated structures, which varies between 1 and 2, depending on the inherent ductility of the superstructure.

8   The foundations should be designed for a shear force F at the level of the isolation system, where

$$F = D_{TD} k_{eff,UB} \tag{13.10}$$

Note that the upper, not lower, bound stiffness has been assumed. In both EC8 and ASCE 7, no structural reduction factor is generally allowed; this is to protect the foundations against the possibility of yield or damage in the design earthquake.

9   The isolation system must be checked for a deflection greater than that corresponding to the ultimate limit state (ULS). In ASCE 7, the design deflection from step 4 must be increased by a factor of around 1.5 when checking the capacity of the isolation system. In EC8, the recommended increase factor for buildings is only 1.2, a significantly lower value, although it rises to 1.5 for bridges. In the author's opinion, an increase factor of less than 1.5 should be used with caution. This particularly applies if ground motions at the site may be subject to 'seismic flings'.

10 In general, the analysis should be carried out for both principal directions of the building.

11 Sensitivity analyses may be required to investigate the effect of variations in the stiffness and damping properties of the isolation system; EC8 recommends this should be done when the system properties may vary by more than $\pm 15\%$ from their mean values.

Usually more sophisticated checks (discussed below) will be needed before finalising the design. However, in almost all cases, the very simple checks will be performed as an essential first stage.

13.5.3   More rigorous analysis

More complex models are required when some or all of the conditions listed at the start of Section 13.5.1 do not apply. This will involve representing the superstructure by a flexible two or three-dimensional model rather than a rigid mass, and modelling the isolators as non-linear springs. A time history analysis will usually be required. Various general dynamic analysis software packages such as ETABS (www.csiberkeley.com/etabs2013) include isolator bearing elements, and the freeware 3D-BASIS (civil.eng.buffalo.edu/3dbasis) was written specifically for analysing isolated structures.

Advice on the more rigorous analysis of seismically isolated systems is given by Christopoulos and Filiatrault (2006) and Kelly et al. (2010).

13.5.4   Ductility and seismically isolated buildings

The procedures discussed so far assume that in an isolated building, the foundations (everything below the isolation level) have a high degree of protection against yielding by designing them for

311

unreduced elastic forces, while the superstructure remains essentially elastic, with limited reduction factors on elastic forces allowing mainly for overstrength rather than ductility.

These procedures clearly result in a building performance during the design earthquake that is superior to that implied by code design for fixed-based buildings, in which considerable excursions into the plastic region are allowed during the ULS earthquake, implying commensurate damage. The possibility therefore exists of designing both foundations and superstructure in an isolated building for reduced lateral strength, and accepting that some damage will occur during the design earthquake. EC8 recognises this situation as 'partial isolation', although no further advice is provided. It is not referred to in ASCE 7. Partial isolation might be attractive when retrofitting an existing building by introducing seismic isolation, particularly when increasing the strength of the existing structural members would be difficult or expensive, but some ductility already exists.

In fact, the option of partial isolation of the superstructure is less straightforward than it seems. The isolation system works by considerably lengthening the period of the motions to which the superstructure is subjected. In these circumstances, very much larger plastic deformations are required in the superstructure to reduce the accelerations in the superstructure to the same extent as in the equivalent fixed-base structure. The effect is the same as that shown in Figure 3.14, which shows that ductility is ineffective in reducing response in very stiff structures. Effectively, the base isolation acts to make the superstructure relatively stiff compared to the motions it is subjected to; this reduces the elastic response, but also limits the capacity for a ductile response to reduce internal forces. Ochiuzzi *et al.* (1994) provide further discussion.

As a result, a simple equivalent linear elastic analysis, using q or R factors appropriate for fixed-base buildings is not a possible option for designing partially isolated buildings. A time history analysis, explicitly accounting for the non-linear behaviour of the isolation system and superstructure is recommended, enabling the local curvature ductility demands in the superstructure to be quantified, and checked as sustainable for the level of seismic detailing provided. As ductility demands may increase rapidly with increasing ground motion intensity after the superstructure starts to yield, a check on a 'maximum considered' as well as a ULS earthquake must also be performed.

13.6. European and US standards for seismically isolated buildings

13.6.1 Structural design standards

EC8 part 1 (CEN, 2004) provides rules for the design of new seismically isolated buildings; however, these are not complete and do not form a fully satisfactory basis for design. The rules are currently under review, and changes are expected when the next major revision of EC8 is issued in about 2020. EC8 part 2 provides more complete rules for seismically isolated bridges. A European product standard, EN 15129: 2009 (anti-seismic devices), covers the testing and specification of anti-seismic devices, including isolation bearings; testing is also addressed by an annexe in EC8 part 2.

The US standard ASCE 7 (ASCE, 2009) provides much more comprehensive rules for the design of seismically isolated buildings.

13.6.2 Testing standards

Bearings for seismically isolated buildings are usually manufactured specially for a particular project, rather than being produced as standard off-the-shelf items. Therefore, they need to be tested to confirm that achieved values of stiffness, damping and deformation capacity accord with design assumptions. One effect that may need investigation is 'scragging', a reversible phenomenon in rubber bearings whereby their initial stiffness reduces after a few cycles of loading but may subsequently recover after a period of time. Another possible effect to investigate is the influence of cycling rate on properties.

In Europe, EN 15129: 2009 (anti-seismic devices) provides product standards for seismic isolation bearings, including testing standards. In the USA, ASCE 7 provides testing requirements for seismic bearings in buildings.

REFERENCES

ASCE (2009) ASCE 7: Minimum design loads for buildings and other structures. American Society of Civil Engineers, Reston, VA, USA.

Beck J and Skinner R (1974) The seismic response of a reinforced concrete bridge pier designed to step. *Earthquake Engineering and Structural Dynamics* **2**: 343–358.

Boardman P, Wood B and Carr A (1983) Union House – a cross-braced structure with energy dissipators. *Bulletin of the New Zealand Society for Earthquake Engineering* **16(2)**: 83–97.

Booth E, Yilmaz C and Sketchley C (2006) Retrofit of Antalya airport international terminal building, Turkey using seismic isolation. In *First European Conference on Earthquake Engineering and Seismology, Geneva.*

Canterbury Earthquakes Royal Commission (2012) *Final Report, Volume 3: Low damage building technologies.* See http://canterbury.royalcommission.govt.nz/Final-Report – Volumes-1-2-and-3.

CEN (2004) EN 1998-1: 2004: Design of structures for earthquake resistance. Part 1: General rules, seismic actions and rules for buildings. European Committee for Standardisation, Brussels, Belgium.

Christopoulos C and Filiatrault A (2006) *Principles of Passive Supplemental Damping and Seismic Isolation.* IUSS Press, Pavia, Italy.

Kasai K, Mita A, Kitamura H *et al.* (2013) Performance of seismic protection technologies during the 2011 Tohoku-Oki earthquake. *Earthquake Spectra* **29(S1)**: S265–S293.

Kelly T, Skinner RI and Robinson WH (2010) *Seismic Isolation for Designers and Structural Engineers.* National Information Centre of Earthquake Engineering, IIT Kanpur, India.

Mayes R and Naeim F (2001) Design of structures with seismic isolation. In *The Seismic Design Handbook* (Naeim F (ed.)). Kluwer Academic, Boston, MA, USA.

Mayes R, Wetzel N, Weaver B *et al.* (2013) Performance based design of buildings to assess damage and downtime and implement a rating system. *Bulletin of the New Zealand Society for Earthquake Engineering* **46(1)**: 40–55.

Moroni M, Sarrazin M and Soto P (2012) Behavior of instrumented base-isolated buildings during the 27 February 2010 Chile earthquake. *Earthquake Spectra* **28(S1)**: S407–S424.

Muir Wood R (1988) Robert Mallet and John Milne – earthquakes incorporated in Victorian Britain. *Earthquake Engineering and Structural Dynamics* **17(1)**: 107–142.

Ochiuzzi A *et al.* (1994) Seismic design of base isolated structures. In *Earthquake Resistant Construction and Design*, Vol. 2 (Savidis S (ed.)). A.A. Balkema, Rotterdam, The Netherlands.

Priestley MJN, Calvi GM and Kowalsky MJ (2007) *Displacement-based Seismic Design of Structures.* IUSS Press, Pavia, Italy.

Sharpe R and Skinner R (1983) The seismic design of an industrial chimney with rocking base. *Bulletin of the New Zealand National Society for Earthquake Engineering* **16(2)**: 98–106.

Smith D (1996) Base isolated structures. In *The Northridge California Earthquake of 1994: a field report by EEFIT* (Blakeborough A *et al.* (eds)). Institution of Structural Engineers, London, UK.

Takayama M and Morita K (2012) Seismic response analysis of seismically isolated buildings using observed records due to 2011 Tohoku earthquake. In *15th World Conference on Earthquake Engineering, Lisbon*.

Earthquake Design Practice for Buildings
ISBN 978-0-7277-5794-4

ICE Publishing: All rights reserved
http://dx.doi.org/10.1680/edpb.57944.315

publishing

Chapter 14
Assessment and strengthening of existing buildings

As the general awareness of earthquake risk increases and standards of protection for new buildings become higher, the safety of older, less earthquake-resistant buildings becomes an increasingly important concern.

Andrew Coburn and Robin Spence. *Earthquake Protection.*
Wiley, Chichester, 2002

This chapter covers the following topics.

- Design strategies for strengthening.
- Assessing the seismic adequacy of existing buildings.
- Analysis methods for existing buildings.
- Methods of strengthening.
- Assessing earthquake-damaged buildings.
- Special considerations for historic buildings.
- Evaluating the seismic performance of large groups of buildings.

Inspection and strengthening of existing buildings requires judgement, experience (and perhaps also a certain amount of luck), and poses earthquake engineers with one of their most testing challenges. This chapter reviews how strengthened buildings have fared in subsequent earthquakes, and describes some of the most common inspection, analysis and strengthening measures.

14.1. Introduction

Even if all future new construction in seismic regions were built to conform to the best current standards (sadly, an unrealistic expectation), the existing stock of substandard construction would continue to pose a large risk for many decades. In mature economies, the rate of new construction is typically only around 1%, and although a much higher rate applies in many developing countries, the existing housing stock does not diminish very rapidly (Coburn and Spence, 2002). It would therefore take many years before all the substandard construction were replaced. In fact, the risk they pose may increase for a number of reasons, including structural deterioration due to poor maintenance, weakening due to the removal of internal partitions to create larger room sizes, and further weakening if earthquakes occur and the resulting damage is not properly repaired.

The existing housing stock in seismic regions is particularly prone to seismic defects. Houses have to support their gravity loads all the time, and the most severe wind load experienced in a 10-year period usually does not vary significantly from one decade to the next. Therefore, fundamental weaknesses generally become evident and local construction practices tend to adjust accordingly, and become (by and large) adequate for gravity and wind loads. A damaging earthquake, however, may occur less than once in a generation. Therefore, as housing styles and requirements change over time, inappropriate practices may develop, which leave the construction vulnerable to earthquake damage. A classic example comes from the Turkish region to the east of Istanbul, which experienced extremely rapid economic expansion in the last decades of the twentieth century. The region was known to be highly seismic, but the seismic construction regulations applying to the area were not enforced. The desire for immediate accommodation was much stronger than any consideration of the threat from possible future earthquakes. Much of the construction was in fact highly unsuitable for earthquake country, and was decimated by the two major earthquakes that affected the region in 1999, resulting in great loss of life, destruction of property and resulting human misery.

A major earthquake often provides an impetus for strengthening substandard buildings in the surrounding area, but in many respects of course this comes too late. In a region assessed as seismic but that has not had an earthquake for many years, it is much harder to persuade building owners of the need for strengthening, particularly if they are non-resident landlords. Strengthening costs are usually a significant proportion of the initial construction cost. Moreover, the process is disruptive, often requiring temporary evacuation of the building. Nevertheless, extensive strengthening programmes have been conducted in many parts of the world, including California, Turkey, New Zealand and Japan. In some places, strengthening is enforced by statute, but more general statutory enforcement is regarded as necessary for seismic risk reduction.

Inspection and strengthening of buildings poses many challenges and indeed pitfalls. A sobering tale of warning is the case of the Pyne Gould Corporation building, described by the Canterbury Earthquakes Royal Commission (2012). This was a five-storey reinforced concrete building in the Christchurch central business district, completed in 1966. Recognising that its construction date predated the introduction of ductile detailing by codes of practice, it was assessed by a reputable firm of consulting engineers in 1997, which performed a site inspection and carried out a detailed non-linear time history analysis. A major weakness was identified in the perimeter columns, and these were strengthened to everyone's satisfaction, but a lesser shortfall identified in the central shear core was not addressed, because it did not fall below the threshold that at that time made strengthening mandatory. Further issues in the reinforcement detailing of the shear core were missed altogether; these subsequently turned out to be important, but they were not recognised by the state of practice at the time. The Darfield earthquake of September 2010 represented the design level shaking for the building; this event and subsequent aftershocks caused some damage and the building was inspected on several occasions by qualified engineers. Probably correctly, it was judged that the capacity of the building had not been significantly reduced and the building continued to be occupied. In February 2011 the Christchurch earthquake occurred, giving rise to ground motions much stronger than those of the previous earthquake; the building suffered a

Figure 14.1 Collapse of the Pyne Gould Corporation building in the 2011 Christchurch earthquake (reproduced from Canterbury Earthquakes Royal Commission, 2012, © Crown Copyright, New Zealand)

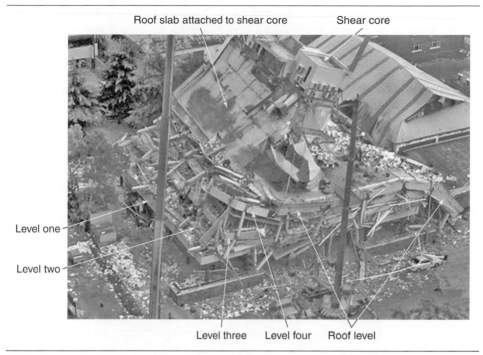

catastrophic collapse, almost certainly originating in the shear core, with the loss of 18 lives (Figure 14.1).

Removing the threat posed by substandard construction in seismic areas poses many complex social, financial and legal problems, as well as technical ones. The rest of this chapter concentrates on the engineering issues; a discussion of the wider issues is provided by Coburn and Spence (2002) and Spence *et al.* (2002).

14.2. Performance of strengthened buildings in earthquakes
14.2.1 Concrete buildings

Reports of the performance of strengthened buildings in strong earthquakes for any form of construction are limited. For concrete buildings, there are a number of reports that buildings retrofitted with concrete shear walls have performed well. One such building was noted in the 1985 Mexican earthquake (EEFIT, 1986; Figure 14.2). Another is Adapazari City Hall in Turkey, which was damaged by an earthquake in 1967. The subsequent retrofit programme involved the addition of new concrete shear walls, and the strengthening of existing beams, columns and shear walls. Adapazari was devastated by the 1999 Kocaeli earthquake, with over 40% of buildings damaged or destroyed. The City Hall, however, performed well; while there was some damage, including cracking of some infill walls and several broken windows, the building remained functional and was heavily used for recovery activities

Figure 14.2 Building in Mexico City after the 1985 earthquake. The building had previously been strengthened by the addition of the external shear walls seen in the photograph

after the earthquake (EERI, 2000). Over the past 20 years, a large number of buildings in highly seismic areas of Turkey, both damaged and undamaged, have been strengthened by the addition of shear walls, and it is likely that much more performance data will emerge as strong earthquakes affect these regions in future.

14.2.2 Steel buildings
Many steel buildings in Japan were retrofitted after the failures in the Kobe earthquake of 1995 discussed in Chapter 9. It is likely that a number of them were severely shaken by the great Tohoku earthquake of 2011, but accounts of how they performed have not been found.

14.2.3 Masonry buildings
The performance of unreinforced masonry (URM) buildings in the Whittier (California) earthquake of 1987 is reported by Deppe (1988). These buildings had been part of an extensive strengthening programme carried out before the earthquake. Losses in the strengthened

buildings were considerably reduced, compared to unstrengthened buildings; however, they still suffered significant damage.

The earthquakes affecting Christchurch, New Zealand in 2010 and 2011 tested a large number of retrofitted URM buildings. Dizhur *et al.* (2010) studied the effects of the first of these earthquakes and reported: 'in general, retrofitted URM buildings performed well, with minor or no earthquake damage observed. Partial or complete collapse of parapets and chimneys were among the most prevalent damage observed [in these buildings]'. The effect of the much more destructive Christchurch earthquake of February 2011 on URM buildings is reported by Dizhur *et al.* (2011); they state the following.

'Common types of retrofit observed in Christchurch URM buildings were:
- Steel moment frames, which increased the lateral capacity of the building
- Steel strong-backs, which helped prevent out-of-plane failure of URM walls
- Application of shotcrete, which increased the in-plane and out-of-plane wall strength
It was concluded that the retrofits which generally performed well were:
- Well-conceived designs which aimed to reduce torsional effects and tied the masonry together
- Well-connected steel strong backs and steel moment frames.'

Dizhur *et al.* (2011) also report on the use of anchors retrofitted to improve the connection between masonry walls and floor diaphragms, as follows

'[There were] numerous cases where anchor connections joining masonry walls or parapets with roof or floor diaphragms failed prematurely, particularly for adhesive anchors. In many cases, these failures were attributed to the low shear strength of masonry, wide anchorage spacing, insufficient embedment depth and/or poor workmanship.'

14.2.4 Historic buildings

Feilden (1982) warns that 'a blind use of reinforced concrete [to strengthen historic buildings] can be disastrous'. Instances were reported after the 1997 earthquake in Umbria-Marche, Italy, of stiff reinforced concrete frames added to strengthen flexible masonry buildings, where the masonry subsequently shook loose, leaving the strengthening elements as the only survivors. D'Ayala and Paganoni (2011a) report similar instances in Italy in the 2009 L'Aquila earthquake. However, other similarly strengthened masonry buildings in the same region performed well, and Booth and Vasavada (2001) noted that strengthening of some historic masonry buildings with concrete diaphragms greatly improved performance in the 2001 Gujarat earthquake. In the 1997 Umbria-Marche earthquake in Italy, damage to the Basilica of St Francis in Assisi was initially attributed by some commentators to the addition in the 1960s of concrete strengthening beams in the roof. Subsequent detailed investigations by Croci (1999) showed that in fact these played no part in the partial collapse of the basilica roof that occurred during the earthquake.

A masonry chapel in Japan built in 1920 and damaged by the 1923 Kanto earthquake was retrofitted with base isolators in 1999 (Seki *et al.*, 2012). The ground motion it experienced from the Tohoku earthquake of 2010 was relatively modest (15%g peak ground acceleration), but the building was fully instrumented and performed well.

14.3. Design strategies for strengthening
14.3.1 Performance objectives for strengthening
There are a wide variety of circumstances under which strengthening may be considered for a building. It may have become unserviceable or unsafe due to earthquake damage. Alternatively, there may be no existing damage, but the owner considers that the building poses an unacceptable threat to life in possible future earthquakes, or that strengthening is a worthwhile investment to protect against future financial losses, both directly from damage to the building fabric and indirectly from business interruption. The strengthening may be a stopgap measure to last a few years until the building is replaced, or it may be intended to preserve a historic building for many future generations. Sometimes, strengthening may be forced on the owner by statute, for example under a municipal programme for improving the seismic safety of a city's building stock.

Performance objectives in the context of the design of new buildings were discussed in Chapter 5 (see Table 5.1). Where strengthening of existing buildings is concerned, the definitions of performance objectives may be similar to those for new construction, but how they are applied will vary greatly. Table 14.1 shows the strengthening objectives for building retrofit defined in the US standard ASCE 41 (ASCE, 2006); Eurocode 8 (EC8) part 3 (CEN, 2005), which also deals with retrofit, has broadly similar definitions. An existing building retrofitted to meet the 'life safety' level of Table 14.1 for a 475-year return earthquake would achieve seismic standards comparable to that for a new building. However, as noted previously, that performance level might be either too high or too low, and adjustments could be made to either the return period of the earthquake considered for design, or to the performance level, or to both. Thus, if preservation of life were the main concern, the 'collapse prevention' level of Table 14.1 might be the governing consideration, while 'immediate occupancy' or 'operational level' would apply to cases in which economic or functional considerations were foremost. In either case, an appropriate return period for the design earthquake would have to be chosen, based on the annual and lifetime risk level to be achieved. EC8 part 3 gives no guidance on the appropriate level of return period to associate with each limit state, although the national annexes to the code of a particular country may provide such guidance for use in that country.

14.3.2 Cost–benefit analysis of seismic strengthening
Coburn and Spence (2002) discuss the application of cost–benefit analysis to the choice of performance level. Achieving higher performance levels implies higher initial costs, because a greater level of upgrading is involved, but it also implies a potential for future savings due to reduced damage in future earthquakes. It also implies lower casualty levels. In fact, it is usually difficult to justify seismic strengthening on purely economic grounds by a cost–benefit analysis, even in highly seismic areas, and Coburn and Spence recommend an estimation of the cost per life saved as a useful additional tool in setting design performance levels. They also suggest that, when planning the upgrading of a large collection of buildings (a city centre, for example), it is most cost-effective to target the most vulnerable buildings for intervention.

Table 14.1 Damage control and building performance levels from ASCE 41 (ASCE, 2006)

	Collapse prevention level	Life safety level	Immediate occupancy level	Operational level
Overall damage	Severe	Moderate	Light	Very light
General	Little residual stiffness and strength, but load-bearing walls function. Some exits blocked. Infills and unbraced parapets failed or at incipient failure. Building near collapse.	Some residual strength and stiffness left in all storeys. Gravity load-bearing elements function. No out-of-plane failure of walls or tipping of parapets. Some permanent drift. Damage to partitions. Building may be beyond economic repair.	No permanent drift. Structure substantially retains original strength and stiffness. Minor cracking of facades, partitions and ceilings as well as structural elements. Elevators can be restarted. Fire protection operable.	No permanent drift. Structure substantially retains original strength and stiffness. Minor cracking of facades, partitions and ceilings as well as structural elements. All systems important to normal operation are functional.
Non-structural components	Extensive damage.	Falling hazards mitigated but many architectural, mechanical and electrical systems are damaged.	Equipment and contents are generally secure, but may not operate due to mechanical failure or lack of utilities.	Negligible damage occurs. Power and other utilities are available, possibly from standby sources.
Comparison with performance intended for (new) buildings, for the design earthquake	Significantly more damage and greater risk.	Somewhat more damage and slightly higher risk.	Less damage and lower risk.	Much less damage and lower risk.

Smith (2003) presents the arguments in a rather different way; he points out that cost–benefit analysis as the basis for strengthening may be misleading if based on the average level of loss, because of the skewed distribution of earthquake ground motions described in Section 2.3. Thus, a low (or zero) level of intervention might give rise to the highest expected financial return, but leave a small but significant risk of a very large loss, if a rare event occurs. Instead, Smith recommends setting a level of unacceptable loss (which could be in financial terms or defined by numbers of lives lost) and then setting a design return period at which this risk must be reduced to an acceptable level. The experience of catastrophic loss in the 2011 Christchurch, New Zealand earthquake lends weight to this view. The intensity of the motions that actually occurred were considered to have a very low probability of exceedence before 2011, but the earthquake shut down the city's central business district, which will take years to reinstate completely.

14.3.3 Strengthening earthquake-damaged buildings

Buildings damaged in an earthquake require a somewhat different approach from undamaged ones, and in some ways the choices are clearer. Some form of intervention is needed, both to achieve an acceptable level of safety and to restore public confidence in the building. At the most basic level, a decision must be taken on whether to demolish and rebuild, or to strengthen. The relative cost of these two options, which depends on the degree of damage and the original deficiencies in the structure, is only one factor; the architectural merits (or demerits) of the building involved and the construction time and disruption associated with each option are also important.

Strengthening may involve merely reinstating the building to its pre-earthquake condition. However, an improved standard usually needs to be achieved. Often, a damaged building will have been evacuated, which makes radical structural intervention easier than for an undamaged, occupied building whose occupants are probably concerned with more immediate, everyday concerns than protecting themselves against a future, hypothetical earthquake. Cosmetic strengthening of a seriously damaged building, which merely hides the evidence of damage without addressing its causes, is of course an option to be avoided, but one that too often has been shown to have been associated with losses in subsequent earthquakes.

14.4. Surveying the seismic adequacy of existing buildings
14.4.1 Undamaged buildings: rapid visual surveys

When a large number of buildings has to be assessed – the building portfolio of a particular client, for example, or all the buildings in a city district – a rapid first pass is usually made, using purely visual methods, so that subsequent more detailed (and time-consuming) surveys can be prioritised. Standard procedures have been developed for US buildings in FEMA 154 (FEMA, 2002) and for New Zealand buildings in a report by that country's earthquake engineering society (NZSEE, 2012). Different building types and design standards in other countries mean these procedures have to be adapted to some extent, but they form a useful basis.

14.4.2 Undamaged buildings: more detailed surveys

As a first essential step, a thorough survey is needed of a building where strengthening is being considered. Guidance on general methods of survey is given in a number of texts, for example Beckmann and Bowles (2004) and the Institution of Structural Engineers (2010). In the first instance, sophisticated equipment is unlikely to be required; a typical checklist is given below. A key aspect is often the standard of workmanship and the degree of corrosion and other deterioration in structural elements. Beautifully finished buildings have been known to hide serious defects behind elaborate plaster work, which the owner may not be very keen for an engineer to chip away. The loft space and service areas are good places to start, because the unadorned structure can often be seen there.

■ Notebook.
■ Pencils.
■ Digital camera with GPS.
■ 300 mm spirit (carpenter) level.
■ Binoculars.
■ 3 m tape.
■ Laser measuring device.
■ Compass.
■ Penknife.
■ Flashlight.
■ A copy of the checklists from ASCE 31 (ASCE, 2003).
■ Ordinary hammer (for tapping surfaces for soundness, and if permitted, knocking off finishes, etc.).
■ Schmidt hammer (for concrete buildings).
■ Covermeter (for concrete buildings).
■ Crack width gauge (for concrete or masonry buildings).
■ Personal safety items: some or all of the following may be needed
    – face mask for entering dusty spaces
    – hard hat
    – site boots
    – mobile phone, to call for help.

Often, the first survey needs to establish if there are any significant areas of seismic weakness. The checklists given in ASCE 31 (ASCE, 2003) are very useful in this respect. In this document, two dozen types of building are identified, classified by their building material and structural form. For each type, a checklist of desirable seismic attributes is given; for example, in concrete moment frame buildings, the list includes the presence of a strong column/weak beam system, and the absence of flat slabs forming part of the lateral load-resisting system. Where the attribute is definitely absent, or cannot be confirmed, a 'concern' is raised, which needs to be further addressed. This could be done by detailed analysis or further testing and inspection. In some instances, it may be possible to rectify the concern directly; for example, the presence of short captive columns created by partial height masonry infill (Figure 8.3) could be addressed by separating the infill from the frame, or building it up to full height. In other cases, more extensive analysis may be needed; for example, a suspected torsional eccentricity (Figure 8.9) would need to be checked out by performing a three-dimensional analysis. ASCE 31 provides checklists not just for the superstructure, but also for foundations and for non-structural elements. They are based on damage observed to typical US buildings, but have been successfully applied to many other parts of the world.

323

More detailed investigations may include soil testing and testing of construction material strength. ASCE 41 (ASCE, 2006) gives advice on the available methods, and further guidance is given by Beckmann and Bowles (2004) and the Institution of Structural Engineers (2010). In the absence of construction drawings, detailed dimensional surveys will be required in order to perform an analysis of the structure; even if there are drawings, their accuracy needs to be confirmed at least by spot checks on site. While overall dimensions are relatively easy to capture, details of structural elements are more problematical. In particular, the reinforcement in concrete members may be very difficult to establish and, in crucial areas like beam–column joints, sometimes almost impossible. A limited amount can be achieved by removing finishes and false ceilings, chipping away concrete cover, using a covermeter and so on, but the confidence with which the properties of the existing structure can be established will be important in the choice of strengthening strategy. The best that can be done may be to infer the likely details from a knowledge of construction practice of the time. When there is large uncertainty about the existing structure, most or all of the seismic resistance will need to be provided by new elements.

14.4.3 Surveying earthquake-damaged buildings

After a damaging earthquake, an urgent need exists to establish which buildings continue to be safe to use, and which should be evacuated. The Applied Technology Council (ATC) in California has published the ATC 20 series of documents containing guidance for rapid and detailed evaluation of earthquake-damaged buildings (of all types) to determine if they can be safety occupied. These documents are all freely available online from www.atcouncil.org. Included are the basic procedures manuals (ATC 20 and ATC 20-2), a field manual (ATC 20-1) and a manual containing case studies of rapid evaluation (ATC 20-3).

14.4.4 Survey methods for earthquake reconnaissance missions

It is vital to gather as much information as possible on building performance from the 'natural laboratory' that is the epicentre of a damaging earthquake affecting populated regions, but the objectives and hence methods involved are necessarily rather different from those discussed in the previous sections. Guidance on conducting earthquake reconnaissance missions is published on the website of the US national earthquake society the Earthquake Engineering Research Institute (EERI; www.eeri.org), and specific advice for non-engineered structures is given by Hughes and Lubkowski (1999), prepared for the UK society for reconnaissance mission, the Earthquake Engineering Field Investigation Team (EEFIT).

14.5. Analysis methods

14.5.1 Approximate initial methods

It is often useful to perform an initial crude analysis to get a rough idea of the likely need for strengthening. The 'tier 1' analysis procedures given in section 3 of ASCE 31 (ASCE, 2003) provide a first estimate of whether or not the strength, ductility and stiffness of a building are likely to be adequate. They are based on approximate equivalent static force procedures.

14.5.2 More detailed analysis

All the forms of analysis discussed in Chapter 3 – equivalent static, response spectrum, non-linear static (pushover) and non-linear time history – may at various times be appropriate for analysing existing buildings. Pushover analysis is particularly useful. This is because it allows the ductility demand to be quantified in the yielding regions of the structure, and checked against the available ductility actually provided. In designing a new building, the adequacy of the ductility supply is usually ensured by following standard detailing rules in codes of practice. Often, however, existing buildings have been designed to standards now considered inadequate, or they may have been substantially altered after construction in ways that reduced their seismic resistance. However, they may still possess some ductility, which (particularly when combined with strengthening measures) may prove adequate, although not complying with current code rules.

For this reason, EC8 part 3 (CEN, 2005), ASCE 41 (ASCE, 2006) and the New Zealand guide (NZSEE, 2012) provide extensive guidance on non-linear static analysis, including data on the plastic rotation of hinges associated with performance levels, such as those defined in Table 14.1. These deformation data have been prepared for structures that do not conform to modern standards, as well as those that do. In ASCE 41, the data are presented in convenient tabular form. In EC8 part 3, they are in the form of equations for concrete, steel and composite construction, which are more versatile, but may be harder to use in practice; they are also acknowledged to be somewhat experimental in nature.

14.6. Methods of strengthening

14.6.1 General

Methods of strengthening are provided by various authors, including Coburn and Spence (2002), Feilden (1982) and Beckmann and Bowles (2004). FEMA 547 (FEMA, 2007) provides extensive information on seismic rehabilitation methods and ASCE 41 (ASCE, 2006) also gives guidance.

14.6.2 Some initial considerations

Initial considerations may include the following.

- Damaged or deteriorated elements are likely to need repair or replacement.
- Irregularities in plan leading to torsional response, caused for example by poorly arranged shear walls or asymmetric masonry infills, should be removed. Adding shear walls or cross-bracing to reduce the torsional eccentricity is an obvious possibility.
- Irregularities in elevation giving rise to soft or weak storeys should be eliminated by the addition of suitable strengthening and stiffening elements.
- Lack of tying together of a building may well be a concern in masonry buildings, particularly at the connections between floors and walls. Local fixings between walls and floors or steel tie rods extending across the entire building may be options to address this.
- Non-structural elements need protection as well as the structure. Cladding elements, chimney stacks, parapets, light fittings, storage units and services may all need to be considered.
- Geotechnical aspects should be addressed, including the possibility of slope instability and liquefaction at or near the site, and foundation movements.

14.6.3 Strengthening options

Among the possible options are the following.

14.6.3.1 Addition of shear walls

Additional concrete shear walls have been widely used to strengthen and stiffen inadequate reinforced concrete moment frame structures. The shear walls reduce the ductility demands in an earthquake on the beam and column frames, which then are much more likely to be able to continue supporting the gravity loads they have to carry. The shear walls also reduce the tendency of a weak or soft storey to form, and their stiffness provides protection to non-structural elements, particularly cladding. The method is particularly suitable for low-rise construction up to about five storeys.

The concrete shear walls are typically formed within an existing concrete frame, with dowelled connections to the surrounding beams and columns (Figure 14.3). Casting each lift of the wall so that it is well compacted up to the soffit of the beam above needs careful placing of the concrete, but is quite easily achieved in practice. Strengthening walls placed outside the frame are also possible (Figure 14.2), and overcome the compaction problems mentioned above, but are more architecturally intrusive, and it is harder to achieve bonding to the existing structure. External buttressing walls have also been used, which may assist

Figure 14.3 Addition of concrete shear walls to a concrete frame building: (a) plan view; and (b) sectional elevation through strengthening shear wall

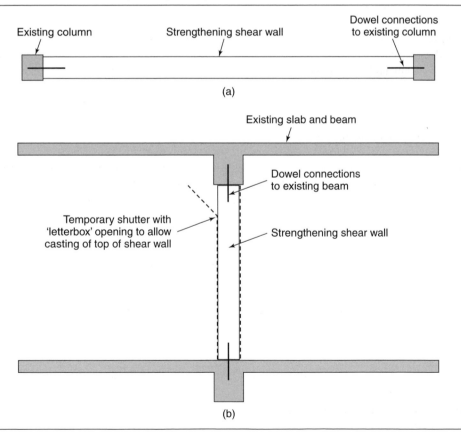

in keeping the building occupied while strengthening works are carried out, but the same architectural and structural difficulties apply.

Instead of reinforced concrete, the strengthening shear walls can take the form of infill masonry, which is particularly suitable if the strength shortfall is low, and the main objective is to remove eccentricities in plan or elevation. The masonry panels may need reinforcement, as described in Section 14.6.3.8. These methods are similar in construction cost to the provisions of additional concrete shear walls, but should prove much less disruptive to the continuing operation of the building.

The additional shear walls need to be provided with foundations that are stiff and strong enough to develop the required moments and shears at their bases, and this can give rise to requirements for substantial new footings, particularly where poor ground conditions occur.

14.6.3.2 Cross-bracing
Adding steel cross-bracing to an inadequate concrete or steel moment frame building is an alternative technique to the addition of concrete shear walls. It has a similar effect of relieving ductility demands on the existing frame, and protecting the non-structural elements. It may be possible to add an additional cross-braced steel frame relatively quickly to an existing building, minimising disruption.

The bracing members may be attached directly to the existing frame, which must then be capable of taking additional axial forces during an earthquake. This is the most efficient and least disruptive solution, but depends on the adequacy of the existing beams and columns. Alternatively, a complete new braced frame can be added, leaving the existing frame to carry only the gravity loads.

14.6.3.3 Passive dampers
Cross-bracing can be added to existing moment-resisting frames by means of passive dampers, which serve to limit the additional forces that the existing frame has to take, while dissipating energy and so reducing response (Figure 14.4). The dampers can take the form of viscous, hysteretic or frictional devices. Further information is given by Christopoulos and Filiatrault (2006).

14.6.3.4 Jacketing of concrete frame elements
Frames with inadequate confinement and shear strength can be surrounded by the elements in a confining jacket. The jacket may be made of steel plate, reinforced concrete or a composite material, such as polymers reinforced with carbon fibre, glass or kevlar. Annexe A of EC8 part 3 (CEN, 2005) provides data quantifying the improvement obtained from these various jacketing methods.

It is usually possible to wrap the jacket around an entire column, producing an efficient confinement. The presence of a slab often makes this difficult or impossible for beams, and so jacketing is mainly useful for increasing the shear strength of beams, with little effect on the flexural ductility. Jacketing of beam–column joints is even more difficult; Weng et al. (2012) describe one approach.

Figure 14.4 Viscous dampers added as retrofit, San Bernardino Justice Center, California, USA (photograph courtesy of Taylor Devices Inc. and SOM (Skidmore Owings & Merrill))

14.6.3.5 Strengthening of steel moment-resisting frame structures

The weld failures at the beam–column joints of steel moment-resisting frames experienced in the Northridge and Kobe earthquakes, discussed in Chapter 9, triggered an extensive research effort to develop upgrading methods to avoid such failures in future. A favoured option is to prevent the formation of plastic hinges at the welded joints. This can be done by welding additional plates to the beams at the joints, so that the plastic hinge forms in the relatively weaker section of beam away from the joint. Alternatively, the same result can be achieved by cutting away the flanges of the beam some distance from the joint (Figure 9.13), although the consequences of this weakening of the beam for the overall lateral strength and stiffness of the building of course need to be checked. FEMA 547 (FEMA, 2007) gives further guidance.

14.6.3.6 Strengthening of floors

Floors play a vital role in seismic resistance by distributing the inertial forces generated in an earthquake back to the lateral resisting elements, and by tying the entire structure together. The strength and stiffness of timber floors can be improved by screwing additional plywood sheets to the floor joists and providing additional blocking elements between joists. Concrete floors may be strengthened by the addition of a concrete screed. The additional gravity loads that the floor bears as a consequence must of course be checked.

14.6.3.7 Reinforcing wall-to-floor connections

For a strengthened floor to be effective, its connection to the surrounding masonry walls must also be adequate. The connection may be improved by the addition of a timber edge beam around the perimeter of the floor, screwed or bolted to the existing floor and attached to the walls with anchor bolts or through-bolts. However, the anchorages were found to be prone to failure in the 2011 Christchurch earthquake (Dizhur *et al.*, 2011).

Steel ties extending through the floors between external masonry walls have also been widely used. D'Ayala and Paganoni (2011b) report that this can be effective, but that once again the anchor points are prone to failure; they discuss the development of dissipative devices at these points to improve performance.

14.6.3.8 Strengthening of masonry walls

Masonry walls can be strengthened by adding a thin layer of mortar to one or both faces. The layer is strengthened with a light mesh reinforcement, and the mortar is usually applied at high pressure to improve compaction and bonding to the masonry, a technique known as guniting. Gluing carbon fibre or polymer reinforced plastic sheets to the walls is an alternative strengthening method.

14.6.3.9 Seismic isolation

Seismic isolation has the potential to protect both structure and non-structure from earthquake damage, while minimising the intervention required to a single isolation plane (see Figure 13.4 in Chapter 13). It is therefore particularly suitable if preservation of the existing architecture and (perhaps) minimising disruption to the continued occupation of a building are important considerations.

14.7. Special considerations for strengthening earthquake-damaged buildings

14.7.1 Categories of damage

Rapid categorisation of buildings as safe or unsafe is needed after an earthquake; Section 14.4.3 discussed the assessment techniques and tools available. In many cases, the most heavily damaged buildings need to be demolished, and retrofitting is only an option for lower grades of damage.

14.7.2 Methods of repair

The performance goal of the strengthening needs to be defined. Is it merely to reinstate the building to its condition before it was damaged by the earthquake, or is it to be upgraded and if so to what level?

Damaged members need to be reinstated at least to the level that they can safely carry their gravity loads; otherwise they must be replaced. Grouting under pressure of cracks in concrete and masonry with epoxy mortar is a well-established technique that is quite reliable in reinstating the concrete to its previous capacity. Jacketing and plating, as previously described, can also be used to reinstate and strengthen damaged elements. Additional elements can then be added to take the demand away from inadequate members under conditions of earthquake loading.

Where foundations have settled during an earthquake, it may be possible to reinstate by jacking up the structure at the points of settlement, but almost certainly this will need to be accompanied by underpinning or other remedial measures to the foundations, particularly if the settlement was caused by liquefaction.

Recommended procedures for upgrading steel moment-resisting frame buildings damaged by earthquakes are provided by FEMA 352 (FEMA, 2000). Advice on the repair of earthquake-damaged concrete and masonry wall buildings is provided by FEMA 308 (FEMA, 1999).

14.8. Upgrading of historic buildings

A number of special considerations apply to historic buildings, including the following.

- Their historical significance implies that they need preserving for many future generations, so the return period of the design needs to be correspondingly long.
- The upgrading needs to pay particular attention to the cultural and architectural values of the original construction. Balancing the need to prevent future damage while preserving the original delight and value of an old building is perhaps as much an art as a science, particularly when the 'original construction' has been built and altered over many years. Figure 14.5 provides an almost ludicrous example of an ancient masonry building insensitively strengthened using external concrete buttresses and steel ring beams.
- Often, it will be difficult to gain full knowledge of the properties of the materials and methods of construction.
- As far as possible, strengthening measures should be reversible, so that they can be removed and modified without damage to the original, if in future more effective measures are developed. In practice, this may be difficult to achieve fully, but the strengthening measures should 'respect, as far as possible, the character and integrity of the original structure' (Feilden, 1987).

Figure 14.5 Concrete buttresses used to retrofit a historic masonry building in Bulgaria (photograph courtesy of Michael Bussell)

Feilden (1982, 1987) has written two classic texts on the restoration of historic buildings. In the second of these two texts (Feilden, 1987), appendices are included giving conclusions and recommendations prepared by the International Centre for the Study and Restoration of Cultural Property, Rome (ICCROM). These contain much practical advice on the issues involved. A case study of the upgrading of a historically important building dating from 1914, using seismic isolation, is provided by Steiner and Elsesser (2004). Marriott (2013) provides an account of the effect of the earthquakes in Christchurch, New Zealand on heritage buildings there, and draws out lessons for retrofitting.

14.9. Assessment of large groups of buildings

The chapter so far has dealt with the assessment and strengthening of individual buildings. Rather different techniques apply when assessing large groups of buildings. Cases include assessment of the building stock of a city centre or an entire region to assist in formulating a seismic strengthening policy, or assessment by an insurance company of the seismic risk associated with its portfolio of buildings. The average annual loss or the 'maximum credible loss' (an ill-defined term) may be of interest. Alternatively, an estimate may be needed of the loss due to a given earthquake 'scenario' – that is, the consequences of an earthquake of a given magnitude at a specified distance that might plausibly be expected to occur.

The technique used is to divide the building stock of interest into a few categories with broadly similar characteristics; for example, low-rise URM, three- to five-storey concrete moment frames and so on. An estimate is then made of the likely distribution of damage within each class of building for a given level of ground shaking. These damage estimates can be based empirically on observations of damage in past earthquakes; usually the ground motion is described in terms of intensity, using the modified Mercalli intensity (MMI) or some other scale. ATC 13 (ATC, 1985) supplemented by Anagnos et al. (1995) gives one Californian source for such data. An alternative approach is provided by HAZUS (FEMA, 2003), which describes the ground motion in terms of a response spectrum rather than by intensity, and uses a non-linear static (pushover) method of analysis. Further discussion of the two methods is provided by Booth et al. (2004). Both approaches have been used to develop more up-to-date and comprehensive vulnerability data as part of the Global Earthquake Model (GEM) programme, which is providing freely accessible data. For information, visit the 'physical vulnerability' page on the GEM website (www. globalquakemodel.org).

REFERENCES

Anagnos T, Rojahn C and Kiremidjian A (1995) NCEER–ATC Joint Study on Fragility of Buildings. Technical Report NCEER-95-0003. National Center for Earthquake Engineering Research, State University of New York, Buffalo, NY, USA.

ASCE (2006) ASCE/SEI 41-06: Seismic rehabilitation of buildings. American Society of Civil Engineers, Reston, VA, USA. NB A revised version is due to be published in 2014 incorporating ASCE 31.

ATC (1985) ATC 13: Earthquake damage evaluation data for California. Applied Technology Council, Redwood City, CA, USA.

Beckmann P and Bowles R (2004) *Structural Aspects of Building Conservation*, 2nd edn. Elsevier, Oxford, UK.

Booth E and Vasavada R (2001) Effect of the Bhuj, India earthquake of 26 January 2001 on heritage buildings. See www.booth-seismic.co.uk/Gujarat%20Intach%20report.pdf.

Booth E, Bird J and Spence R (2004) Building vulnerability assessment using pushover methods – a Turkish case study. *Proceedings of an International Workshop on Performance-based Seismic Design, Bled, Slovenia*, 28 June to 1 July.

Canterbury Earthquakes Royal Commission (2012) Final Report Volume 2: *The performance of Christchurch CBD buildings*. http://canterbury.royalcommission.govt.nz/Final-Report – Volumes-1-2-and-3.

CEN (2005) EN 1998-3: 2005: Design of structures for earthquake resistance. Part 3: Assessment and retrofitting of buildings. European Committee for Standardisation, Brussels, Belgium.

Christopoulos C and Filiatrault A (2006) *Principles of Passive Supplemental Damping and Seismic Isolation*. IUSS Press, Pavia, Italy.

Coburn A and Spence R (2002) *Earthquake Protection*. Wiley, Chichester, UK.

Croci G (1999) Strengthening of the Basilica of St Francis at Assisi. Reported by Booth E. *SECED Newsletter* **13(4)**: 8–9. Society for Earthquake and Civil Engineering Dynamics, UK.

D'Ayala D and Paganoni S (2011a) Assessment and analysis of damage in L'Aquila historic city centre after the 6th April 2009 earthquake. *Bulletin of Earthquake Engineering* **2(2)**: 81–104.

D'Ayala D and Paganoni S (2011b) Performance based earthquake damage assessment and development of dissipative retrofitting devices for masonry buildings. *SECED Newsletter* **23(1)**: Society for Earthquake and Civil Engineering Dynamics, UK.

Deppe K (1988) Whittier Narrows, California earthquake of October 1, 1987: evaluation of strengthened and unstrengthened unreinforced masonry in Los Angeles city. *Earthquake Spectra* **4(1)**: 157–180.

Dizhur D, Ismail N, Knox C, Lumantarna R and Ingham J (2010) Performance of unreinforced and retrofitted masonry buildings during the 2010 Darfield earthquake. *Bulletin of the New Zealand Society for Earthquake Engineering* **43(4)**: 321–339.

Dizhur D, Ingham J, Moon L *et al.* (2011) Performance of masonry buildings and churches in the 22 February 2011 Christchurch earthquake. *Bulletin of the New Zealand Society for Earthquake Engineering* **44(4)**: 279–296.

Earthquake Engineering Field Investigation Team (EEFIT) (1986) *The Mexican Earthquake of 19 September 1985*. SECED, Institution of Civil Engineers, London, UK.

Earthquake Engineering Research Institute (EERI) (2000) Kocaeli, Turkey, earthquake of August 17, 1999: reconnaissance report. *Earthquake Spectra* **16 (Suppl. A)**.

Feilden B (1982) *Conservation of Historic Buildings*. Architectural Press, Oxford, UK.

Feilden B (1987) Between two earthquakes: cultural property in seismic zones. *Proceedings of an International Conference on Conservation Rome (ICCROM)*, Rome, Italy.

FEMA (Federal Emergency Management Agency) (1999) FEMA 308: Repair of earthquake damaged concrete and masonry wall buildings. FEMA, Washington, DC, USA.

FEMA (2000) FEMA 352, 2nd edn: Post earthquake evaluation and repair criteria for welded steel moment-frame buildings. FEMA, Washington, DC, USA.

FEMA (2002) FEMA 154: Rapid visual screening of buildings for potential seismic hazards: A handbook, 2nd edn. Federal Emergency Management Agency, Washington, DC, USA. See https://www.fema.gov/resource-document.library.

FEMA (2003) HAZUS technical manual. FEMA, Washington, DC, USA.

FEMA (2007) FEMA 547: Techniques for the Seismic Rehabilitation of Existing Buildings. See www.fema.gov/library/viewRecord.do?id=2393.

Hughes R and Lubkowski Z (1999) *The survey of earthquake damaged non-engineered structures: a field guide by EEFIT*. Earthquake Engineering Field Investigation Team, see www.eefit.org.uk.

Institution of Structural Engineers (2010) *Appraisal of Existing Structures*, 3rd edn) Institution of Structural Engineers, London, UK.

Marriott A (2013) The performance of heritage buildings in the 2010/2011 Christchurch earthquake swarm. *SECED Newsletter* **24(2)**: May. Society for Earthquake and Civil Engineering Dynamics, UK.

NZSEE (New Zealand Society for Earthquake Engineering) (2012) *Assessment and improvement of the structural performance of buildings in earthquakes*: June 2006 edition with corrigenda 1 and 2. NZSEE, Wellington, New Zealand. Free download from NZSEE website: www.nzsee.org.nz.

Seki M, Yoshida O and H. Katsumata H (2012) *Behavior of Rikkyo Univ. Chapel Building Retrofitted by Seismic Isolation in the 3.11, 2011 Earthquake, Japan*. Paper 0830, 15WCEE, Lisbon, Portugal.

Smith W (2003) Criteria for strengthening buildings: cost–benefit analysis is misleading. *Bulletin of the New Zealand Society for Earthquake Engineering* **36(4)**: 260–262.

Spence R, Peterken O, Gulkan P *et al.* (2002) Earthquake risk mitigation: lessons from recent experience in Turkey. *Proceedings of the 12th European Conference on Earthquake Engineering*. Elsevier Science, Oxford, UK.

Steiner F and Elsesser E (2004) *Oakland City Hall Repair and Upgrades*. Earthquake Engineering Research Institute, Oakland, CA, USA.

Weng D, Zhang J, Xia D, Lu X and Zhang S (2012) *Experimental and Numerical Study on RC Frame Joint Strengthened with Enveloped Steel Plates*. Paper 1653, 15WCEE, Lisbon, Portugal.

Earthquake Design Practice for Buildings
ISBN 978-0-7277-5794-4

ICE Publishing: All rights reserved
http://dx.doi.org/10.1680/edpb.57944.335

Index

Page numbers in *italics* denote figures separate from the corresponding text.